PRACTICAL RESIDUAL STRESS MEASUREMENT METHODS

PRACTICAL RESIDUAL STRESS MEASUREMENT METHODS

Edited by

Gary S. Schajer
University of British Columbia, Vancouver, Canada

A John Wiley & Sons, Ltd., Publication

This edition first published 2013
© 2013 John Wiley & Sons Ltd

Registered office

John Wiley & Sons Ltd, The Atrium, Southern Gate, Chichester, West Sussex, PO19 8SQ, United Kingdom

For details of our global editorial offices, for customer services and for information about how to apply for permission to reuse the copyright material in this book please see our website at www.wiley.com.

The right of the author to be identified as the author of this work has been asserted in accordance with the Copyright, Designs and Patents Act 1988.

All rights reserved. No part of this publication may be reproduced, stored in a retrieval system, or transmitted, in any form or by any means, electronic, mechanical, photocopying, recording or otherwise, except as permitted by the UK Copyright, Designs and Patents Act 1988, without the prior permission of the publisher.

Wiley also publishes its books in a variety of electronic formats. Some content that appears in print may not be available in electronic books.

Designations used by companies to distinguish their products are often claimed as trademarks. All brand names and product names used in this book are trade names, service marks, trademarks or registered trademarks of their respective owners. The publisher is not associated with any product or vendor mentioned in this book.

Limit of Liability/Disclaimer of Warranty: While the publisher and author have used their best efforts in preparing this book, they make no representations or warranties with respect to the accuracy or completeness of the contents of this book and specifically disclaim any implied warranties of merchantability or fitness for a particular purpose. It is sold on the understanding that the publisher is not engaged in rendering professional services and neither the publisher nor the author shall be liable for damages arising herefrom. If professional advice or other expert assistance is required, the services of a competent professional should be sought.

Library of Congress Cataloging-in-Publication Data
Practical residual stress measurement methods / edited by Gary S. Schajer.
    pages cm
  Includes bibliographical references and index.
  ISBN 978-1-118-34237-4 (hardback)
  1. Residual stresses. I. Schajer, Gary S., editor of compilation.
  TA648.3.P73 2013
  620.1'123–dc23
                                                                                    2013017380

A catalogue record for this book is available from the British Library.

ISBN 9781118342374

Set in 10/12 Times by Laserwords Private Limited, Chennai, India
Printed and bound in Singapore by Markono Print Media Pte Ltd

This book is dedicated to the memory of

Iain Finnie

late Professor of Mechanical Engineering at the University of California, Berkeley, a pioneer developer of the Slitting Method for measuring residual stresses.

Respectfully dedicated in appreciation of his encouragement, teaching, mentorship and personal friendship.

The royalties from the sale of this book have been directed to the Leonard and Lilly Schajer Memorial Bursary at the University of British Columbia, to provide bursaries to Mechanical Engineering students on the basis of financial need.

Contents

List of Contributors		xv
Preface		xvii
1	**Overview of Residual Stresses and Their Measurement**	1
	Gary S. Schajer and Clayton O. Ruud	
1.1	Introduction	1
	1.1.1 Character and Origin of Residual Stresses	1
	1.1.2 Effects of Residual Stresses	3
	1.1.3 Residual Stress Gradients	4
	1.1.4 Deformation Effects of Residual Stresses	5
	1.1.5 Challenges of Measuring Residual Stresses	6
	1.1.6 Contribution of Modern Measurement Technologies	7
1.2	Relaxation Measurement Methods	7
	1.2.1 Operating Principle	7
1.3	Diffraction Methods	13
	1.3.1 Measurement Concept	13
	1.3.2 X-ray Diffraction	14
	1.3.3 Synchrotron X-ray	15
	1.3.4 Neutron Diffraction	15
1.4	Other Methods	16
	1.4.1 Magnetic	16
	1.4.2 Ultrasonic	17
	1.4.3 Thermoelastic	17
	1.4.4 Photoelastic	18
	1.4.5 Indentation	18
1.5	Performance and Limitations of Methods	18
	1.5.1 General Considerations	18
	1.5.2 Performance and Limitations of Methods	19
1.6	Strategies for Measurement Method Choice	19
	1.6.1 Factors to be Considered	19
	1.6.2 Characteristics of Methods	24
	References	24

2	**Hole Drilling and Ring Coring**	29
	Gary S. Schajer and Philip S. Whitehead	
2.1	Introduction	29
	2.1.1 Introduction and Context	29
	2.1.2 History	30
	2.1.3 Deep Hole Drilling	31
2.2	Data Acquisition Methods	31
	2.2.1 Strain Gages	31
	2.2.2 Optical Measurement Techniques	33
2.3	Specimen Preparation	35
	2.3.1 Specimen Geometry and Strain Gage Selection	35
	2.3.2 Surface Preparation	38
	2.3.3 Strain Gage Installation	40
	2.3.4 Strain Gage Wiring	40
	2.3.5 Instrumentation and Data Acquisition	41
2.4	Hole Drilling Procedure	42
	2.4.1 Drilling Cutter Selection	42
	2.4.2 Drilling Machines	43
	2.4.3 Orbital Drilling	44
	2.4.4 Incremental Measurements	45
	2.4.5 Post-drilling Examination of Hole and Cutter	46
2.5	Computation of Uniform Stresses	47
	2.5.1 Mathematical Background	47
	2.5.2 Data Averaging	49
	2.5.3 Plasticity Effects	50
	2.5.4 Ring Core Measurements	50
	2.5.5 Optical Measurements	50
	2.5.6 Orthotropic Materials	50
2.6	Computation of Profile Stresses	51
	2.6.1 Mathematical Background	51
2.7	Example Applications	54
	2.7.1 Shot-peened Alloy Steel Plate – Application of the Integral Method	54
	2.7.2 Nickel Alloy Disc – Fine Increment Drilling	54
	2.7.3 Titanium Test-pieces – Surface Processes	56
	2.7.4 Coated Cylinder Bore – Adaptation of the Integral Method	57
2.8	Performance and Limitations of Methods	57
	2.8.1 Practical Considerations	57
	2.8.2 Common Uncertainty Sources	58
	2.8.3 Typical Measurement Uncertainties	59
	References	61
3	**Deep Hole Drilling**	65
	David J. Smith	
3.1	Introduction and Background	65

3.2	Basic Principles	68
	3.2.1 Elastic Analysis	68
	3.2.2 Effects of Plasticity	71
3.3	Experimental Technique	72
3.4	Validation of DHD Methods	75
	3.4.1 Tensile Loading	75
	3.4.2 Shrink Fitted Assembly	77
	3.4.3 Prior Elastic–plastic Bending	78
	3.4.4 Quenched Solid Cylinder	79
3.5	Case Studies	80
	3.5.1 Welded Nuclear Components	80
	3.5.2 Components for the Steel Rolling Industry	82
	3.5.3 Fibre Composites	82
3.6	Summary and Future Developments	83
	Acknowledgments	84
	References	85

4 The Slitting Method — 89
Michael R. Hill

4.1	Measurement Principle	89
4.2	Residual Stress Profile Calculation	90
4.3	Stress Intensity Factor Determination	96
4.4	Practical Measurement Procedures	96
4.5	Example Applications	99
4.6	Performance and Limitations of Method	101
4.7	Summary	106
	References	106

5 The Contour Method — 109
Michael B. Prime and Adrian T. DeWald

5.1	Introduction	109
	5.1.1 Contour Method Overview	109
	5.1.2 Bueckner's Principle	110
5.2	Measurement Principle	110
	5.2.1 Ideal Theoretical Implementation	110
	5.2.2 Practical Implementation	110
	5.2.3 Assumptions and Approximations	112
5.3	Practical Measurement Procedures	114
	5.3.1 Planning the Measurement	114
	5.3.2 Fixturing	114
	5.3.3 Cutting the Part	115
	5.3.4 Measuring the Surfaces	116
5.4	Residual Stress Evaluation	117
	5.4.1 Basic Data Processing	117
	5.4.2 Additional Issues	120

5.5	Example Applications		121
	5.5.1	Experimental Validation and Verification	121
	5.5.2	Unique Measurements	127
5.6	Performance and Limitations of Methods		130
	5.6.1	Near Surface (Edge) Uncertainties	130
	5.6.2	Size Dependence	131
	5.6.3	Systematic Errors	131
5.7	Further Reading On Advanced Contour Method Topics		133
	5.7.1	Superposition For Additional Stresses	133
	5.7.2	Cylindrical Parts	134
	5.7.3	Miscellaneous	134
	5.7.4	Patent	134
	Acknowledgments		134
	References		135
6	**Applied and Residual Stress Determination Using X-ray Diffraction**		**139**
	Conal E. Murray and I. Cevdet Noyan		
6.1	Introduction		139
6.2	Measurement of Lattice Strain		141
6.3	Analysis of Regular $d_{\phi\psi}$ vs. $\sin^2\psi$ Data		143
	6.3.1	Dölle-Hauk Method	143
	6.3.2	Winholtz-Cohen Least-squares Analysis	143
6.4	Calculation of Stresses		145
6.5	Effect of Sample Microstructure		146
6.6	X-ray Elastic Constants (XEC)		149
	6.6.1	Constitutive Equation	150
	6.6.2	Grain Interaction	151
6.7	Examples		153
	6.7.1	Isotropic, Biaxial Stress	153
	6.7.2	Triaxial Stress	154
	6.7.3	Single-crystal Strain	156
6.8	Experimental Considerations		159
	6.8.1	Instrumental Errors	159
	6.8.2	Errors Due to Counting Statistics and Peak-fitting	159
	6.8.3	Errors Due to Sampling Statistics	159
6.9	Summary		160
	Acknowledgments		160
	References		160
7	**Synchrotron X-ray Diffraction**		**163**
	Philip Withers		
7.1	Basic Concepts and Considerations		163
	7.1.1	Introduction	163
	7.1.2	Production of X-rays; Undulators, Wigglers, and Bending Magnets	166
	7.1.3	The Historical Development of Synchrotron Sources	167

		7.1.4 Penetrating Capability of Synchrotron X-rays	169
7.2	Practical Measurement Procedures and Considerations		169
	7.2.1	Defining the Strain Measurement Volume and Measurement Spacing	170
	7.2.2	From Diffraction Peak to Lattice Spacing	173
	7.2.3	From Lattice Spacing to Elastic Strain	173
	7.2.4	From Elastic Strain to Stress	178
	7.2.5	The Precision of Diffraction Peak Measurement	179
	7.2.6	Reliability, Systematic Errors and Standardization	180
7.3	Angle-dispersive Diffraction		184
	7.3.1	Experimental Set-up, Detectors, and Data Analysis	184
	7.3.2	Exemplar: Mapping Stresses Around Foreign Object Damage	186
	7.3.3	Exemplar: Fast Strain Measurements	187
7.4	Energy-dispersive Diffraction		188
	7.4.1	Experimental Set-up, Detectors, and Data Analysis	189
	7.4.2	Exemplar: Crack Tip Strain Mapping at High Spatial Resolution	189
	7.4.3	Exemplar: Mapping Stresses in Thin Coatings and Surface Layers	190
7.5	New Directions		191
7.6	Concluding Remarks		192
	References		193
8	**Neutron Diffraction**		**195**
	Thomas M. Holden		
8.1	Introduction		195
	8.1.1	Measurement Concept	195
	8.1.2	Neutron Technique	196
	8.1.3	Neutron Diffraction	196
	8.1.4	3-Dimensional Stresses	198
	8.1.5	Neutron Path Length	198
8.2	Formulation		199
	8.2.1	Determination of the Elastic Strains from the Lattice Spacings	199
	8.2.2	Relationship between the Measured Macroscopic Strain in a given Direction and the Elements of the Strain Tensor	199
	8.2.3	Relationship between the Stress $\sigma_{i,j}$ and Strain $\varepsilon_{i,j}$ Tensors	200
8.3	Neutron Diffraction		201
	8.3.1	Properties of the Neutron	201
	8.3.2	The Strength of the Diffracted Intensity	202
	8.3.3	Cross Sections for the Elements	203
	8.3.4	Alloys	204
	8.3.5	Differences with Respect to X-rays	205
	8.3.6	Calculation of Transmission	205

8.4	Neutron Diffractometers		206
	8.4.1	*Elements of an Engineering Diffractometer*	206
	8.4.2	*Monochromatic Beam Diffraction*	206
	8.4.3	*Time-of-flight Diffractometers*	209
8.5	Setting up an Experiment		210
	8.5.1	*Choosing the Beam-defining Slits or Radial Collimators*	210
	8.5.2	*Calibration of the Wavelength and Effective Zero of the Angle Scale, $2\theta_0$*	210
	8.5.3	*Calibration of a Time-of-flight Diffractometer*	210
	8.5.4	*Positioning the Sample on the Table*	211
	8.5.5	*Measuring Reference Samples*	211
8.6	Analysis of Data		211
	8.6.1	*Monochromatic Beam Diffraction*	211
	8.6.2	*Analysis of Time-of-flight Diffraction*	212
	8.6.3	*Precision of the Measurements*	213
8.7	Systematic Errors in Strain Measurements		213
	8.7.1	*Partly Filled Gage Volumes*	213
	8.7.2	*Large Grain Effects*	214
	8.7.3	*Incorrect Use of Slits*	214
	8.7.4	*Intergranular Effects*	215
8.8	Test Cases		215
	8.8.1	*Stresses in Indented Discs; Neutrons, Contour Method and Finite Element Modeling*	215
	8.8.2	*Residual Stress in a Three-pass Bead-in-slot Weld*	218
	Acknowledgments		221
	References		221
9	**Magnetic Methods**		**225**
	David J. Buttle		
9.1	Principles		225
	9.1.1	*Introduction*	225
	9.1.2	*Ferromagnetism*	226
	9.1.3	*Magnetostriction*	226
	9.1.4	*Magnetostatic and Magneto-elastic Energy*	227
	9.1.5	*The Hysteresis Loop*	228
	9.1.6	*An Introduction to Magnetic Measurement Methods*	228
9.2	Magnetic Barkhausen Noise (MBN) and Acoustic Barkhausen Emission (ABE)		229
	9.2.1	*Introduction*	229
	9.2.2	*Measurement Depth and Spatial Resolution*	230
	9.2.3	*Measurement*	232
	9.2.4	*Measurement Probes and Positioning*	233
	9.2.5	*Calibration*	233
9.3	The MAPS Technique		235
	9.3.1	*Introduction*	235
	9.3.2	*Measurement Depth and Spatial Resolution*	237

	9.3.3	MAPS Measurement	238
	9.3.4	Measurement Probes and Positioning	239
	9.3.5	Calibration	240
9.4	Access and Geometry		243
	9.4.1	Space	243
	9.4.2	Edges, Abutments and Small Samples	244
	9.4.3	Weld Caps	244
	9.4.4	Stranded Wires	244
9.5	Surface Condition and Coatings		244
9.6	Issues of Accuracy and Reliability		245
	9.6.1	Magnetic and Stress History	245
	9.6.2	Materials and Microstructure	246
	9.6.3	Magnetic Field Variability	248
	9.6.4	Probe Stand-off and Tilt	248
	9.6.5	Temperature	249
	9.6.6	Electric Currents	250
9.7	Examples of Measurement Accuracy		250
9.8	Example Measurement Approaches for MAPS		252
	9.8.1	Pipes and Small Positive and Negative Radii Curvatures	252
	9.8.2	Rapid Measurement from Vehicles	252
	9.8.3	Dealing with 'Poor' Surfaces in the Field	253
9.9	Example Applications with ABE and MAPS		253
	9.9.1	Residual Stress in a Welded Plate	253
	9.9.2	Residual Stress Evolution During Fatigue in Rails	253
	9.9.3	Depth Profiling in Laser Peened Spring Steel	254
	9.9.4	Profiling and Mapping in Ring and Plug Test Sample	254
	9.9.5	Measuring Multi-stranded Structure for Wire Integrity	255
9.10	Summary and Conclusions		256
	References		257
10	**Ultrasonics**		**259**
	Don E. Bray		
10.1	Principles of Ultrasonic Stress Measurement		259
10.2	History		264
10.3	Sources of Uncertainty in Travel-time Measurements		265
	10.3.1	Surface Roughness	265
	10.3.2	Couplant	265
	10.3.3	Material Variations	265
	10.3.4	Temperature	265
10.4	Instrumentation		266
10.5	Methods for Collecting Travel-time		266
	10.5.1	Fixed Probes with Viscous Couplant	267
	10.5.2	Fixed Probes with Immersion	267
	10.5.3	Fixed Probes with Pressurization	270
	10.5.4	Contact with Freely Rotating Probes	270
10.6	System Uncertainties in Stress Measurement		270

10.7	Typical Applications		271
	10.7.1	Weld Stresses	271
	10.7.2	Measure Stresses in Pressure Vessels and Other Structures	272
	10.7.3	Stresses in Ductile Cast Iron	273
	10.7.4	Evaluate Stress Induced by Peening	273
	10.7.5	Measuring Stress Gradient	273
	10.7.6	Detecting Reversible Hydrogen Attack	273
10.8	Challenges and Opportunities for Future Application		274
	10.8.1	Personnel Qualifications	274
	10.8.2	Establish Acoustoelastic Coefficients (L_{11}) for Wider Range of Materials	274
	10.8.3	Develop Automated Integrated Data Collecting and Analyzing System	274
	10.8.4	Develop Calibration Standard	274
	10.8.5	Opportunities for L_{CR} Applications in Engineering Structures	274
	References		275
11	**Optical Methods**		**279**
	Drew V. Nelson		
11.1	Holographic and Electronic Speckle Interferometric Methods		279
	11.1.1	Holographic Interferometry and ESPI Overview	279
	11.1.2	Hole Drilling	282
	11.1.3	Deflection	285
	11.1.4	Micro-ESPI and Holographic Interferometry	286
11.2	Moiré Interferometry		286
	11.2.1	Moiré Interferometry Overview	286
	11.2.2	Hole Drilling	287
	11.2.3	Other Approaches	289
	11.2.4	Micro-Moiré	289
11.3	Digital Image Correlation		290
	11.3.1	Digital Image Correlation Overview	290
	11.3.2	Hole Drilling	291
	11.3.3	Micro/Nano-DIC Slotting, Hole Drilling and Ring Coring	292
	11.3.4	Deflection	293
11.4	Other Interferometric Approaches		294
	11.4.1	Shearography	294
	11.4.2	Interferometric Strain Rosette	294
11.5	Photoelasticity		294
11.6	Examples and Applications		295
11.7	Performance and Limitations		295
	References		298
	Further Reading		302
Index			**303**

List of Contributors

Don E. Bray, Don E. Bray, Inc., Texas, USA
David J. Buttle, MAPS Technology Ltd., GE Oil & Gas, Oxford, UK
Adrian T. DeWald, Hill Engineering, LLC, California, USA
Michael R. Hill, Department of Mechanical and Aerospace Engineering, University of California, Davis, California, USA
Thomas M. Holden, National Research Council of Canada, Ontario, Canada (Retired)
Conal E. Murray, IBM T.J. Watson Research Center, New York, USA
Drew V. Nelson, Stanford University, Stanford, California, USA
I. Cevdet Noyan, Columbia University, New York, USA
Michael B. Prime, Los Alamos National Laboratory, New Mexico, USA
Clayton O. Ruud, Pennsylvania State University, Washington, USA (Retired)
David J. Smith, University of Bristol, Bristol, UK
Philip S. Whitehead, Stresscraft Ltd., Shepshed, Leicestershire, UK
Philip J. Withers, University of Manchester, Manchester, UK

Preface

Residual stresses are created by almost every manufacturing process, notably by casting, welding and forming. But despite their widespread occurrence, the fact that residual stresses occur without any external loads makes them easy to overlook and ignore. This neglect can cause great design peril because residual stresses can have profound influences on material strength, dimensional stability and fatigue life. Sometimes alone and sometimes in combination with other factors, unaccounted for residual stresses have caused the failure of major bridges, aircraft, ships and numerous smaller structures and devices, often with substantial loss of life. At other times, residual stresses are deliberately introduced to provide beneficial effects, such as in pre-stressed concrete, shot-peening and cold hole-expansion.

Starting from early curiosities such as "Rupert's Drops," understanding of the character and mechanics of residual stresses grew with the rise in the use of cast metals during the Industrial Revolution. The famous crack in the Liberty Bell is due to the action of residual stresses created during casting. Early methods for identifying the presence of residual stresses involved cutting the material and observing the dimension changes. With the passage of time, these methods became more sophisticated and quantitative. Complementary non-destructive methods using X-rays, magnetism and ultrasonics were simultaneously developed.

Modern residual stress measurement practice is largely based on the early historical roots. However, the modern techniques bear the same relationship to their predecessors as modern jet planes to early biplanes: they share similar conceptual bases, but in operational terms the current measurement techniques are effectively "new." They have attained a very high degree of sophistication due to greatly increased conceptual understanding, practical experience and much more advanced measurement/computation capabilities. All these factors join to give substantial new life into established ideas and indeed to produce "new lamps for old."

Conceptual and technological progress has been a collective endeavor by a large group of people. The list of names is a long and distinguished one. To paraphrase Isaac Newton's words, the present Residual Stress community indeed "stands on the shoulders of giants." A particular one of these giants that several of the contributors to this book were privileged to know and learn from, was Iain Finnie, late Professor of Mechanical Engineering at the University of California, Berkeley. Professor Finnie was a pioneer of the Slitting Method, described in detail in Chapter 4 of this book. I join with the other authors in dedicating this book to him as a sign of respect and of appreciation for his encouragement, teaching, mentorship and personal friendship. Those of us who aspire to be researchers and teachers can do no better than look to him for example.

On a personal note, I would like to express my sincere gratitude and appreciation to all the chapter authors of this book. The depth of their knowledge and experience of their various specialties and their generous willingness to share their expertise makes them a true "dream team." They have been extraordinarily patient with all my editorial requests, both large and small, and have worked with me with grace and patience. Thank you, you have been good friends!

I also would like to thank the staff at John Wiley & Sons for the support and encouragement of this project, and for the careful way they have carried forward every step in the production process.

And finally, more personally, I would like to acknowledge my late parents, Leonard and Lilly Schajer, whose fingerprints are to be found on these pages. They followed the biblical proverb "Train up a child in the way he should go: and when he is old, he will not depart from it." In keeping with their philosophy, the royalties from the sale of this book have been directed to support students in financial need through the Leonard and Lilly Schajer Memorial Bursary at the University of British Columbia. All book contributors have graciously supported this endeavor and in this way hope to add to the available shoulder-space on which the next generation may stand.

Gary Schajer
Vancouver, Canada
April 2013

1

Overview of Residual Stresses and Their Measurement

Gary S. Schajer[1] and Clayton O. Ruud[2]
[1]*University of British Columbia, Vancouver, Canada*
[2]*Pennsylvania State University, Washington, USA (Retired)*

1.1 Introduction

1.1.1 Character and Origin of Residual Stresses

Residual stresses are "locked-in" stresses that exist in materials and structures, independent of the presence of any external loads [1]. The stresses are self-equilibrating, that is, local areas of tensile and compressive stresses sum to create zero force and moment resultants within the whole volume of the material or structure. For example, Figure 1.1 schematically illustrates how a residual stress distribution through the thickness of a sheet of toughened glass can exist without an external load. The tensile stresses in the central region balance the compressive stresses at the surfaces.

Almost all manufacturing processes create residual stresses. Further, stresses can also develop during the service life of the manufactured component. These stresses develop as an elastic response to incompatible local strains within the component, for example, due to non-uniform plastic deformations. The surrounding material must then deform elastically to preserve dimensional continuity, thereby creating residual stresses. The mechanisms for creating residual stresses include:

1. Non-uniform plastic deformation. Examples occur in manufacturing processes that change the shape of a material including forging, rolling, bending, drawing and extrusion, and in service during surface deformation, as in ball bearings and railway rails.
2. Surface modification. Examples occur in manufacture during machining, grinding, plating, peening, and carburizing, and in service by corrosion or oxidation.

Practical Residual Stress Measurement Methods, First Edition. Edited by Gary S. Schajer.
© 2013 John Wiley & Sons, Ltd. Published 2013 by John Wiley & Sons, Ltd.

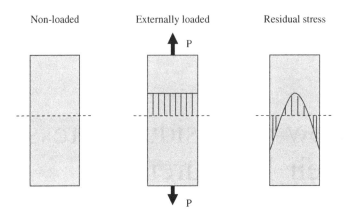

Figure 1.1 Schematic diagram of the cross-section of a sheet of toughened glass showing how residual stresses can exist in the absence of an external load

3. Material phase and/or density changes, often in the presence of large thermal gradients. Examples occur in manufacture during welding, casting, quenching, phase transformation in metals and ceramics, precipitation hardening in alloys and polymerization in plastics, as well as in service from radiation damage in nuclear reactor components and moisture changes in wood.

Residual stresses are sometimes categorized by the length scale over which they equilibrate [2]. Type I are macro residual stresses that extend over distances from mm upwards. These are the "macro stresses" that appear in manufactured components. Type II are micro residual stresses that extend over distances in the micron range, for example, between grains in metals. Type I macro-stress, whether residual or applied, is one cause of Type II micro-stresses. Finally, Type III are residual stresses that occur at the atomic scale around dislocations and crystal interfaces. The Type I macro stresses are the target of most of the measurement techniques described in this book. Several of the techniques can be scaled down and used also to measure Type II and possibly Type III stresses. However, for some of the diffraction methods, the presence of Type II stresses can impair attempts to measure Type I stresses.

Figure 1.2 schematically illustrates examples of some typical ways in which residual stresses are created in engineering materials. The diagrams illustrate how localized dimension changes require the surrounding material to deform elastically to preserve dimensional continuity, thereby creating residual stresses. For example, the upper left panel illustrates shot peening, where the surface layer of a material is compressed vertically by impacting it with small hard balls [8]. In response, the plastically deformed layer seeks to expand horizontally, but is constrained by the material layers below. That constraint creates compressive surface stresses balanced by tensile interior stresses, as schematically shown in the graph. A similar mechanism occurs with plastic deformation created in cold hole expansion and bending, although with completely different geometry. Phase transformations, such as martensitic transformations in steel, can also cause the dimensions of a part of material to change relative to the surrounding areas, also resulting in residual stresses.

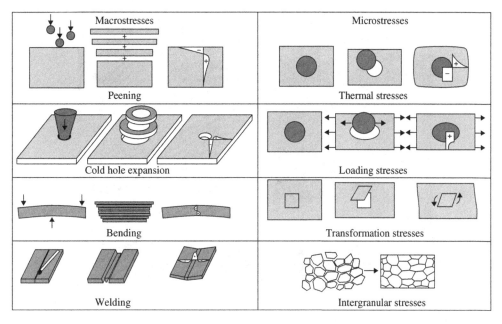

Figure 1.2 Examples of some typical ways in which residual stresses are created in engineering materials. Reproduced with permission from [2], Copyright 2001 Maney

Solidification and differential shrinkage cause large tensile and compressive residual stresses in welds. The weld metal is stress-free while molten, and can support residual stresses only after solidification. The very hot weld metal and heat-affected zone (HAZ) cool over a larger temperature range than the surrounding cooler material and therefore shrinks more. Thus, to maintain dimensional continuity through compatible longitudinal strains, large longitudinal tensile residual stresses are created in the weld metal and HAZ balanced by compressive stresses in the surrounding material.

1.1.2 Effects of Residual Stresses

Because of their self-equilibrating character, the presence of residual stresses may not be readily apparent and so they may be overlooked or ignored during engineering design. However, they are stresses and must be considered in the same way as stresses due to external loading [6].

In terms of material strength, the main effect of residual stresses is as an addition to the loading stresses. The contribution of the residual stresses can be beneficial or harmful, dependent on the sign and location of the residual stresses. For example, the surface compressive residual stresses in the toughened glass shown in Figure 1.1 strengthen the overall structure because glass is brittle and has low tensile strength. The failure mechanism is by crack growth, but most cracks (scratches) are at the surfaces. Thus, the compressive residual stresses act to bias the loading stresses towards compression in the areas of the tension-sensitive surface cracks. There are few if any cracks in the central region and so

Figure 1.3 Cracking in a cast aluminum ingot due to excessive residual stresses. Courtesy of Alcoa Inc.

the material there can tolerate the elevated local tensile stresses. The resultant effect of the combined stresses is an increased capacity of the glass component to support external loads. A similar concept applies to shot peening, where impacting a surface with small hard balls induces surface compressive stresses. An increased fatigue life is achieved by biasing the mean of the varying stresses at the surface towards compression, where fatigue cracks usually initiate.

Residual stresses can also be harmful and significantly reduce material strength and cause premature fracture. Figure 1.3 shows longitudinal fractures in an aluminum alloy direct chill cast ingot, a precursor to hot rolling. The fractures are caused by residual stresses induced by inhomogeneous cooling after solidification during casting.

Some further examples of harmful effects of residual stresses are:

- Corrosion fatigue fracture of heart valves caused nearly 200 fatalities due to residual stresses induced by the bending of retainer struts during fabrication [9].
- Fatigue fractures enhanced by circumferential tensile residual stresses on rivet holes in a Boeing 737 caused the top half of the fuselage to be torn away with the loss of a flight attendant and injury to 65 passengers [10].
- Stress corrosion cracking of heat exchanger tubes in nuclear reactors caused loss of power production [11].

1.1.3 Residual Stress Gradients

Because residual stresses are non-zero but have zero force resultant, they must be non-uniform, sometimes quite substantially so, with large stress gradients. Figure 1.4

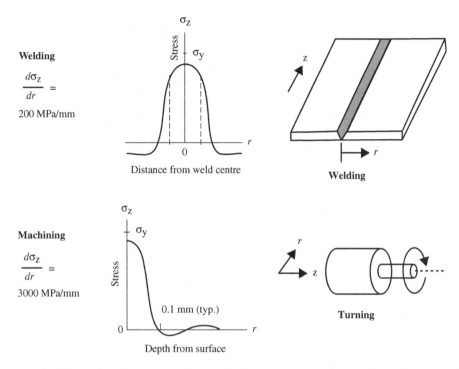

Figure 1.4 Schematics illustrating typical residual stress gradients induced by various manufacturing processes

shows two examples of typical stress gradients found in manufactured components. The first shows welding residual stresses and indicates a stress gradient of ∼200 MPa/mm adjacent and parallel to the weld. The second example shows a machining stress gradient of ∼3000 MPa/mm from the surface to about 0.1 mm in depth.

Because of concerns for premature failure through fatigue and stress corrosion cracking, and because of the high stress gradients and the uncertainty of the area of highest stresses, it is often necessary to make many stress measurements on as small a number of elements of the component as possible. Thus, the spatial resolution and thickness of the measurement volume is an important consideration in most residual stress investigations, as is measurement speed and cost.

1.1.4 Deformation Effects of Residual Stresses

If a component containing residual stresses is cut in some way, the stresses with force components acting on the cut surface will relieve and the stresses within the remaining material will redistribute to maintain interior force equilibrium. The strains associated with the stress redistribution cause the component to distort, sometimes quite substantially [6,7]. Figure 1.5 shows an example of an aircraft cargo ramp that had a major fraction of material removed to reduce structural weight. The particular forging contained residual

Figure 1.5 C-17 cargo ramp warped by the release of residual stresses from material removed during the manufacturing process. Courtesy of D. Bowden (Boeing Company)

stresses that were excessively large and/or very widespread, whose relief during machining caused the dramatic deformation shown in the photo. This deformation became apparent after the component was detached from the milling machine worktable.

Deformation of machined components due to release of residual stresses can be a serious problem, particularly when high dimensional precision is required. The most direct solution is to reduce the size of the residual stresses present either during material manufacture or by subsequent heat treatment. A further approach is to machine components incrementally, preferably symmetrically, and gradually converge on the desired dimensions.

The deformation caused by the residual stress redistribution after material cutting provides the basis of a major class of residual stress measurement methods, commonly called "relaxation" methods or "destructive" methods [2,3]. By measuring the deformations after the material has been cut in some way, the originally existing residual stresses can be mathematically determined. Chapters 2–4 describe some well-established "relaxation" type residual stress measurement techniques. Although seemingly less desirable because they typically damage or destroy the measured specimen, the relaxation methods are very versatile and so are often the method of choice. The non-destructive residual stress measurement techniques described in Chapters 6–10 provide further approaches, particularly useful when specimen damage is not acceptable.

1.1.5 Challenges of Measuring Residual Stresses

The "locked-in" character of residual stress makes them very challenging to evaluate, independent of the measurement technique used. Even with stresses caused by external loads, measurements are indirect; a proxy such as strain or displacement is measured, from which stresses are subsequently interpreted. The typical procedure is to make comparative measurements on the structure without and with the external load applied and then evaluate stresses based on the difference of the measurements. However, residual stresses cannot simply be removed and applied. When using the relaxation measurement methods described in Section 1.2, residual stresses are "removed" by physically cutting away the material containing those stresses. A complication introduced by this approach is that the stress-containing material is destroyed and measurements must therefore be made on the adjacent remaining material. This separation of stress and measurement locations creates mathematical challenges that require specialized stress evaluation methods [44,45]. The non-destructive measurement techniques described in Sections 1.3 and 1.4 typically avoid any material removal and some must use some identification of a "stress-free" reference state when interpreting measurements made with intact residual stresses. Achieving such reference states can be quite challenging to do reliably. A consequence of all these challenges is that measurements of residual stresses do not typically reach the accuracy or reliability possible when working with applied stresses. However, the various residual stress measurement methods are now quite mature and the accuracy gap is often not very large.

1.1.6 Contribution of Modern Measurement Technologies

Most of the residual stress measurement methods described in the subsequent chapters are well established and have long histories. However, high-precision machinery and modern instrumentation have enabled such substantial advances in experimental technique and measurement quality that the modern procedures are essentially "new" methods when compared with the early versions. Modern computer-based computation methods have similarly revolutionized residual stress computation capabilities, allowing stress evaluations that were far beyond reach in earlier times. In the subsequent sections of this chapter and in the following chapters, various residual stress measurement and computation techniques are considered. The features and applications of each method are described, also their expected evaluation accuracy and potential concerns. Figure 1.6 summarizes several of the methods in terms of their spatial resolution and their ability to make residual stress measurements deep within a specimen, the "penetration." It is evident that several factors need to be carefully considered and balanced to make an appropriate choice of a residual stress measurement method for a given application.

1.2 Relaxation Measurement Methods

1.2.1 Operating Principle

Figures 1.3 and 1.4 directly show the structural deformations that accompany the stress redistribution that occurs when residual stresses are released by cutting or material removal. These deformations ("relaxations") are typically elastic in character, and so there is a linear relationship between the deformation size and the released residual stresses. This observation provides the basis for the "relaxation" methods for measuring residual stresses [3,4,7]. While many different measurement technologies, specimen and cutting geometries are used, all methods seek to identify residual stresses from the

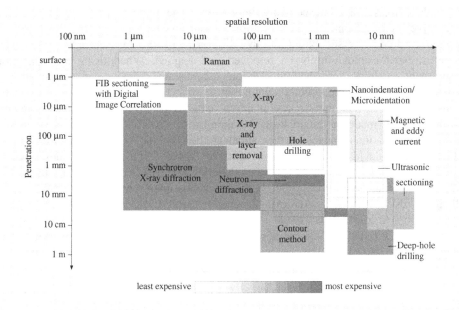

Figure 1.6 Measurement penetration vs. spatial resolution for various residual stress measurement methods. Courtesy of Michael Fitzpatrick, Open University, UK

measured deformations caused by material cutting or removal, hence the alternative name, "destructive" methods. For some specimen geometries the deformation/stress relationship can be determined analytically, other times finite element calibrations are needed. In almost all cases the deformation/stress relationship is made complicated by the characteristic that the stress is removed from one region of the specimen while the measurements are made on a different region where only partial stress relief occurs. Chapter 12 describes some mathematical approaches to handling this situation.

Many "relaxation" methods for measuring residual stresses have been developed over the years for both general and specific types of specimens. Despite their large differences in geometry and experimental technique, all methods share the concept of measurement of deformation caused by local cutting of stressed material. This section gives a brief overview of a range of relaxation methods for measuring residual stress. Chapters 2–5 give more extended details of the most commonly used of the general-purpose methods.

The splitting method [12,13] mimics the deformations seen in material cracking due to excessive residual stresses, such as seen in Figure 1.3. A deep cut is sawn into a specimen such as in Figure 1.7(a) and the opening or closing of the adjacent material indicates the sign and the approximate size of the residual stresses present. This method is commonly used as a quick comparative test for quality control during material production. The same testing geometry is used for the "prong" test for assessing stresses in dried lumber [14].

Figure 1.7(b) shows another variant of the splitting method, used to assess the circumferential residual stresses in thin-walled heat exchanger tubes. This procedure is also a generalization of Stoney's Method [15], sometimes called the curvature method. It

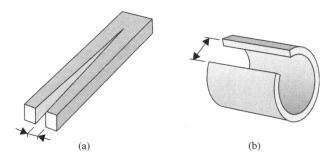

Figure 1.7 The splitting method (a) for rods and (b) for tubes

involves measuring the deflection or curvature of a thin plate caused by the addition or removal of material containing residual stresses. The method was developed for evaluating the stresses in electroplated materials, and is also useful for assessing the stresses induced by shot-peening.

The sectioning method [16,17] combines several other methods to evaluate residual stresses within a given specimen. It typically involves attachment of strain gages, or sometimes the use of diffraction measurements (see Sections 1.2 and 1.3) and sequentially cutting out parts of the specimen. The strain relaxations measured as the various parts are cut out provide a valuable source of data from which both the size and location of the original residual stresses can be determined. Figure 1.8 shows an example where a sequence of cuts was made to evaluate the residual stresses in an I-beam [17].

The layer removal method [18] involves observing the deformation caused by the removal of a sequence of layers of material. The method is suited to flat plate and cylindrical specimens where the residual stresses vary with depth from the surface but are uniform parallel to the surface. Figure 1.9 illustrates examples of the layer removal method, (a) on a flat plate, and (b) on a cylinder. The method involves measuring deformations on one surface, for example using strain gages, as parallel layers of material are removed from the opposite surface. In the case of a cylindrical specimen, deformation measurements can be made on either the outside or inside surface (if hollow), while annular layers are removed from the opposite surface. When applied to cylindrical specimens, the layer removal method is commonly called "Sachs' Method" [19].

The hole-drilling method [20], described in detail in Chapter 2, is probably the most widely used relaxation method for measuring residual stresses. It involves drilling a small hole in the surface of the specimen and measuring the deformations of the surrounding surface, traditionally using strain gages, and more recently using full-field optical techniques, (see Chapter 11). Figure 1.10(a) illustrates the process. The hole-drilling method is popular because it can give reliable and rapid results with many specimen types, and creates only localized and often tolerable damage. The measurement procedure is well developed [21,22] and can identify the through-depth profile of the in-plane residual stresses to a depth approximately equal to the hole radius. It is now standardized as ASTM E837 [20].

The ring-core method [23,24] is an "inside-out" variant of the hole-drilling method where the "hole" is around the outside and the measurements on the inside. Figure 1.10(b) illustrates the geometry. The ring-core method has the advantage over the hole-drilling

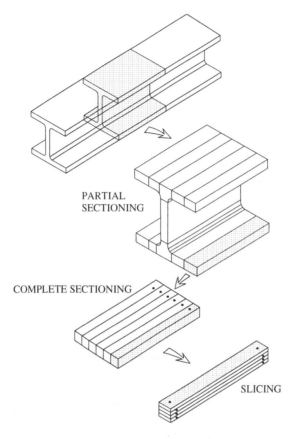

Figure 1.8 Sectioning method. Reproduced with permission from [17], Copyright 1973 Springer

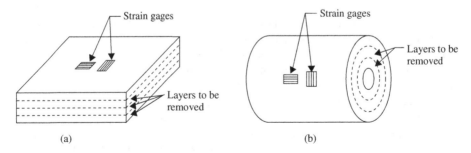

Figure 1.9 Layer removal method (a) flat plate and (b) cylinder. Reproduced with permission from [7], Copyright 2010 Springer

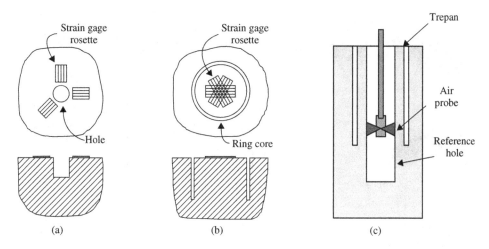

Figure 1.10 Hole-drilling methods: (a) conventional hole-drilling method, (b) ring-core method and (c) deep-hole method. Reproduced with permission from [7], Copyright 2010 Springer

method that it provides much larger surface strains and can identify larger residual stresses. However, it creates much greater specimen damage and it is more difficult to implement in practice.

The deep-hole method [25,26] is a further variant procedure that combines elements of both the hole drilling and ring-core methods. It involves drilling a hole deep into the specimen, and then measuring the diameter change as the surrounding material is overcored. Figure 1.10(c) illustrates the geometry. The main feature of the method is that it enables the measurement of deep interior stresses. The specimens can be quite large, for example, steel and aluminum castings weighing several tons. On a yet larger scale, the deep-hole method is often used to measure stresses in large rock masses. Chapters 2 and 3 describe the hole-drilling, ring-core and deep-hole methods in more detail.

The slitting method [27,28], illustrated in Figure 1.11, is also conceptually similar to the hole-drilling method, but using a long slit rather than a hole. Alternative names are the crack compliance method, the sawcut method or the slotting method. Strain gages are attached on the front or back surfaces, or both, and the relieved strains are measured as the slit is incrementally increased in depth using a thin saw, milling cutter or wire EDM.

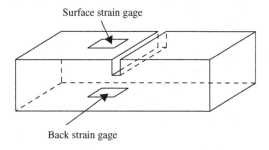

Figure 1.11 Slitting method. Reproduced with permission from [7], Copyright 2010 Springer

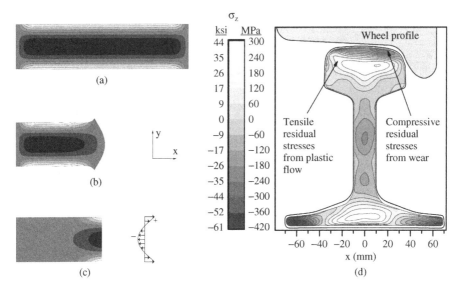

Figure 1.12 Contour Method (a) original stresses, (b) stress-free after cutting, (c) stresses to restore flat surface and (d) measured stress profile of a railway rail. Diagrams courtesy of Michael Prime, Los Alamos National Labs, USA

The slitting method has the advantage over the hole-drilling method that it can evaluate the stress profile over the entire specimen depth. However, it provides only the residual stresses normal to the cut surface. Chapter 4 describes the slitting method in more detail.

The Contour Method [29,30], illustrated in Figure 1.12(a–c) is a newly developed technique for making full-field residual stress measurements. It involves cutting through the specimen cross-section using a wire EDM, and measuring the surface height profiles of the cut surfaces using a coordinate measuring machine or a laser profilometer. The residual stresses shown in Figure 1.12(a) are released by the cut and cause the material surface to deform (pull inwards for tensile stresses, bulge outward for compressive stresses), as shown in Figure 1.12(b). The originally existing residual stresses normal to the cut can be evaluated from finite element calculations by determining the stresses required to return the deformed surface shape to a flat plane. In practice, to avoid any effects of measurement asymmetry, the surfaces on both sides of the cut are measured and the average surface height map is used. The contour method is remarkable because it gives a 2D map of the residual stress distribution over the entire material cross-section. Figure 1.12(d) shows an example measurement of the axial residual stress profile within the cross-section of a railway rail [31]. In comparison, other techniques such as layer removal and hole-drilling give one-dimensional profiles. Chapter 5 describes the contour method in more detail.

The various relaxation techniques differ greatly in their characteristics, for example their applicable specimen geometry, their cutting procedure, measurement procedure, residual stress components identified, spatial resolution, and so on. Sometimes the nature of the specimen dictates a specific test procedure, but often a judgment needs to be made to select an advantageous measurement method. Section 1.7 of this chapter describes some practical strategies for measurement method choice.

1.3 Diffraction Methods

The diffraction methods provide the possibility for non-destructive procedures to measure residual stresses. Section 1.4 describes some further methods that can be non-destructive. "Non-destructive" implies that the component may be returned to service after the residual stresses are measured and the stress fields evaluated. Thus, either the measuring instrument must be portable and sufficiently compact to be brought to the component, or the component must be brought to the instrument, intact and without sectioning. In addition, most of the methods described in Sections 1.3 and 1.4 provide the high spatial resolution needed to resolve high stress gradients.

1.3.1 Measurement Concept

Diffraction methods exploit the ability of electromagnetic radiation to measure the distance between atomic planes in crystalline or polycrystalline, materials. When any external mechanical or thermal load is applied or incompatible strains occur, the material deforms in response. This deformation is linear when the response is in the elastic range. The diffraction methods effectively measure a crystal inter-planar dimension that can be related to the magnitude and direction of the stress state existing within the material. These measurements are independent of whether that stress is residual or applied.

Diffraction of electromagnetic radiation occurs when the radiation, typically X-rays and neutrons for residual stress measurements, interact with atoms or crystallites that are arranged in a regular array, for example atoms in crystals. The radiation is absorbed and then reradiated with the same frequency such that strong emissions occur at certain orientations and minimal emissions at other orientations. The angles at which the strong emissions occur are described by Bragg's Law:

$$n\lambda = 2d \sin\theta \qquad (1.1)$$

where n is an integer, λ is the wavelength of the electromagnetic radiation, d is the distance between the diffracting planes (inter-atomic lattice spacing) and θ is the Bragg angle. Figure 1.13 illustrates these quantities. It can be seen that Equation (1.1) describes the condition where the additional path length of diffracted radiation (the three line segments shown in Figure 1.13) from each crystal plane is an integer number of radiation wavelengths. Thus, the radiation components diffracted by the various lattice planes emerge in phase.

For stress measurement using X-ray and neutron diffraction a range of θ angles are scanned and the angle at which the most intense radiation is detected is established as the Bragg angle. Small changes in the corresponding d-spacing that tend to broaden the diffracted peak reflect Type II and Type III stresses. For synchrotron diffraction, θ is sometimes held constant and the detector scans a range of energies to determine the λ that meets the Bragg condition. The measured lattice strains are "absolute" quantities, that is, relative to a zero-strain datum. This is a significant feature of diffraction methods because it allows residual stresses to be measured as well as applied stresses. In contrast, strain gages can only measure the differential strains associated with applied stresses, that is, the strain difference between the initial condition when the strain gage was attached and some subsequent condition.

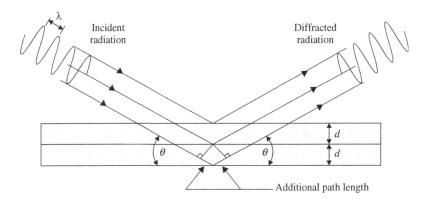

Figure 1.13 Radiation diffraction within a crystal structure $d =$ spacing between lattice planes, $\theta =$ Bragg angle, and $\lambda =$ wavelength of the radiation

1.3.2 X-ray Diffraction

The X-ray diffraction (XRD) techniques are capable of measuring the inter-atomic lattice spacing, which is indicative of the strain in the irradiated area. The SAE International has published an excellent handbook supplement on XRD stress measurement [32].

The most commonly used X-ray wavelengths applied in stress measurement are not capable of penetrating deeply into most materials. Usually, characteristic X-rays from a specific anode or target of X-ray tube, for example, copper, chromium and iron are used, with wavelengths ranging from 0.7 to 2 Ångstroms (18 to 5 keV in energy). Typically, X-ray penetration is of the order of 0.025 mm and thus in most cases is considered a surface stress measurement method. X-ray techniques have become the most widely used techniques for evaluating these stresses [32].

In the XRD method, the X-radiation only penetrates a few microns. This shallow depth accommodates the assumption that the stress normal to the surface is zero. Thus, the irradiated volume or gage volume, under investigation is considered to be in a state of plane stress. This condition allows for simplification of the stress-strain equations and avoids the need for precise determination of the unstressed lattice plane dimension. Thus, simplified equations using the difference between specific lattice planes at several angles to the surface plane are used to extrapolate the strain condition to a vector in the plane of the surface [32].

Despite the fact that X-rays provide stress evaluations only to a depth of about 0.025 mm, the non-contact X-ray diffraction techniques are presently the only generally applicable, truly non-destructive techniques for measuring surface residual stresses. Their reliability has been extensively demonstrated and documented; see, for example, the handbook supplement published by SAE [32].

There are three ASTM Standards relating to XRD stress measurement. These are E915-10 for verifying the alignment of an instrument, E1426-98 for determining the effective elastic parameter of the material, and E2860-12 for XRD stress measurements in bearing materials.

Instrumentation for bringing XRD measurements from the laboratory environment to field applications has advanced rapidly in the last two decades, especially toward increased portability, compactness and speed of operation [32,33]. These instruments are referred to as X-ray diffraction stress instrumentation and are instruments specifically designed for X-ray stress measurements and are not to be confused with conventional X-ray diffraction instruments modified with stress measurement attachments. Measurement times of a few seconds and spatial resolutions of less than one square mm are possible with X-ray diffraction, as well as automated stress mapping [34].

1.3.3 Synchrotron X-ray

This method uses X-rays as does the traditional XRD method, however, the X-rays are much more intense and of much higher energy, and because of their high energy they penetrate much deeper, on the order of mm, into materials. This deep penetration allows for the measurement of bulk stresses of which traditional XRD is incapable. The source of the X-rays are synchrotron radiation facilities, and "hard" X-rays greater than 50 keV in energy (less than 0.25 Ångstrom wavelength) are commonly used. In comparison, traditional XRD uses "soft" X-rays ranging from 18 to 5 keV in energy (0.7 to 2 Ångstroms in wavelength). The strain measurement is also based upon Equation (1.1), but sometimes the Bragg angle is held constant and the energy of the diffracted X-rays is measured.

The energy that is diffracted changes with the strain (change in lattice spacing) and is measured with an energy dispersive X-ray detector. The synchrotron method has become possible through advances in detector technology allowing better precision in energy measurement [35]. Low Bragg angles on the order of 10 degrees are used.

Because the higher energy radiation from synchrotron sources penetrate many tens or even hundreds of mm into a material, the stress condition on the irradiated area, gage volume, cannot be considered under plane stress and the full three dimensional stress condition must be considered. Thus, strain is determined by subtracting the unstressed lattice parameter from the parameter under stress. This results in the necessity for the unstressed lattice parameter to be precisely known, and this requirement can introduce significant uncertainty into the strain measurement. The unstressed lattice parameter can be affected by composition (alloy content), phase composition, and other factors and thus is not often precisely known. Thus, the uncertainty of the unstressed lattice parameter is a major source of error in synchrotron diffraction residual stress measurement. The synchrotron X-ray method has most of the same limitations as neutron diffraction (see Section 1.3.4). Measurement times as small as msec and resolution of stresses in gage volumes in the cubic micrometer range are possible.

1.3.4 Neutron Diffraction

Neutron diffraction (ND) uses penetrating radiation as do the previous two diffraction methods, however, neutrons interact directly with the nucleus of the atom, and the contribution to the diffracted intensity is different than for X-rays which interact with electrons. It is also often the case that light (low atomic number, Z) atoms contribute just as strongly to the neutron diffracted intensity as do large Z atoms, and thus neutrons penetrate equally well into high Z material as low Z. The scattering characteristic varies irregularly from isotope to isotope rather than linearly with the atomic number. An element like iron is a strong scatterer of X-rays, but relative to X-rays it scatters less and thus neutrons can penetrate several cm into the material. This penetration allows for the measurement of bulk stresses. Neutrons for ND may be generated by fission or spallation. Neutron energies that provide wavelengths in the range from 0.7 to 3 Ångstroms are commonly used in ND stress measurement.

Neutron diffraction is capable of measuring the elastic strains induced by residual stresses throughout the volume of components of thickness in the 0.1–1.5 m range with a spatial resolution less than one mm. Such capabilities provide for the measurement of residual stress inside of components without the necessity of sectioning or layer removal. ND methods, as with XRD methods, measure the Bragg angle of the scattered radiation; which is related to the spacing between crystallographic planes. And this spacing is affected by residual and applied stress. However, unlike traditional XRD techniques, the accuracy of ND techniques requires that the unstressed lattice spacing of the measured crystallographic planes be precisely known since the stress state is, in principle, tri-axial at depth.

As with the synchrotron method the unstressed lattice spacing at the point of strain measurement must be known but is not easily measured. This problem is aggravated by the fact that the elemental composition, and thus the lattice spacing, vary considerably within a component made from an alloy. Additional limitations are that the component must be brought to a neutron source, each strain measurement requires several minutes to over an hour, a single stress determination in one small gage volume of the component (on the order of one cubic mm) requires at least three strain measurements, and the measurements are costly.

Nevertheless, the ND methods have been applied to residual stress measurements in weldments, rolled rods, plastically deformed plate [36], rocket case forgings, [37], and many other types of components.

1.4 Other Methods

The category "Other Methods" includes non-destructive and semi-destructive techniques that rely on the measurement of some property affected by strain.

1.4.1 Magnetic

The most widely applied magnetic residual stress measurement method uses Magnetic Barkhausen Noise (MBN) [38]. MBN analysis involves measuring the number and magnitude of abrupt magnetic re-orientations made by the magnetic domains in a ferromagnetic material during magnetization reversal. These reorientations are observed as pulses that are random in amplitude, duration and temporal separation, and therefore roughly described as noise. Many engineering materials are not ferromagnetic, so MBN is not applicable to them. Nevertheless, most steels and some ceramics are ferromagnetic and will produce MBN.

The MBN technique has a somewhat limited range of stress sensitivity, around ±300 MPa, and a few mm depth of measurement. The latter condition might be relieved by using Acoustic Barkhausen Emission (ABE), an ultrasound analog to MBN. However, the sensitivity of either of these techniques to other properties and characteristics of metallic components and the consequent need for calibration with a nearly identical specimen severely compromises and restricts MBN's reliability, accuracy, and applicability. Measurement times of a few seconds and spatial resolutions in the sub-mm range are possible with magnetic methods.

1.4.2 Ultrasonic

The ultrasound acoustic methods, described in more detail in Chapter 10, offer many techniques that have the potential to determine material properties such as crystallographic orientation, grain size, phase composition, and stress (residual and applied). Fundamentally, an ultrasonic (acoustic) wave is induced in the material and the reflected, transmitted, or scattered wave is measured. For stresses, the velocity of some mode of ultrasonic wave is measured. The forms include longitudinal, transverse and surface waves.

Ultrasonic technology offers a number of wave modes in which to probe materials. These include bulk waves such as longitudinal and shear, and surface waves typically of the Rayleigh type. Each mode offers many unique parameters for extracting information. The primary effect of stress-induced strain on ultrasonic propagation in metals is on velocity. This may be detected in a number of ways, including measurements of wave velocity, shear wave birefringence, and dispersion.

The uncertainty in the application of ultrasound to residual stress measurement is that the velocity of ultrasound is not only affected by stress, but also is affected by many other micro-structural characteristics such as grain size, second phases, crystallographic texture, and so on. The effect of these other characteristics compromises the accuracy of ultrasonic stress measurement method. Nevertheless, in spite of the effect of micro-structural variations in manufactured products, success in the application of ultrasonic methods to residual stress measurement has been achieved in several specific cases [39,40]. Measurement times of a few seconds and volume resolutions of several cubic mm are possible with the ultrasonic method.

1.4.3 Thermoelastic

Thermoelastic residual stress measurement derives from a coupling between mechanical elastic strain and changes in thermal energy of an elastic material. The rate of change in temperature of a dynamically loaded body depends on the rate of change of the principal stress sum under adiabatic conditions. A SPATE (Stress Pattern Analysis by measurement of Thermal Emission) method was developed to detect changes in infrared emission due to minute changes in the temperature of dynamically stressed material. These temperature changes are also sensitive to static stresses, including residual [41]. However, the very small change in the thermal effect due to residual stress presently restricts the application of the method. Thus, the thermoelastic method tends to be qualitative and thus most suited to comparing components to establish whether they differ from one another significantly.

1.4.4 Photoelastic

Photoelastic methods are also referred to as birefringent methods. Under the action of stress, many transparent materials become birefringent, that is, their refractive index changes in a directional manner. If polarized light is transmitted through the stressed material, a colored fringe pattern is produced illustrating the shear stresses within the material [42]. The method is limited to transparent materials. It is commonly used for stress analysis within transparent model structures, but can also be used for residual stress measurements. A common use is in high-quality lenses, where the birefringence caused by residual stresses can significantly impair image quality. With calibration, quantitative measurement of shear stresses can be achieved. A variation on the method applicable to opaque materials is to coat the component with a photoelastic polymer. Then, when stress changes are induced in the component, for example, by the hole-drilling method, strain is induced in the coating and the resulting birefringence can be observed and measured with a reflection polariscope.

1.4.5 Indentation

The presence of residual stresses slightly influences the material hardness values measured using indentation methods, where tensile stresses tend to decrease apparent hardness and compressive stresses increase it. This influence provides a possible means to measure residual stresses. Many studies over the years have sought to refine the indentation technique to obtain more quantitative results [43]. However, the measured effect is small and the accuracy of the indentation method is at present relatively low compared to XRD and hole drilling. Measurement times of a few minutes and spatial resolutions of less than one square mm are possible with indentation methods.

1.5 Performance and Limitations of Methods

1.5.1 General Considerations

Most stress measurement methods do not directly measure stress; instead they measure some proxy for stress such as strain or displacement. A mathematical conversion from the proxy data then needs to be done to complete the desired stress identification. In the case of evaluation of load-induced stresses, the stress identification can be as simple as multiplying measured strains by Young's modulus. However, for residual stresses, their locked-in, self-equilibrating presence in the absence of external loads can significantly complicate the needed mathematical calculations [44,45].

The indirect stress evaluation approach imposed by the locked-in, self-equilibrating character of residual stresses causes residual stress measurements to be less precise than load-induced stress measurements. In the case of the relaxation measurement methods, a further degree of indirectness is produced by the loss of the stressed material that is cut or removed, and the consequent need to infer those stresses from the resulting changes in the surrounding material. The diffraction and other methods have their own challenges, notably that the proxy measurements they use can also be indirect and so require detailed and extensive calibrations. For example, ultrasonic and magnetic methods require detailed calibration very closely tailored to the specific specimen material, and even small material variations may have serious adverse effects. In summary, measurements of residual stresses will not in general reach the level of precision of measurements of load-induced stresses and should not be expected to do so.

A further consideration is the balance between measurement accuracy and spatial resolution. Often, the two desirable features are mutually conflicting, with greater spatial resolution coming at the cost of reduced precision, and vice versa. Accuracy is typically improved by data averaging, but this tends to reduce spatial resolution. Conversely, the desire for improved spatial resolution requires measurement of small differences between data relating to nearby measurement points. Small errors in the absolute measurements cause large relative errors in their difference and hence large stress evaluation errors. Making many measurements and using averaging/smoothing techniques can achieve simultaneous advances in both desired features. In this respect, the non-destructive methods have an advantage because the measurements can be repeated at will and further data acquired.

1.5.2 Performance and Limitations of Methods

Table 1.1 summarizes some general comments about common residual stress measurement methods. This table should be interpreted as an outline guide only because of the wide variability of practice and application of the various methods listed. For many methods, few if any formal accuracy evaluations have been conducted, so most quoted numerical values are just personal estimates. The lower end of the precision estimates in the second column of Table 1.1 indicates the likely measurement precision of the various listed methods. An experienced specialist who has made many measurements over an extended period may be expected to achieve precision results at the lower end of the scale, while generalists with less experience may expect precisions in the middle to upper range. This is not to suggest a lack of skill or care on the part of the generalist, but just that many of the methods, although appearing straightforward in concept, have subtle details that must be learned from experience to achieve the most refined results.

1.6 Strategies for Measurement Method Choice

1.6.1 Factors to be Considered

The preceding sections summarize many different residual stress measurement methods, and several more exist for specialized measurement needs. The question then arises as to how one should choose an appropriate measurement technique for a given measurement need. The answer is often non-unique and depends on several different factors, some of

Table 1.1 Characteristics of various residual stress measurement methods. Percent accuracy estimates are for measurements of large stresses (~50% of yield stress), percentage numbers will be greater when measuring small stresses. For the most part the precision values given are for steel alloys

	Precision	Depth Penetration	Works Best For	Has Limitations With
Splitting	20–50%	Specimen thickness	Routine comparative quality control	Non-uniform and untypical stresses
Sectioning	10–30%	Specimen thickness	Specimens with more regular-shaped geometry	Challenging calculations for multiple sectioning
Stoney	5–20%	Layer thickness	Thin layers on flexible beams	Determining layer thickness accurately
Layer Removal	10–30%	Specimen thickness	Flat plates and cylinders of uniform thickness	Time consuming procedure subject to measurement drift
Hole Drilling (uniform stress)	5–20%	Up to 2 mm typically	Near surface measurements of in-plane uniform stresses	Stresses often are not uniform, max. stress = 70% of yield
Hole Drilling (stress profile)	10–30%	Up to 2 mm typically	Near surface measurements of in-plane stress profiles	Sensitivity to noisy data, max. stress = 70% of yield
Deep Hole	5–15%	Specimen thickness	Large components	Now done only by specialists and compromised by plasticity
Slitting	5–20%	Specimen thickness	1-D perpendicular stress in prismatic shaped specimens	Stresses that are non-uniform across width

(*continued*)

Table 1.1 (*continued*)

Contouring	5–20%	Specimen X-section	2-D perpendicular stress in prismatic shaped specimens	Requires very accurate cutting, not good for near-surface
X-ray Diffraction	~20 MPa	<0.03 mm	Near surface measurements on crystalline materials	Variations in grain structure and surface texture
Synchrotron Diffraction	~50 MPa	>5 mm	Deeper non-destructive measurements than X-ray	Requires synchrotron radiation source and zero stress reference
Neutron Diffraction	~50 MPa	25 mm steel 100 mm Al.	Deeper non-destructive meas. than synchrotron	Requires neutron radiation source and zero stress reference
Magnetic BNA	>25 MPa	1 mm	Ferromagnetic materials only	Requires material-specific calibration
Ultrasonic	>25 MPa	1–20 mm	Low-cost comparative measurements	Requires material-specific calibration
Thermoelastic	Qualitative	Varies	Low-cost comparative measurements	Results are not quantitative
Photoelastic	10–30%	Specimen thickness	Full-field measurements in transparent materials	Transparent materials, Results are not quantitative
Indentation	Qualitative	<1 mm	Comparative measurements	Results are not quantitative

which may be conflicting. In some cases, more than one method may be suitable, and in other cases, no method may be entirely satisfactory. Some major factors to consider are:

1. **Measurement Objective?** Is the measurement to investigate the propensity to distortion or premature fracture? The measurement method must be able to provide the residual stresses in the needed directions, in the needed places and at the needed quantitative and spatial resolution.
2. **Specimen damage: acceptable, not acceptable?** If either partial or complete damage to the specimen is acceptable, then the relaxation methods are often an attractive choice because they are very adaptable and give good results with a wide range of materials and specimen geometries. Diffraction, ultrasonic or magnetic methods are also good candidates, notably when specimen damage is unacceptable.
3. **Specimen shape: simple geometry, complex shape?** Simple specimen geometries often lend themselves to specific measurement methods, for example, tube splitting for thin-wall tubes, and layer removal for flat plates. Other general-purpose methods such as hole drilling and X-ray diffraction can be used with non-specific specimen shapes.
4. **Specimen dimensions: bench top size, very large, very small?** Most residual stress measurements are done where the specimen and/or the measurement equipment can fit on a laboratory bench top. Most measurement methods fit this category. Many measurement methods can be scaled both up and down substantially, likely with adaptations to the measurement technique used. Synchrotron and neutron diffraction are particularly suited to measuring residual stresses in the interior of large metal specimens (10–100 mm). Deep-hole drilling, slitting, contour and sectioning are suited to even larger specimens (100–1000 mm).
5. **Measurement environment: lab or field use?** Almost all measurement methods were originally developed as laboratory techniques. Some require carefully controlled conditions and may only be done in a laboratory, for example, the contour method and measurements involving laser interferometry or nuclear reactors. However, other measurement methods can be adapted for field use, for example, strain gage hole-drilling, X-ray diffraction, ultrasonic and magnetic measurements. Several manufacturers make portable equipment specifically designed for field applications and for stress measurements on moving parts.
6. **Availability of required equipment and experience in doing the measurements.** This is often a deciding factor. It is certainly advantageous to use a measurement method that is familiar and whose characteristics and limitations are well understood. These features can easily outweigh the potential advantages of a nominally superior but unfamiliar method. However, taking a "one size fits all" approach can be a dangerous strategy, so favoring method familiarity should not be taken too far.
7. **Nature of the residual stresses: uniform, rapidly varying, surface, interior?** Several measurement methods require residual stresses to be uniform over significant distances, for example layer removal and slitting, while others are quite localized, for example X-ray diffraction and hole-drilling. XRD is also useful for surface measurements and hole-drilling for near-surface measurements. Conversely, the contour method, slitting, and neutron diffraction can evaluate interior stresses.
8. **Specimen material: metal, crystalline, amorphous, high and low yield strength?** The relaxation type measurements can be used with most materials, although

Table 1.2 Characteristics of various residual stress measurement methods. X = feature is somewhat present, XX = feature is substantially present

	Low Damage	Field Use	General Geometry	General Material	Quick, Low Cost	Quantitative	Stress Profile	Near Surface	Deep Interior
Splitting				XX	XX				
Stoney	X	XX		XX	XX	X		XX	
Sectioning			X	XX		X	XX	X	XX
Layer Removal			X	XX		XX	XX	X	XX
Hole Drilling	X	XX	XX	XX	X	XX	X	XX	
Deep Hole	X		XX	XX		XX	XX		XX
Slitting			XX	XX		XX		XX	XX
Contouring			XX	XX		XX		X	XX
X-ray Diffraction	XX	XX	XX	X	X	XX	XX	XX	
Synchrotron Diffraction	XX		XX	X		X	X		XX
Neutron Diffraction	XX		XX	X		X	X		XX
Magnetic BNA	XX	XX	XX		XX	X	XX	XX	
Ultrasonic	XX	XX	XX	XX	XX	X	XX	XX	
Thermoelastic	XX	X	X	X	X		X	X	
Photoelastic	XX	X	X		X		X		X
Indentation	X	XX	X	XX	XX		X	XX	

sometimes materials with very high or low yield strength can present a challenge for making stress-free cuts. Crystalline materials are required for the diffraction type of measurements, and ferromagnetic materials are required for the magnetic methods. Coated components require careful consideration to achieve effective results.

9. **Accuracy and spatial resolution: detailed or approximate results needed?** Some methods are designed for rapid quality control purposes, for example, tube splitting. Others such as magnetic and ultrasonic can give quantitative results only after calibration for the specific material. Accuracy and spatial resolution are often conflicting needs, more demand on one often tends to diminish the other. Simultaneous advances in both can be achieved by making many measurements and using averaging/smoothing techniques. References 44 and 45 describe some mathematical techniques.
10. **Cost and duration of test procedure: specimen value, number of evaluations?** A costly measurement such as neutron diffraction can be justified when a small number of measurements are required on a valuable specimen. At the other end of the scale, low cost measurements such as ultrasonic or magnetic and sometimes XRD are appropriate for production line use.
11. **Other important features: radioactive, high temperature, and so on**. These factors must be carefully considered according to the particular circumstances and may force the use of an otherwise non-optimal measurement method.

1.6.2 Characteristics of Methods

Table 1.2 summarizes some significant characteristics of the residual stress measurement methods discussed in this chapter. It can provide a useful starting point for more detailed investigations of measurement method choice.

References

Residual Stresses

[1] Totten, G., Howes, M., Inoue, T. (eds) (2002) *Handbook of Residual Stress and Deformation of Steel*. ASM International, Materials Park, OH.
[2] Withers, P. J., Bhadeshia, H. K. (2001) "Residual Stress, Part 1: Measurement Techniques." *Materials Science and Technology* 17(4):355–365.
[3] Lu, J. (ed.) (1996) *Handbook of Measurement of Residual Stresses*. Fairmont Press, Lilburn, USA.
[4] Schajer, G. S. (2001) "Residual Stresses: Measurement by Destructive Methods." In: Buschow, K. H. J. et al. (eds) *Encyclopedia of Materials: Science and Technology*, Section 5a, Elsevier Science, Oxford.
[5] Ruud, C. O. (2000) "Residual Stress Measurements." In: Kuhn, H., Medlin, D. (eds) *ASM Handbook*, ASM International, 8:886–904.

Influence of Residual Stress

[6] Withers, P. J. (2007) "Residual stress and its Role in Failure." *Reports on Progress in Physics* 70(12):2211–2264.
[7] Schajer, G. S. (2010) "Relaxation Methods for Measuring Residual Stresses: Techniques and Opportunities." *Experimental Mechanics* 50(8):1117–1127.
[8] Kobayashia, M., Matsuia, T., Murakamib, Y. (1998) "Mechanism of Creation of Compressive Residual Stress by Shotpeening." *International Journal of Fatigue* 20(5):351–357.

[9] de Mol, A. J. M., Overkamp, P. J., van Gaalen, G. L., Becker, A. E. (1997) "Non-destructive Assessment of 62 Dutch Bjork-Shiley Convexo-Concave Heart Valves." *European Journal of Cardio-Thoracic Surgery*, 11(4):703–709.

[10] McIlree. A. R., Ruud, C. O., Jacobs, M. E. (1993) *"The Residual Stresses in Stress Corrosion Performance of Roller Expanded Inconel 600 Stream Generator Tubing."* Proc. International Conference on Expanded and Rolled Joint Technology, Canadian Nuclear Society, Toronto, Canada, 139–148.

[11] Masubuchi, K., Blodgett, O. W., Matsui, S., Ruud, C. O., Tsai, C. C. (2001) "Residual Stresses and Distortion." Chapter 7, *Welding Handbook* 9ed, American Welding Society.

Splitting

[12] Walton, H. W. (2002) "Deflection Methods to Estimate Residual Stress." In: Totten, G. E., Howes, M. A. H., Inoue, T. (eds) *Handbook of Residual Stress and Deformation of Steel*. ASM International, Materials Park, OH, 89–98.

[13] ASTM (2007) *Standard Practice for Estimating the Approximate Residual Circumferential Stress in Straight Thin-walled Tubing, Standard Test Method E1928-07*, American Society for Testing and Materials, West Conshohocken PA.

[14] Fuller, J. (1995) *Conditioning stress development and factors that influence the prong test*. USDA Forest Products Laboratory, Research Paper FPL–RP–537.

Stoney's Method

[15] Stoney, G. G. (1909) *The Tension of Thin Metallic Films Deposited by Electrolysis*. Proc. Royal Society of London, Series A 82(553):172–175.

Sectioning

[16] Shadley, J. R., Rybicki, E. F., Shealy, W. S. (1987) "Application Guidelines for the Parting out in a Through Thickness Residual Stress Measurement Procedure." *Strain* 23(4):157–166.

[17] Tebedge, N., Alpsten, G. and Tall, L. (1973) "Residual-stress Measurement by the Sectioning Method." *Experimental Mechanics* 13(2):88–96.

Layer Removal

[18] Treuting, R. G., Read, W. T. (1951) "A Mechanical Determination of Biaxial Residual Stress in Sheet Materials." *Journal of Applied Physics* 22(2):130–134.

[19] Sachs, G., Espey, G. (1941) The Measurement of Residual Stresses in Metal. *The Iron Age*, Sept 18, 63–71.

Hole Drilling

[20] ASTM (2008) *Standard Test Method for Determining Residual Stresses by the Hole-Drilling Strain-Gage Method, Standard Test Method E837-08*, American Society for Testing and Materials, West Conshohocken, PA.

[21] Grant, P. V., Lord, J. D., Whitehead, P. S. (2002) The Measurement of Residual Stresses by the Incremental Hole Drilling Technique. *Measurement Good Practice Guide*, No.53. National Physical Laboratory, UK.

[22] Measurements Group (2001) *Measurement of Residual Stresses by Hole-Drilling Strain Gage Method*. Tech Note TN-503-6, Vishay Measurements Group, Raleigh, NC.

Ring Coring

[23] Kiel, S. (1992) "Experimental Determination of Residual Stresses with the Ring-Core Method and an On-Line Measuring System." *Experimental Techniques* 16(5):17–24.

[24] Ajovalasit, A., Petrucci, G., Zuccarello, B. (1996) "Determination of Non-Uniform Residual Stresses using the Ring-Core Method." *Journal of Engineering Materials and Technology* 118(2):224–228.

Deep-Hole Drilling

[25] Leggatt, R. H., Smith, D. J., Smith, S. D., Faure, F. (1996) "Development and Experimental Validation of the Deep Hole Method for Residual Stress Measurement." *Journal of Strain Analysis* 31(3):177–186.
[26] DeWald, A. T., Hill, M. R. (2003) "Improved Data Reduction for the Deep-Hole Method of Residual Stress Measurement." *Journal of Strain Analysis* 38(1):65–78.

Slitting

[27] Prime, M. B. (1999) "Residual Stress Measurement by Successive Extension of a Slot: The Crack Compliance Method." *Applied Mechanics Reviews* 52(2):75–96.
[28] Cheng, W., Finnie, I. (2007) *Residual Stress Measurement and the Slitting Method*. Springer, New York.

Contour Method

[29] Prime, M. B. (2001) "Cross-Sectional Mapping of Residual Stresses by Measuring the Surface Contour After a Cut." *Journal of Engineering Materials and Technology* 123(2):162–168.
[30] DeWald, A. T., Hill, M. R. (2006) "Multi-Axial Contour Method for Mapping Residual Stresses in Continuously Processed Bodies." *Experimental Mechanics* 46(4):473–490.
[31] Kelleher, J., Prime, M. B., Buttle, D., Mummery, P. M., Webster, P. J., Shackleton, J., Withers, P. J. (2003) "The Measurement of Residual Stress in Railway Rails by Diffraction and other Methods." *Journal of Neutron Research* 11(4):187–193.

X-Ray Diffraction

[32] SAE (2003) *Residual Stress Measurement by X-ray Diffraction*. SAE J784, Society of Automotive Engineers Handbook Supplement, Warrendale, PA.
[33] Ruud, C. O., DiMascio, P. S., Snoha, D. J. (1984) "A Miniature Instrument for Residual Stress Measurement." *Advances in X-Ray Analysis*, 27:273–283.
[34] Pineault, J., Brauss, M. E. (1994) *Automated Stress Mapping – A New Tool for the Characterization of Residual Stress and Stress Gradients*. Proc. 4th International Conference on Residual Stresses, Society for Experimental Mechanics, Bethel CT, 40–44.

Synchrotron Diffraction

[35] Reimers, W., Pyzalla, A., Broda, M., Brusch, G., Dantz, D., Schmackers, T., Liss, K-D., Tschentscher, T. (1999) "The Use of High-Energy Synchrotron Diffraction for Residual Stress Analyses." *Journal of Materials Science Letters*, 18(7):581–583.

Neutron Diffraction

[36] Hayaski, M., Ohkido, S., Minakawa, N., and Murii, Y. (1997) *Residual Stress Distribution Measurement in Plastically Bent Carbon Steel by Neutron Diffraction*. 5th International Conference on Residual Stresses, Linköping University, Sweden, 2:762–769.
[37] Root, J. H., Hosbaus, R. R., Holden, T. M. (1991) 'Neutron Diffraction Measurements of Residual Stresses Near a Pin Hole in a Solid-Fuel Booster Rocket Casing'. In: Ruud, C. O. (ed) *Practical Applications of Residual Stress Technology*, ASM International 83–93.

Magnetic Methods

[38] Jiles, D. C. (1988) Review of Magnetic Methods for Nondestructive Evaluation. *NDT International* 21(5):311–319.

Ultrasonic

[39] Leon-Salamanca, T., Bray, D. E. (1996) "Residual stress measurement in steel plates and welds using critically refracted longitudinal (LCR) waves." *Research in Nondestructive Evaluation* 7(4):169–184.

[40] Santos, A. A., Bray, D. E. (2000) Application of Longitudinal Critically Refracted Waves to Evaluate Stresses in Railroad Wheels. *Topics on Nondestructive Testing*, 5, The American Society for Nondestructive Testing.

Thermoelastic

[41] Wong, A. K., Dunn, S. A., Sparrow, J. G. (1988) "Residual Stress Measurement by Means of the Thermoelastic Effect." *Nature* 332(6165):613–615.

Photoelastic

[42] Hetenyi, M. (ed) (1950) *Handbook of Experimental Stress Analysis*, Wiley, New York.

Indentation

[43] Chiang, S. S., Marshall, D. B., Evans, A. G. (1982) "The Response of Solids to Elastic/Plastic Indentations I: Stresses and Residual Stresses." *J. Appl. Phys.*, 50(1):298–311.

Residual Stress Computations

[44] DeWald, A. T., Hill, M. R. (2009) "Eigenstrain-based Model for Prediction of Laser Peening Residual Stresses in Arbitrary Three-Dimensional Bodies." *Journal of Strain Analysis for Engineering Design* 44(1) Part 1:1–11, Part 2:13–27.

[45] Schajer, G. S., Prime, M. B. (2006) "Use of Inverse Solutions for Residual Stress Measurements." *Journal of Engineering Materials and Technology* 128(3):375–382.

2

Hole Drilling and Ring Coring

Gary S. Schajer[1] and Philip S. Whitehead[2]
[1]*University of British Columbia, Vancouver, Canada*
[2]*Stresscraft Ltd., Shepshed, Leicestershire, UK*

2.1 Introduction

2.1.1 Introduction and Context

The hole-drilling method is the most widely used general-purpose technique for measuring residual stresses in materials. It is convenient to use, has standardized procedures and it has good accuracy and reliability. The test procedure involves some damage to the specimen but the damage is often tolerable or repairable. For this reason, the method is sometimes called "semi-destructive."

The hole-drilling method involves drilling a small hole in the test specimen at the place where the residual stresses are to be evaluated. This removal of stressed material causes a redistribution of the residual stresses in the remaining material around the hole and associated localized deformations. Figure 2.1 schematically illustrates the deformations around a hole drilled into material with tensile residual stresses. The consequent stress release causes elastic springback that slightly expands the hole edge, with a small local surface rise due to Poisson strain. The reverse happens with compressive stresses. For experimental evaluations, strain gage or optical techniques are available to quantify the surface deformations of the surrounding material, from which the residual stresses originally existing within the hole can be determined.

The ring-core method is an "inside-out" variant of the hole-drilling method, where the measurement area is in the middle and the "hole" takes the form of a surrounding annular groove. Figure 2.2 compares the geometry of the hole drilling and ring-core methods. The two methods are identical mathematically, and differ only in the numerical constants used for the residual stress evaluations. The ring-core method has the advantage of producing

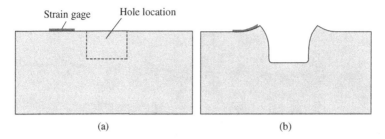

Figure 2.1 Schematic cross-sections around a hole drilled into tensile residual stresses. (a) Before hole drilling and (b) after hole drilling

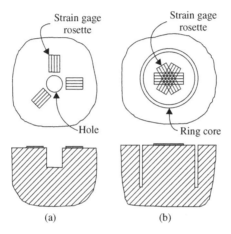

Figure 2.2 Residual stress measurement methods. (a) Hole drilling and (b) ring-core. Reproduced with permission from [13], Copyright 2010 Springer

larger relieved strains and has superior capability to measure very large residual stresses close to the material yield stress. However, the hole-drilling method is the more commonly used procedure because of its much greater ease of use and lesser specimen damage.

2.1.2 History

The hole-drilling method derives from the pioneering work of Mathar in the 1930s [1]. Since that time, the method has grown and developed remarkably, with contributions from many researchers. The hole-drilling method is now well-established, with an ASTM Standard Test Procedure [2] and extensive instructional literature [3–6]. It is a tribute to the fertility of Mathar's original concept that, after over 75 years, interest in hole drilling continues to grow, with frequent new developments.

From an early stage, the mechanical extensometer used by Mathar was recognized as a major factor limiting the accuracy and reliability of hole-drilling residual stress measurements. The development of strain gages in the 1940s provided an opportunity for substantial improvements in deformation measurement quality. In 1950, Soete [7] introduced the use of strain gages for hole-drilling measurements, greatly improving measurement accuracy and reliability, and allowing smaller holes to be used. The use of strain gages was further investigated by Kelsey who explored the incremental drilling technique to estimate stress vs. depth profile [8]. In the same period, Milbradt [14] introduced the ring-core method, with subsequent developments by Gunnert [15] and Hast [16].

The modern application of the strain gage hole-drilling method dates from the work of Rendler and Vigness [9] in 1966. They established a standardized strain gage geometry for residual stress measurements and developed the hole-drilling method into a systematic and repeatable procedure. Their work provided the basis for the establishment of ASTM Standard Test Method E837 in 1981, updated several times since then [2]. Early hole-drilling measurements were used to identify uniform stresses, then approximate methods were used to identify stress profile with depth [8]. The later availability of finite element calculations enabled accurate stress profile measurements to be achieved [10,11], and the procedure has now been standardized in E837. A large literature on strain gage hole-drilling measurements has also developed, with descriptive information [4,5], a good practice guide [3], and measurement accuracy analysis [6].

2.1.3 Deep Hole Drilling

A further variant approach is the deep-hole method, described in detail in Chapter 3. It is useful for determining the residual stresses within the deep interior of large specimens. The method was initially developed as a means of measuring geological stresses within large rock masses [17], and was later extended to the measurement of residual stresses in large metal components such as castings [18,19]. The method involves drilling a deep hole into the test material and then measuring the change in diameter as the surrounding material is overcored. The method combines some mechanical elements of the hole drilling and ring-core methods, but it differs significantly in that the measurements are made in the interior of the hole rather than at the surface. This is an important feature because the location of the measurements controls the location of the measured stresses. Conventional hole drilling and ring coring involve measurements at the surface, so they are mostly sensitive to the residual stresses at the surface, with some diminishing sensitivity to stresses within a depth approximately equal to the hole radius. In contrast, deep-hole measurements indicate the residual stresses in the deep interior.

2.2 Data Acquisition Methods

2.2.1 Strain Gages

Strain gages have, over an extended period, proven to be a robust and reliable means for measuring the surface deformations that occur during hole-drilling residual stress measurements. Following the work of Rendler and Vigness in the 1960s, specialized strain gage rosettes have been manufactured commercially for hole-drilling measurements. The design of these rosettes takes advantage of the photographic production method of strain gages to ensure that the individual gages of the rosettes are accurately oriented in space. This feature significantly reduces alignment error sources and greatly enhances the

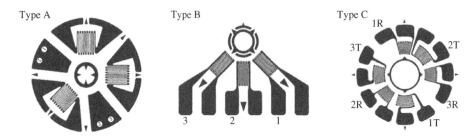

Figure 2.3 Standardized hole-drilling strain gage rosettes. Reprinted with permission from ASTM E837-08 [2]

quality of the measured data. Hole-drilling rosettes typically contain three radial strain gages arranged in rectangular format ($0°-135°-270°$ or $0°-45°-90°$) to identify the three in-plane stress components σ_x, σ_y and τ_{xy}.

Modern strain gages and associated electronic instrumentation can make very accurate and stable strain measurements, which is an essential feature because hole-drilling strains tend to be small, typically low hundreds of microstrain, sometimes less than one hundred. The compact and portable character of strain gage equipment enables effective field use. This ability to make successful hole-drilling measurements within a wide range of outside-lab measurement environments is a major factor in the wide acceptance of the strain gage hole drilling method.

A further important advantage of standardizing hole-drilling rosette geometry is that the calibration constants that relate the measured strains to the residual stress results also become standardized. This feature greatly simplifies the stress computations and allows documents such as ASTM E837 to give explicit stress calculation instructions.

E837 describes the use of three different rosette types to suit a range of measurement needs. Figure 2.3 illustrates the rosette geometries. Type A, which follows the Rendler and Vigness geometry, is a general-purpose design appropriate for most measurement needs. Type B has all three strain gages placed on the same side of the hole location and is useful for making measurements adjacent to obstacles. However, this rosette pattern should be used only for this purpose because the single-sided geometry increases its sensitivity to hole eccentricity errors.

The Type C rosette is a specialized design suited to measurement of small residual stresses and to measurements on materials with low thermal conductivity such as plastics. The design comprises three radial strain gages and three circumferential gages, connected in three half-bridge circuits. This arrangement increases the effective strain sensitivity of the rosette and also provides compensation for thermal strains, both very useful features when measuring small strains. The thermal strain compensation also greatly stabilizes measurements on low-conductivity materials that do not provide adequate heat dissipation for strain gages when connected within quarter-bridges. Again, this rosette pattern should be used only for these purposes because the half-bridges are costly and time-consuming to assemble.

2.2.2 Optical Measurement Techniques

Starting in the 1980s and 1990s, several optical techniques have been introduced as alternative surface deformation measurement techniques when evaluating residual stresses by the hole-drilling method. Chapter 11 describes these techniques in detail. Camera-based optical techniques have the advantage of providing full-field data, which enable the possibility for data averaging, error checking and extraction of detailed information. In contrast, strain gages provide strain measurements in just three discrete directions. Figure 2.4 compares the localized information provided by strain gages (area within the three squares) with the much richer information available from full-field displacement measurements.

Three full-field techniques have been applied so far to hole-drilling residual stress measurements: Moiré Interferometry, Holographic Interferometry and Digital Image Correlation. When using Moiré interferometry [20–23], a diffraction grating consisting of finely ruled lines, typically 600–1200 lines/mm, is attached or made directly on the specimen surface. This area is illuminated by two symmetric light beams that are derived from a single coherent laser source. Diffraction of the light beams creates a "virtual grating," giving interference fringes consisting of light and dark lines that are imaged by a video camera. The fringe lines represent contours of in-plane surface displacement at intervals of about 0.5 μm.

Holographic interferometry [24–26] provides a further important method for measuring the surface displacements around a drilled hole. A modern variant, Electronic Speckle Pattern Interferometry (ESPI) has become popular because its use of a video camera allows "live" fringe patterns to be produced by image subtraction [26,27]. Figure 2.4 shows an example of ESPI fringe pattern created by hole drilling. In-plane, out-of-plane or surface slope ESPI measurements [28] are possible, depending on the optical configuration

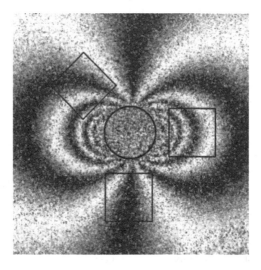

Figure 2.4 Comparison of strain gage and full-field ESPI data

used. A significant feature of ESPI is that it can work with a plain specimen surface, without attachment of the diffraction grating needed for Moiré measurements. This makes it possible to do ESPI measurements rapidly, and potentially to use the method as an industrial quality control tool.

Digital Image Correlation (DIC) is another optical technique that can be used for hole-drilling residual stress measurements [29–31]. The 2D technique involves painting a textured pattern on the specimen surface and imaging the region of interest using a high-resolution digital camera. The camera, set perpendicular to the surface, records images of the textured surface before and after deformation. The local details within the two images are then mathematically correlated and their relative displacements in the two in-plane directions determined. The algorithms used for doing this have become quite sophisticated, so that a well-calibrated optical system can resolve displacements of ± 0.02 pixel.

Three-dimensional displacement measurements can also be made by stereoscopic imaging using two cameras. The additional out-of-plane deformation data created could potentially improve the accuracy of residual stress evaluations from hole-drilling measurements. However, the effect is likely to be modest because the out-of-plane displacements are much smaller and therefore less influential than the in-plane displacements.

The full-field optical techniques are complementary to the strain gage technique, each approach having generally opposite advantages and disadvantages. Table 2.1 lists some of their features. The optical techniques have the advantage that they can provide dramatically larger data sets. The availability of "excess" data creates the possibility to improve stress evaluation accuracy and reliability by data averaging, and to be able to identify errors, outliers or additional features. However, the optical methods generally require fairly controlled conditions, while strain gages are much better suited to field use. The two measurement approaches also have opposite cost characteristics, with strain gages having relatively low equipment cost but high per-measurement cost. Optical apparatus has high equipment cost but relatively low per-measurement cost.

Table 2.1 Features of strain age and optical measurements

Strain Gage Measurements	Optical Measurements
• Moderate equipment cost, high per-measurement cost	• High equipment cost, moderate per-measurement cost
• Significant preparation and measurement time	• Preparation and measurement time can be short
• Small number of very accurate and reliable measurements	• Large number of moderately accurate measurements available for averaging
• Stress calculations are relatively compact	• Stress calculations often quite large
• Modest capabilities for data averaging and self-consistency checking	• Extensive capabilities for data averaging and self-consistency checking
• Relatively rugged, suitable for field use	• Less rugged, more suited to lab use
• Sensitive to hole-eccentricity errors	• Hole center can be identified accurately

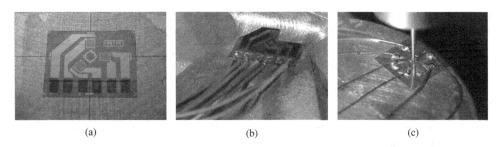

Figure 2.5 Hole-drilling rosette examples. (a) Hole-drilling rosette, (b) rosette with leadwires and (c) drilling operation. Images courtesy of Stresscraft

2.3 Specimen Preparation

2.3.1 Specimen Geometry and Strain Gage Selection

The ASTM Standard Test Method E837 for hole drilling [2] describes the required characteristics of an ideal test specimen. The measurement location on an ideal specimen has a plane smooth surface that is distant from any other surfaces, edges, or discontinuities such as holes or steps. In addition, the specimen material is linear-elastic, isotropic and homogeneous.

Figure 2.5 shows some example hole-drilling rosette installations. Figure 2.5(a) illustrates the "ideal" case where a rosette can be placed on a smooth, flat surface, remote from any obstacles or discontinuities. Figure 2.5(b) shows another rosette with leadwires attached. Here, the rosette backing has been trimmed to accommodate some features on the specimen surface. Figure 2.5(c) shows the drilling operation in progress on another style of rosette.

Table 2.2 summarizes the geometrical requirements for specimen thickness, distance from adjacent features and surface shape. The main specifications in the table derive from the ASTM Standard Test Method, with some additional suggestions based on the practical experience of the authors.

The prime factor to be considered for measurement planning is the gage size because it determines the maximum depth to which residual stresses can be detected. However, the maximum gage size may be limited by the specimen thickness and the proximity of the proposed gage location to any nearby specimen features.

The three types of ASTM strain gage rosettes identified in section 2.1 (Figure 2.3) are manufactured in a number of sizes and configurations [2]. Type A rosettes are manufactured in 031, 062 and 125-sizes as "open" gages (with an exposed metal surface, pattern RE) and in the 062-size as an encapsulated gage (with the metal surface covered by an insulating film, pattern UL) with large copper solder pads for direct soldering of leadwires. The Type B rosette is produced as an encapsulated 062-size gage only (pattern UM) while the Type C rosette is produced as an open 030-size gage and is identified as pattern RR.

Table 2.2 Geometrical specifications for hole-drilling rosette use, adapted and expanded from ASTM E837 [2]. Dimensions in mm. * indicates author's suggestions

	Symbol	E837 Rosette Type					
		A			B		C
Vishay Pattern	–	031RE	062RE	062UL	125RE	062UM	030RR
Gage mean diameter	D	2.57	5.13	5.13	10.26	5.13	4.32
Nominal hole diameter	D_o	1.0	2.0	2.0	4.0	2.0	2.0
Max drilled hole depth	zh_{max}	0.7	1.4	1.4	2.8	1.4	1.7
Max. stress data depth	zs_{max}	0.5	1.0	1.0	2.0	1.0	1.25
Min. specimen thickness	ts_{min}	2.5	5.0	5.0	10.0	5.0	6.5
Min. distance to edge feature	de_{min}	2.5	5.0	5.0	10.0	4.0	5.0
Min. distance to step feature*	ds_{min}	2.0	4.0	4.0	8.0	2.0	5.0
Min. radius of curvature*	rc_{min}	6.0	12.0	12.0	24.0	12.0	12.0
Min. gage-to-gage distance*	dg_{min}	6.0	12.0	12.0	24.0	12.0	12.0

Additional hole-drilling rosette patterns are commercially available from several different manufacturers and are suitable for use providing that the associated calibration constants are used when computing the residual stresses corresponding to the measured strain data. The manufacturers typically provide these calibration constants either explicitly or within available computer software.

Practical specimens are often less than fully ideal because of their complex geometry and the presence of local geometric features. The photographs in Figure 2.6 illustrate the dimensions defined in Table 2.2 and show ways in which the standard rosette patterns can be used and possibly adapted to accommodate various practical circumstances. The extreme gage position dimensions (minimum or maximum) listed in Table 2.2 for ts, de, ds, rc are values for which the published Integral Method coefficient values [2] remain valid within a range of ca. $\pm 4\%$. However, coefficients for combinations of extreme installation dimensions (for example near-edge and small radius of curvature) may lie further from the ASTM E837 values.

Specimen and gage position dimensions are established as follows:

- *Nominal hole diameter:* Typically $D_o = 0.4\,D$ with a recommended range of diameters of $\pm 0.04\,D$. The use of smaller diameter holes leads to small strain outputs, in particular at depths close to the surface. This reduced strain response increases uncertainties in the calculated residual stress values. Conversely, the use of a large hole diameter is a concern because it can lead to damage to the gage and bond close to the innermost parts of the gage elements.
- *Drilled hole depth:* The maximum depth zh_{max} to which holes are drilled is defined by the size and geometry of the gage pattern. It equals $0.28\,D$ for rosette types A and B and $0.4\,D$ for type C. Drilling beyond these depths produces no useful additional strain data.
- *Stress data depth:* The maximum depth zs_{max} to which residual stresses can be evaluated is also defined by the size and geometry of the gage pattern. This depth is less

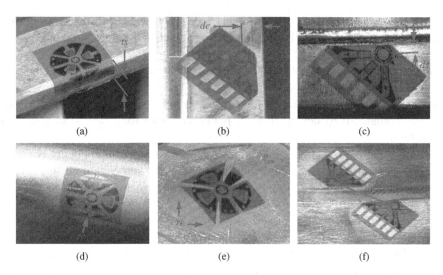

Figure 2.6 Dimensions relating to the specimen and gage rosette position. (a) Specimen thickness, (b) distance to edge, (c) distance to step feature, (d) surface curvature (single), (e) surface curvature (double) and (f) gage spacing. Images courtesy of Stresscraft

than zh_{max}. Because the computed stress within each hole depth increment is associated with the center of the increment, the effective stress data depth is reduced below zs_{max} by half the thickness of the final calculation increment. Taking this into account, the stress data depth is typically 0.2 D to 0.25 D for rosette types A and B and 0.3 D to 0.35 D for type C.

- *Specimen thickness:* For measurements on a thin specimen (Figure 2.6(a)), the drilling process causes significant bending of the specimen. For rosette types A and B, ASTM E837 [2] proposes a minimum specimen thickness $ts_{min} = 1.2\,\text{D}$. Rosette type C includes circumferential elements for which deeper hole depths cause significantly greater changes in stiffness. Accordingly, the minimum specimen thickness is somewhat greater.
- *Distance to an edge feature:* The proximity of a free edge to the drilled hole causes changes in stiffness and departures from the conditions used to calculate the published Integral Method coefficients. This is a particular concern when using rosette type B (Figure 2.6(b), where all the gage elements are contained within a single quadrant. In extreme cases, the drilled hole could be positioned very close to an edge. Where the distance between the gage center and specimen edge de_{min} is less than 0.8 D, the validity of published Integral Method coefficients may lie outside the range $\pm 4\%$. The presence of chamfers at the edge may further influence the validity of the coefficients.
- *Distance to a step feature:* The proximity of a step feature in the region of the gage results in changes to the stiffness of material around the drilled hole that are usually less severe than changes caused by an edge feature at the same distance. It is unlikely that the presence of small steps, such as weld beads (Figure 2.6(c)), will affect the validity of published Integral Method coefficients for practical ranges of dimension

'*ds*' which can be achieved using available rosettes. However, distances from drilled holes to larger steps should be greater than the tabulated values to avoid excessive uncertainties.

- *Radius of curvature:* Curvature of the gage installation surface rc_{min} (Figures 2.6(d) and 2.6(e)) causes two significant concerns:
  - the required Integral Method coefficients for a curved surface (cylindrical or spherical) will progressively deviate from the published values for a flat surface as curvature increases.
  - drilling a flat bottomed hole into a curved surface results in ambiguity concerning the selection of the hole datum depth (from which all subsequent depth increments are measured). This uncertainty is most significant at shallow drilling increments at surfaces with a small radius of curvature.

- *Gage spacing:* The relaxation effects of hole drilling extend beyond the boundaries of the rosette. Where it is required to install a number of rosettes on a specimen surface (Figure 2.6(f)) potential interference between adjacent holes must be considered. For a hole of diameter D_o drilled to a depth of $0.7 D_o$ in an equi-biaxial stress field, the stress relaxation at a distance of $6 D_o$ is less than 1% of the stress originally at the hole [3]. Accordingly, it is recommended that the minimum distance between adjacent holes dg_{min} should be at least six hole diameters.

One further important factor to be considered when planning measurements concerns the accuracy of the drilling machine. The hole drilling method requires relaxed strains to be recorded at a number of hole depth increments. The accuracy of subsequent residual stress calculations directly depends on the dimensional accuracy of the depth increments, which in turn depends on the quality of the cutter control during the drilling process. This issue becomes a significant concern when using many small depth increments (10–25 are typical) and when using very small rosettes. In the latter case the required hole diameter and depth increments vary in proportion to the rosette size, and so a greater absolute accuracy is required to maintain a reasonable level of relative accuracy of the drill depth increments and the concentricity of the hole within the rosette. Thus, while the smallest "031" rosettes may appear attractive to fit in sites of restricted area or on thin, curved specimens, they can be less attractive because of their more demanding drilling accuracy, handling and soldering requirements.

Where the installation of the gage on the specimen surface differs significantly from the above specifications, detailed finite element models of the specimen (incorporating the drilled hole) can be used to quantify the strain response of the residual stresses [12]. However, useful comparative information can often still be gained from processing non-compliant strain data using the standard Integral Method coefficients. For example, comparison of results from two or more differently processed specimens can indicate large differences in residual stresses even though the results are not strictly quantitative.

2.3.2 Surface Preparation

The way in which the surface is treated prior to gage installation forms an important part of the hole drilling procedure. In the worst cases, inappropriate preparation practices can lead to poor gage bonding and contamination of near-surface material with spurious residual stresses. The three main aims of surface preparation are to provide:

- a target site with sufficiently smooth surface that the bonded rosette can function well,
- a target site of adequate microscopic roughness to promote secure gage bonding,
- layout lines for accurate location and orientation of the gage.

For field measurements, it may first be necessary to clean any loose debris, corrosion or paint from the specimen so that the surface may be examined in detail. The surface profile should be sufficiently smooth to enable the gage backing material to be in intimate contact with the specimen surface, with the minimum possible adhesive layer thickness and with no significant irregularities in the region of the hole drilling and gage element areas.

If the specimen has been machined, die-cast or processed in some way to provide a surface that is sufficiently regular in profile, then no preliminary re-profiling will be required; the surface will only require treatment to ensure that the roughness is sufficient for bonding. In such cases, it will be possible to perform hole drilling with fine near-surface increments to provide detailed evaluations of near-surface residual stresses.

Un-machined sand castings, welded joints or coarsely machined components may present surface irregularities to which it is not possible to bond the target gage rosette satisfactorily. In such cases, there is no opportunity to obtain detailed near-surface stress distributions. It is then usual to remove material from the higher surface asperities to produce the required profile. This can be done using a number of readily available methods that include silicon carbide abrasive paper, abrasive paper flap wheels, files (conventional or diamond) and mounted points (revolving abrasive stones). The penetration of spurious stresses resulting from these processes can range from a few microns (fine abrasive paper with light hand pressure and a supply of water coolant) to a few hundred microns (a coarse mounted point). These methods can be used singly or in combination in the appropriate order to remove the required amount of material while limiting the depth of spurious stress penetration at the finished surface.

Less readily available surface profiling methods may include ECM/super-polishing and EDM. It is to be noted that EDM processes produce a thin, highly stressed recast layer; in steel specimens, such a layer may also be very hard and require removal to prevent drilling cutter damage. Some hole drilling or milling machines may be configured to machine a spot-face using a hole-drilling cutter. The machining process can be controlled so that machining-induced stresses are limited to a few microns in depth.

Any material removal at a stressed specimen surface will lead to a local redistribution of stresses. This must be considered prior to any material removal. Where near-surface stress gradients are large, removal of surface material may lead to unacceptable changes in stresses.

When a satisfactory specimen surface profile has been produced, the surface is treated to produce a suitable level of surface roughness to provide a suitable "key" for the gage adhesive. The Vishay surface preparation publication [32] proposes that the surface roughness for satisfactory bonding should be in the range Ra 1.6 to 3.2 μm. Two readily available methods that can achieve this are:

- *Swab-etching:* A suitable acid is applied to the specimen surface. To promote the etching process, the acid can be agitated and the specimen gently warmed. The acid can be removed from the specimen to allow inspection of the surface. The aim is to remove the minimum amount of material (say between 1 μm or 3 μm) while providing a surface of matte appearance. The operator should follow all health and safety recommendations when dealing with acid compounds. Following etching, a neutralizing solution is applied to return the surface pH to a value suitable for the bonding surface [32]. Where it is required to provide high quality near-surface stress data, trials should be performed on the specimen material for control of the amount of material removed during etching.

- *Abrasion:* The etchants required for some materials (for example titanium) are extremely hazardous and an alternative method of producing a matte surface may be required. Several passes of fine silicon carbide paper using light hand pressure can achieve this. A new piece of paper should be used for each site to "cut" a light cross-hatched pattern into the specimen, rather than to polish the surface.

Gage alignment layout lines can then be burnished onto the surface using a tungsten carbide ball tip [32], for example, a ball-point pen. These can be readily identified against the matte etched or cross-hatched abraded surface. It is usual to make four lines around the target center to align with features around the edge of the gage pattern so that no marks are present close to the drilled hole or gage elements (Figure 2.5(a)).

Where it is required to install the strain gage rosette at a particular position with respect to another feature on the specimen surface (e.g., a weld, hole or edge), then gage placement can be made while viewing the target area using the optical head provided with the drilling machine.

2.3.3 Strain Gage Installation

The aim when attaching a strain gage to a specimen surface is to achieve a thin, high-strength, creep-free bond that will reliably transfer the surface deformations by shear loading to the strain gage backing film and hence to the metal film resistance grid. Cyanoacrylate adhesives are the most widely used for strain gage work [33] because they are easy to handle and have a relatively short curing time. Other adhesive types are available, for example, two-part epoxy resins, but require longer curing times with more elaborate arrangements for applying pressure to the gage/specimen joint, possibly at an elevated temperature. The instructions given by strain gage and/or adhesive suppliers should be followed carefully.

For installations on irregular specimens, "open" strain gage types should be selected in preference to encapsulated gages because their lesser thickness enables them to follow surface contours without creating unacceptable bond thickness. In addition, gage backing material may be trimmed or slit so that the active parts of the gage can be installed with the thinnest possible bonds, see Figure 2.6 for some examples.

2.3.4 Strain Gage Wiring

In most cases, the requirements for wiring strain gage rosettes are straightforward. Protection of the wiring is not usually needed because the hole drilling procedure is completed within a short period of time. It is recommended that the three-wire arrangement be used for the connection to each gage element to reduce the effect of leadwire temperature changes [34]. This contributes significantly to the stability of strain readings when measuring small strain changes. Gages with leadwires are shown in Figures 2.5(b) (encapsulated) and 2.5(c) (open). Leadwires should be as short as practicable to minimize the effect of leadwire resistance. Vishay publications also describe practical strain gage soldering techniques [35].

2.3.5 Instrumentation and Data Acquisition

The stability, resolution and accuracy of the strain measurement electronic instrumentation are important to reduce uncertainties and scatter in residual stress results. These features are of particular importance where small changes in strain outputs are to be measured during the drilling of fine increments. The instrumentation used for strain measurements must be calibrated and the operator must be familiar with all necessary features for satisfactory operation. As a general guide, instrumentation for the strain measurement should include the following:

- *Resolution:* $\pm 1\,\mu\varepsilon$. ASTM E837 [2] requires that strain measurement instrumentation should have a resolution of $\pm 1\,\mu\varepsilon$. Most modern instrumentation can achieve this.
- *Stability:* $\pm 2\,\mu\varepsilon$ over the duration of the hole-drilling test; typically 20 to 40 minutes.
- *Accuracy:* $\pm 0.1\%$. This can readily be achieved by modern instrumentation.
- *Range:* $\pm 10,000\,\mu\varepsilon$. Strain readings seldom go outside the range $\pm 2,500\,\mu\varepsilon$ and are typically much smaller. However, additional range capability is required to accommodate any offsets produced when wrapping gages around curved surfaces.
- *Excitation:* A low excitation voltage is desirable because it reduces the "thermal output" of strain gages caused by ohmic heating. Many modern strain measurement devices incorporate low voltage DC bridge excitation, typically 2v or less, and are sufficiently stable and sensitive to accommodate the resulting small signal sizes. Thermal output is generally not significant when working with metallic materials because they are good thermal conductors and therefore provide a good heat-sink effect. However, it is significant when working with poor thermal conductors such as plastics or ceramics. In the latter case, the use of Type C rosettes can be additionally helpful because their half-bridge design provides thermal compensation.
- *Data acquisition:* Strain data are required on completion of each drilling increment. Simple instrumentation provides strain readings that must be recorded manually for entry into a data reduction program. More sophisticated instrumentation provides for electronic recording of strain data. This feature may be utilized by recording strains discretely after each hole depth increment, or continuously throughout the test, with subsequent selection of strain values. The latter method can produce a significant volume of data for a long drilling procedure. However, the data will not only provide the strain readings for subsequent data reduction, but will also indicate the strain variations due to transient thermal loading throughout the procedure. These additional data can be useful for detecting drilling problems and for establishing the "settling time" required between the end of drilling and strain recording.

2.4 Hole Drilling Procedure

2.4.1 Drilling Cutter Selection

Table 2.2 in the previous section lists the nominal hole diameters for ASTM rosette types A, B and C. In practice, a slightly smaller size drill is required to produce a target hole diameter because of vibration, clearances, and so on. Where drilling is to be performed using an orbital motion, a significantly smaller diameter drill is required, with orbit eccentricity making a significant contribution to the finished hole diameter. For conventional plunge drilling, a 1.8 mm diameter drilling cutter will typically produce a hole diameter around 2 mm suitable for a 062-size rosette. For orbital drilling, a 1.2 mm diameter drill with an orbit eccentricity set to 0.35 mm will produce a suitable hole for a 062-size rosette. Figure 2.7 shows a selection of cutters; some experimentation may be required to achieve the required results.

The following features are desirable for cutters used for strain gage hole drilling:

- A tungsten carbide head. This is suitable for drilling most aluminium, bronze, steel and nickel alloys, in addition to plastics. Cutters may be made from solid tungsten carbide or a tungsten carbide head bonded to a steel shank, typically 1.6 mm in diameter.
- An inverted cone shape with typically 5° relief on each side. This feature provides clearance for chip removal and eliminates rubbing between the cutter and specimen.
- A flat leading edge across the cutter end so that the initial contact between the cutter and specimen occurs over the entire hole area and that the drilled hole has a flat end. In practice, a slightly concave end profile is acceptable, but not a convex shape because of the difficulty to detect the drill zero datum depth.

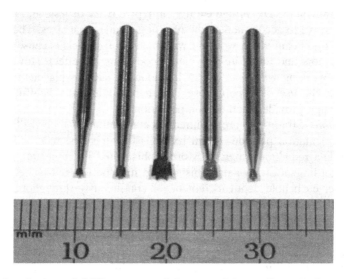

Figure 2.7 A selection of drilling cutters, 0.6 mm to 2.4 mm diameter. Images courtesy of Stresscraft

- A sharp corner between the end and side flanks. This is required to produce a sharp corner in the hole so that the hole profile matches the model used to produce the Integral Method coefficients. A corner profile with a chamfer or radius causes smearing of strain data and resultant residual stress results.

For extremely hard materials, for example, carburized or nitrided steel, ceramics, glass filled plastics, and so on, cutters coated with CBN or diamond particles can be used to drill holes. Because of the method of construction, coated cutters do not have sharp corners and so are suitable only for incremental drilling using a small number of coarse increments. Furthermore, this type of cutter cannot cut at the center of rotation in a satisfactory manner and should be used only when drilling with an orbital motion. It is to be noted that diamond coated drills are not suitable for use with steel specimens.

Prior to use, it is recommended that each cutter is carefully inspected to confirm that the diameter is correct and that the cutting edges are not damaged. For drilling in all but the softest materials, a fresh cutter is recommended for each new hole.

2.4.2 Drilling Machines

A specialized drilling machine is typically used to cut the hole into the specimen at the center of the strain gage rosette. The features required to achieve this include a:

- motor with a chuck or collet to hold the drilling cutter,
- system to adjust the position of the machine to align with the gage,
- drill feed and depth control,
- means to measure the drilled hole diameter,
- mounting or clamping provision for the specimen.

Figure 2.8 shows examples of commercial machines that are designed specifically for strain gage hole drilling:

- *SINT MTS3000 (RESTAN)*: This machine uses an air turbine drilling motor and a stepper motor to control the drill depth. A computer controls the air turbine, drill depth and, when linked to a suitable strain indicator, records the relaxed strains.
- *Micro-measurements RS-200* (Vishay Precision Group): This machine also uses an air turbine drilling motor. The drill depth is controlled manually using a large-scale micrometer head located around the drill barrel.

Both machines use a replaceable optical head to align with the rosette center and to measure the hole diameter after drilling. Each machine mounts on the specimen using three pads that can be cemented to the surface.

The methods of operation of the two machines are quite different and each shows distinct advantages in certain areas of operation. The MTS3000 machine can control the drill depth with a very fine resolution, and offers a high level of integration of drilling, strain measurement and calculation of residual stresses. However, the RS-200 drilling machine allows eccentric mounting of the drilling motor within its holder so that the machine can be configured to drill with an orbital cutting motion.

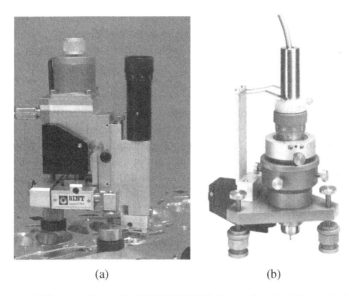

(a)   (b)

Figure 2.8 Hole drilling machines (a) SINT MTS3000 (Reproduced with permission from SINT Technology). (b) Micro-Measurements RS-200. Courtesy of Micro-Measurements, Raleigh, NC, a brand of Vishay Precision Group

2.4.3 Orbital Drilling

Early hole drilling measurements were carried out using conventional low-speed drilling [9]. In recent years it has become typical to use high-speed drilling at speeds 20,000–200,000 rpm using air-turbine or electric motors [12]. While high-speed drilling can provide a satisfactory solution in many materials, plunge drilling in hard or tough materials can rapidly damage the tips of the cutting tool. Such damage distorts the hole profile (non-circularity and loss of the sharp corner at the bottom of the hole) in addition to the creation of spurious stresses at the hole surface. Unfortunately, because of the difficulty in resetting the cutter and controlling the hole depth and diameter, the replacement of a drilling cutter at any stage during drilling is not a realistic option for holes of 2 mm diameter or less.

Practical experience shows that machining with the side of the drilling cutter (orbital drilling or circular milling) offers several advantages over machining with the front of the cutter (plunge drilling):

- The drilling cutter is significantly smaller than the drilled hole; this provides a large area for the exit of drilling debris as shown in Figure 2.9.
- The tangential and axial forces resulting from orbital drilling are significantly smaller than the axial forces at the equivalent plunge-drilling cutter.
- There is less heat input to the specimen, causing faster strain settling times (between motor switch-off and strain reading) and smaller temperature increases at the critical inner parts of the gage/specimen bond.
- The deflections/distortions between the drill and specimen are reduced.
- The wear of the drilling cutter and motor bearings is also reduced.

Figure 2.9 Schematic view of orbital drilling. Image courtesy of Stresscraft

- The tendency for an air-turbine motor to stall is significantly decreased, particularly when drilling into tough materials.

For orbital drilling, the drill size and orbit eccentricity must be selected to produce the required completed hole diameter. In practice, each drilling increment can be divided into a number of sub-increments, each of which may be only a few microns in depth. An axial feed of the required depth is applied to the drill and the orbital movement is then made for at least one complete revolution (until the motor tone indicates that the cutting process is complete). The drilling of sub-increments is repeated in this way until the required increment depth is achieved. Where required to reduce cutter vibration and prevent distortion of the hole shape, the starting position of the drill around the orbit may be varied so that the axial feed does not occur at the same circumferential position.

2.4.4 Incremental Measurements

The sequence of incremental hole drilling proceeds as follows (some details may vary depending on the machine type in use):

- With the target strain gage installed, checked and connected to the strain indicator, the drilling machine is fixed to the specimen (with the optical head in position) so that the center of the drill axis is located close to the gage center and, most importantly, the drill axis is perpendicular to the specimen surface at the center of the gage.
- Fine adjustments are made to the drill position so that the optical head cross-hairs lie over the target alignment markers at the center of the gage. The drill is locked in this position.
- The drilling cutter is inserted into the drilling motor collet and locked in position.
- The drill motor holder is fitted into the drilling guide and its height adjusted so that it is located above the gage surface.

Figure 2.10 Removal of gage backing material to reach the specimen surface. The photographs are taken at 0.001″ (0.025 mm) drill advance intervals. Images courtesy of Stresscraft

- In order to establish the drill datum depth, the drill motor is switched on and the cutter slowly advanced to cut through the gage backing material and adhesive bond layer until contact is made between the drill and specimen. The drilling datum may be detected by electrical contact (between the cutter and specimen) or by visual inspection (using a magnifying eyepiece after the cutter has been withdrawn from the surface). In practice, specimen surfaces are rarely completely flat nor the drill axis set precisely perpendicular to the surface and some visual inspection is required to determine when the drill has made contact over approximately 50% of the periphery of the hole. Figure 2.10 shows a series of images of a target gage taken at 0.001″ (0.025 mm) intervals of drill advance. The fourth image illustrates the partial removal of the strain gage backing material and the slight scratching of the specimen surface that indicate the axial position of the drill at the zero hole depth datum.
- The depth gage is reset to zero and the strain indicator bridge is re-balanced.
- The first increment is drilled in a controlled manner with a low feed rate. The motor tone can be a useful indicator to confirm that the cutting process has been completed. After drilling is complete the drill is backed out of the hole and the motor stopped. The strain indicator output is examined and the strain levels are monitored until a steady state is reached. The three strain readings are then recorded.
- Drilling and strain reading is repeated for each depth increment to the final hole depth.
- The drilling cutter is completely withdrawn from the hole and the drill holder removed from the drilling machine.

2.4.5 Post-drilling Examination of Hole and Cutter

On completion of the drilling process, the following examinations are made:

- The strain gage rosette is viewed using the optical head to check for any de-bonding around the drilled hole. This can be a serious concern if debonding has extended under the gage elements. The concentricity of the drilled hole and gage alignment markers is also examined at this time.

- The leadwires are unsoldered and the gage rosette carefully removed from the specimen. The initial stage of gage removal is achieved by inserting a sharp blade under the edge of the gage. The gage is then peeled away from the specimen providing the operator with an opportunity to judge the strength and quality of the bond. The under side of the gage rosette should have a uniform matte appearance with disturbances only at the solder terminals and at the drilled hole. Any irregular markings at the bond surface in the region of the individual gage elements are causes for concern. The gage may be stored as part of the record of the drilling process.
- The final hole depth can be checked using a depth gage. This will indicate any large discrepancies in setting the zero depth datum.
- The edge of the hole can be examined using the optical head to determine whether any burring has occurred. If present, burs indicate plastic deformation around the hole and the inclusion of any spurious stresses in the indicated results.
- The focus of the optical head can be adjusted to the bottom of the hole to show whether a significant radius has been created at the hole lower corner. This should be sharp.
- The hole diameter is measured across two perpendicular diameters using the graticule scale within the optical head. The average of the two readings is subsequently input into the Integral Method calculation. Significant diameter differences (>2%) indicate an irregular hole shape and may be a cause for concern.
- After removal, the used drilling cutter is carefully examined to detect the pattern of wear and any damage. Important features may include rounding and breakage of the tooth tips, and flats on the cutting surfaces. The cutter may also be stored as part of the record of the drilling process. Except when drilling into very soft materials such as plastics, it is prudent to use a fresh cutter for each new hole.

The results of the above examinations can provide valuable information for monitoring the performance of the strain gage rosette and hole drilling equipment and for identifying any needed changes in future supplies of consumable items (adhesives, drilling cutters, etc.).

2.5 Computation of Uniform Stresses

2.5.1 Mathematical Background

Two types of residual stress calculations are possible with the hole-drilling method, "uniform stresses" when the in-plane stresses can be assumed not to vary with depth from the specimen surface, and "stress profiling" when they do vary significantly with depth. The first case, considered in this section, is the simpler one because there are only three unknown in-plane residual stresses to be determined.

The conventional strain gage procedure involves measuring the strains as the hole is drilled within a strain gage rosette of the type shown in Figure 2.3. For a "thick" material, the hole reaches a depth approximately equal to the hole diameter, and for a "thin" material, the hole goes all the way through. The relationship between the measured strains and the in-plane residual stresses is [2]:

$$\varepsilon = \frac{\sigma_x + \sigma_y}{2} \frac{(1+\nu)\bar{a}}{E} + \frac{\sigma_x - \sigma_y}{2} \frac{\bar{b}}{E} \cos 2\theta + \tau_{xy} \frac{\bar{b}}{E} \sin 2\theta \qquad (2.1)$$

or

$$\varepsilon = P\frac{(1+v)\bar{a}}{E} - Q\frac{\bar{b}}{E}\cos 2\theta + T\frac{\bar{b}}{E}\sin 2\theta \qquad (2.2)$$

where

$$P = \frac{\sigma_y + \sigma_x}{2}, \qquad Q = \frac{\sigma_y - \sigma_x}{2}, \qquad T = \tau_{xy} \qquad (2.3)$$

In Equations (2.1) and (2.3), σ_x, σ_y and τ_{xy} are the "uniform" in-plane stresses and θ is the angle between the strain gage axis and the x-direction. $\bar{a}$ and $\bar{b}$ are calibration constants that define the strain/stress sensitivity of the measurement. Their numerical values depend on hole diameter and depth. Combination stresses P, Q and T respectively represent the isotropic, 45° shear and axial shear stresses. The factor $(1+v)$ appears with the $\bar{a}$ term to account for the Poisson ratio dependence of the isotropic strains. The Poisson effects on the shear strains associated with the $\bar{b}$ terms are negligible and so no extra factors are required.

Figure 2.11 shows the variation of $\bar{a}$ and $\bar{b}$ with hole depth. The graphs show that the calibration constants approach limiting values at hole depths beyond about 0.4 D where D is the diameter of the circle containing the centers of the strain gages. For typical hole sizes, this depth approximately equals the hole diameter D_o. No significant additional strain response occurs by drilling to greater depths. To a first approximation, the calibration constants are proportional to the square of the hole diameter. Thus, for greater strain sensitivity, it is helpful to use a hole diameter near the upper end of the allowable range, around 0.4–0.45 D. The maximum allowable hole diameter is limited by the need not to cut into the strain gage grids.

Figure 2.11 illustrates how the strain gage circle diameter D controls the strain response of the strain gage rosette. All the curves have similar shapes when normalized this way. This feature indicates that it is the diameter of the rosette that controls the shape of

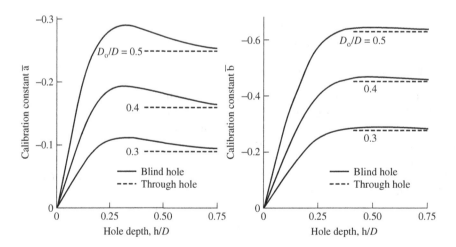

Figure 2.11 Strain gage calibration constants $\bar{a}$ and $\bar{b}$ for a Type A strain gage rosette. Adapted from [11]

the strain response, which is to be expected because the strain gage is a source of the measurement. The diameter of the hole controls the sizes of the curves but not their shape.

Equation (2.2) can be inverted to evaluate the axial and principal residual stresses from the three measured strains ε_1, ε_2 and ε_3:

$$P = \frac{E}{(1+\nu)\bar{a}} p, \qquad Q = \frac{E}{\bar{b}} q, \qquad T = \frac{E}{\bar{b}} t \qquad (2.4)$$

$$\sigma_{max}, \sigma_{min} = P \pm \sqrt{Q^2 + T^2} \qquad \beta = \frac{1}{2} \text{atan}\left(\frac{-T}{-Q}\right) \qquad (2.5)$$

where

$$p = \frac{\varepsilon_3 + \varepsilon_1}{2}, \qquad q = \frac{\varepsilon_3 - \varepsilon_1}{2}, \qquad t = \frac{\varepsilon_3 - 2\varepsilon_2 + \varepsilon_1}{2} \qquad (2.6)$$

and

$$\sigma_x = P - Q, \qquad \sigma_y = P + Q, \qquad \tau_{xy} = T \qquad (2.7)$$

The combination strains p, q and t represent the isotropic, 45° shear and axial shear strains corresponding to the combination stresses P, Q and T. The equations above can also be written directly in terms of the axial stresses σ and axial strains ε. The form using the isotropic and shear stresses and strains is used here because it corresponds with the format used for the stress profiling calculations described in the next section.

In Equation (2.5), σ_{max} and σ_{min} are respectively the more and less tensile (less and more compressive) principal stresses, and β is the clockwise angle from the gage 1 axial direction to the σ_{max} direction. The minus signs are retained in the β equation so that if the angle is determined using the conventional rules for the two-argument arctangent function, it will be placed in the correct quadrant and will refer specifically to the more tensile principal stress σ_{max}.

As shown in Figure 2.11, the calibration constants $\bar{a}$ and $\bar{b}$ have negative numerical values because they describe the strain changes that occur when the residual stresses are *removed* by hole drilling. For convenience of tabulation, E837-08 [2] quotes these constants as positive quantities and places minus signs in Equation (2.4).

2.5.2 Data Averaging

Theoretically, it is sufficient to drill a hole to its final depth, typically 0.4 D, and directly evaluate the residual stresses using Equations (2.1–2.7). However, residual stress evaluation accuracy can be significantly increased by averaging a set of strain measurements made at a series of small depth increments as the hole is drilled from zero to the final depth [36]. ASTM E837 [2] specifies the use of eight equal hole increments, each 0.05 D deep. The averaging can be done by replacing Equation (2.3) with:

$$P = \frac{E}{1+\nu} \frac{\sum(\bar{a} \cdot p)}{\sum(\bar{a}^2)} \qquad Q = E \frac{\sum(\bar{b} \cdot q)}{\sum(\bar{b}^2)} \qquad T = E \frac{\sum(\bar{b} \cdot t)}{\sum(\bar{b}^2)} \qquad (2.8)$$

2.5.3 Plasticity Effects

A significant limitation of the hole-drilling method is that the hole creates a stress concentration that can cause localized plastic deformations if the nearby residual stresses are high. Typical residual stress computation methods, such as described by Equations (2.1–2.8) rely on material linearity. The localized yielding near the hole boundary caused by stress concentrations starts to cause noticeable deviations from linearity for residual stresses greater than 60% of the material yield stress. If the hole depth is limited to 0.2 D, the range of linear response can be extended to 70% of the material yield stress. Fortunately, in most cases, the effect of local yielding is to overestimate the size of the residual stress, often to values significantly above the material yield stress. Thus, the existence of problematic results is readily apparent and the errors are conservative. For strain gage measurements, correction procedures have been developed to allow accurate measurement of residual stresses up to 90% of the material yield stress [37].

2.5.4 Ring Core Measurements

Equations (2.1–2.8) also apply to the ring-core method, but with different numerical values of $\bar{a}$ and $\bar{b}$. Since ring-coring relieves all the residual stress in the central island, the corresponding $\bar{a}$ and $\bar{b}$ values are relatively high, giving the ring-core method a high strain/stress sensitivity. By comparison, the strains measured in the hole drilling method are only partially relieved, and so are much smaller. Because the stress concentrations mostly occur in the material around the milled annulus and with only small effect on the central island, their influence on the measurements is modest. Thus, the ring-core method can directly measure residual stresses close to the material yield stress.

2.5.5 Optical Measurements

Optical measurements differ from strain gage measurements in that they measure surface displacements rather than strains, and that they give "full-field" results with displacement data at hundreds of thousands, even millions of independent pixels. Figure 2.4 shows an example measurement. Equation (2.1) still applies, with surface displacement replacing surface strain on the left side. The large quantity of available data can improve measurement accuracy through data averaging and can give opportunities for error checking and for extraction of detailed information. The challenge is to be able to do the needed calculations to take advantage of the rich data source while minimizing the computational burden. Early residual stress evaluation methods sought to extract the data corresponding to strain gage measurements and then complete the calculations using Equations (2.1–2.8). While effective, Figure 2.4 shows that this approach uses only a small fraction of the available data, and many valuable measurements are lost. More recent computation methods [26,31] use a least-squares approach that uses the large majority of the available data.

2.5.6 Orthotropic Materials

The trigonometric relationship in Equation (2.1) occurs only for hole-drilling measurements in an isotropic material. However, for orthotropic materials such as fiber composites and wood the strain/stress relationship is not trigonometric. Residual stresses can still be evaluated, but in a more complex matrix format using seven calibration constants [38,39].

2.6 Computation of Profile Stresses

2.6.1 Mathematical Background

In the general case, residual stresses vary with depth from the specimen surface. This feature introduces a significant challenge into the residual stress calculation because the surface deformations depend not on just a single uniform stress, but on the combination of all the stresses through the specimen depth. For example, for the isotropic stresses and strains:

$$p(h) = \frac{1+\nu}{E} \int_0^h \hat{a}(H, h) \, P(H) \, dH \quad (2.9)$$

where $p(h)$ is the isotropic strain combination defined in Equation (2.6) that is relieved when the hole reaches a depth h, $P(H)$ is the isotropic stress combination defined in Equation (2.3) that exists at depth H from the measured surface, and kernel function $\hat{a}(H, h)$ is a generalization of the uniform stress calibration constant $\bar{a}(h)$ defined in Equation (2.4). (For compactness, Equations (2.1–2.8) omit explicitly showing the dependence on h of $\bar{a}$, $\bar{b}$ and the associated strain quantities). The shear stress and strain parts of Equation (2.4) can similarly be generalized in the format of Equation (2.9) using the kernel function $\hat{b}(H,h)$.

The form of Equation (2.9) is classified mathematically as a Volterra equation of the first kind. It describes an "inverse problem" because the stress quantity to be determined occurs contained within the integral on right side, rather than openly on the left, as in Equation (2.4). Solution of Equation (2.9) can be achieved using so-called "inverse methods," specifically designed for equations of this type.

The computational approach most commonly used to solve Equation (2.9) is called the "Integral" or "unit pulse" method [40], which is a generalization of Equation (2.4). The associated measurement technique involves hole drilling in a sequence of small depth increments, with strain measurements after each increment. ASTM E837-08 specifies $n = 20$ equal steps to a final hole depth 0.2 D. The interior stresses are assumed to be locally constant within each of the hole depth increments, giving rise to the stepwise stress profile representation shown in Figure 2.12. With this approach, the scalar stress quantities P, Q and T in Equation (2.4) become vector quantities P_j, Q_j and T_j where $j = 1, n$ is an index for the n stress depth increments corresponding to the hole depth increments. The strain quantities p, q and t similarly become vectors p_i, q_i and t_i, where $i = 1, n$.

In an analogous way, the calibration constants $\bar{a}$ and $\bar{b}$ become matrix quantities $\bar{a}_{ij}$ and $\bar{b}_{ij}$ whose elements relate the stresses within depth increment j to the strains measured with a hole i increments deep.

$$\bar{a}_{ij} = \int_{h_{j-1}}^{h_j} \hat{a}(H, h_i) \, dH \qquad \bar{b}_{ij} = \int_{h_{j-1}}^{h_j} \hat{b}(H, h_i) \, dH \quad (2.10)$$

Figure 2.13 shows a physical interpretation of matrix $\bar{a}$ [40]. Quantity $\bar{a}_{32}$ represents the strain caused by a unit stress within increment two of a hole three increments deep. The matrix is lower triangular because only stresses that exist within the hole contribute to the measured strains. The numerical values of the various calibration constants can be found using finite element calculations [11,40].

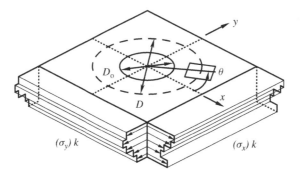

Figure 2.12 Stepwise variation of residual stress with depth used by the Integral Method. Reprinted with permission from ASTM E837-08 [2]

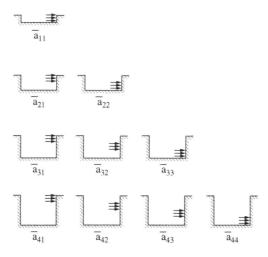

Figure 2.13 Physical interpretation of matrix coefficients $\bar{a}$ for the hole-drilling method. Reprinted with permission from ASTM E837-08 [2]

With the above generalizations, Equation (2.4) becomes:

$$\bar{a}\,P = \frac{E}{1+\nu}\,p \qquad \bar{b}\,Q = E\,q \qquad \bar{b}\,T = E\,t \qquad (2.11)$$

where symbols in **bold** font represent matrix and vector quantities. For example, with $n = 4$, the first equation would appear in expanded form as:

$$\begin{bmatrix} \bar{a}_{11} & & & \\ \bar{a}_{12} & \bar{a}_{22} & & \\ \bar{a}_{13} & \bar{a}_{23} & \bar{a}_{33} & \\ \bar{a}_{14} & \bar{a}_{24} & \bar{a}_{34} & \bar{a}_{44} \end{bmatrix} \begin{bmatrix} P_1 \\ P_2 \\ P_3 \\ P_4 \end{bmatrix} = \frac{E}{1+\nu} \begin{bmatrix} p_1 \\ p_2 \\ p_3 \\ p_4 \end{bmatrix} \qquad (2.12)$$

and similarly for the other two. Conveniently, all three Equations (2.11) are independent of each other and can be solved individually. The stress results can then be combined as in Equations (2.7) and (2.5) to determine the Cartesian and principal stresses within each depth increment.

There is a physical limit on how deep stresses with a specimen material can be evaluated. As described by St. Venant's Principle, the further the stresses are away from the measured surface, the less the influence will be on the measurements. This effect can be visualized from Figure 2.13 where the matrix elements along the diagonal are expected to have a small size because they represent the surface strains caused by stresses situated adjacent to the bottom of the hole. The material at the bottom of the hole provides substantial support to those stresses, so only a small fraction of their effect appears at the surface. This effect becomes more extreme for greater hole depths, giving ever smaller diagonal elements in the matrix. When a diagonal element approaches zero, the matrix becomes singular and no further stress evaluation is possible. In practice, stresses can be evaluated to a depth of about 0.2 D [2]. In the mathematical literature it is described that inverse equations such as Equation (2.9) are ill-conditioned, with consequent ill-conditioning in Equations (2.11) and (2.12). Thus, it can easily happen that substantial noise in the form of sharp local oscillations in the computed stresses can occur. However, it should be clearly understood that the behavior has a physical cause associated with the action of St. Venant's Principle. The problem is not of mathematical origin and therefore cannot be "solved" through the use of a different mathematical approach. The first practical response to the ill-conditioning problem is strict attention to meticulous experimental technique. Reduced measurement errors directly reduce stress evaluation errors.

After experimental quality has been given priority, some mathematical techniques can further be used to ameliorate the effects of the remaining experimental imperfections. One approach is to choose to use only a small number of hole depth increments, with sizes that become larger as the hole depth increases [41]. In this way, each increment has greater area within which the stresses can act and thus have a larger force resultant. In addition, the area of each successive increment gets increasingly larger to offset its increasing remoteness from the measured surface. This technique is effective but tends to give rather coarse spatial resolution.

An alternative approach is to make measurements at many small hole depth increments and to increase the quantity of data used for the calculation. Used by itself, this approach will produce substantial oscillations in the stress solution because the stress solution is driven by the differences in successive strain measurements, which in this case are very small and therefore subject to large relative errors. The mathematical approach to deal with this is to seek a best-fit solution that smoothly approximates the measured data. An effective way of doing this is by using Tikhonov Regularization [42]. This involves slightly modifying Equations (2.11) to penalize the noisy component of the stress solution:

$$(\bar{\mathbf{a}}^T \bar{\mathbf{a}} + \alpha_P \mathbf{c}^T \mathbf{c}) \mathbf{P} = \frac{E}{1+\nu} \bar{\mathbf{a}}^T \mathbf{p} \tag{2.13}$$

and similarly for stress quantities $\mathbf{Q}$ and $\mathbf{T}$. The matrix c contains the second-derivative operator, for example with $n = 4$:

$$\mathbf{c} = \begin{bmatrix} 0 & 0 & & \\ -1 & 2 & -1 & \\ & -1 & 2 & -1 \\ & & 0 & 0 \end{bmatrix} \tag{2.14}$$

which contains the number sequence [−1 2 −1] in each row, centered on the main diagonal, excluding the first and last rows. The coefficient α is the regularization parameter that controls the amount of regularization used. For $\alpha = 0$, Equation (2.13) reduces to Equation (2.11), and there is no regularization effect. With increasing α values, the amount of regularization increases. Too small an α value insufficiently smoothes the stress solution and leaves too much noise. Too large an α value excessively smoothes the stress solution and eliminates local details. Optimal regularization eliminates most of the measurement noise while preserving the spatial details of the stress solution. The optimal regularization occurs then the misfit, for example:

$$\mathbf{p}_{misfit} = \mathbf{p} - \frac{1+\nu}{E} \bar{\mathbf{a}} \mathbf{P} \qquad (2.15)$$

has a 2-norm (root-mean square value) equal to the standard error of the measurements. This is called the Morozov Criterion. The Integral Method with Tikhonov regularization [42] works well in practice because it averages a large quantity of measured data. In addition, the smooth trends that are expected in the measured strains make isolated erroneous measurements easier to identify. The procedure has now been standardized in the ASTM Standard Test Method E837 [2], which gives further details of the mathematical procedure.

2.7 Example Applications

2.7.1 Shot-peened Alloy Steel Plate – Application of the Integral Method

The application of the Integral Method to evaluate residual stress variation with depth was investigated by making measurements on a weld made along the center of a 150 mm long × 50 mm wide × 4 mm thick steel plate. The welding procedure created longitudinal tensile stresses adjacent to the weld. Subsequently, parts of the plate surfaces were masked while other parts were shot-peened. Several gages (type EA-031RE-120) were installed on the plates and were drilled at 11 × 0.064 mm depth increments. The strain data from one pair of gages in the (1) un-peened and (2) shot-peened state were first used to calculate residual stresses using the Integral Method. The gage 2 results in Figure 2.14 show the intensity of near-surface shot-peening stresses, with a sub-surface tensile peak of over 400 MPa (greater than the un-peened stress) occurs at depth 0.2 mm. Sets of both longitudinal and transverse sub-surface stresses from both gages at depth 0.5 mm are very similar.

2.7.2 Nickel Alloy Disc – Fine Increment Drilling

Figure 2.15 shows residual stresses calculated from strain data obtained from hole drilling on the side of a machined (turned) Inconel 718 disc. Hole drilling was done using very fine (16 μm) increments close to the surface, increasing to 32 μm at greater depths. Residual stresses shown in Figure 2.15(a) have been calculated (Integral Method) using only relaxed strains at 128 μm increments. While some gradient of hoop stresses is seen to occur between the first two increments, radial stresses are seen to be uniformly distributed at all

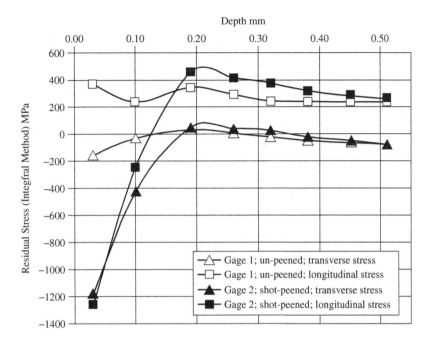

Figure 2.14 Residual stress distributions calculated using the Integral Method. Images courtesy of Stresscraft

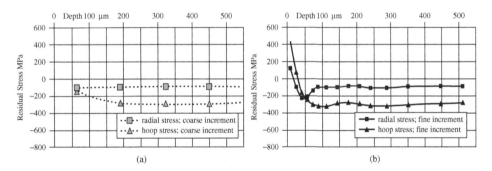

Figure 2.15 Residual stress profiles calculated using strain data from coarse and fine drilling increments. (a) Stresses calculated using data from coarse drilling increments and (b) Stresses calculated using data from fine drilling increments. Images courtesy of Stresscraft

depths to 0.5 mm. The calculated residual stresses are compressive at all depths; the sub-surface radial stress is -100 MPa while the corresponding hoop stresses is approximately -300 MPa. These are the residual stress distributions that would be evaluated following incremental drilling at 128 μm (approximately 0.005 inch) depth increments.

Figure 2.15(b) shows the distributions of residual stresses calculated using all the available (16 μm increment) strain data. In this case, near-surface residual stresses are shown to be tensile; in particular, the first increment hoop stress (the stress in the machining

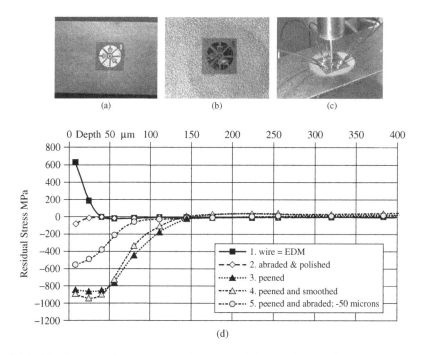

Figure 2.16 Titanium test-piece: assessment of residual stresses from surface processes. (a) Gage installation on abraded surface, (b) Gage installation on shot-peened surface, (c) Incremental hole drilling on smoothed surface and (d) Distributions of near-surface residual stresses for 5 surface conditions. Images courtesy of Stresscraft

direction) exceeds 400 MPa. Both radial and hoop stresses show typical machining stress profiles comprising near-surface tensile peaks immediately followed by large negative gradients and local sub-surface stress minima. While the use of smaller near-surface drilling and calculation increments results in greater uncertainties in stress levels in individual increments, the nature of near-surface stress distributions is more clearly revealed than by the coarse increment results.

2.7.3 Titanium Test-pieces – Surface Processes

Figure 2.16 shows some of the specimens and a summary of results from a series of hole drilling tests carried out on a series of titanium Ti6/4 alloy test-pieces. Vishay 031-size rosettes were applied to the test-pieces and drilled using 16 μm near-surface increments. The conditions of the material surface for the five gages are:

1. Wire-EDM cut surface.
2. EDM surface; recast layer removed using a fine abrasive stone (Figure 2.16(a)).
3. Shot-peened (Figure 2.16(b)).
4. Shot peened (as 3) followed by smoothing using a fine stone to remove peening indentations (Figure 2.16(c)).
5. Shot-peened and smoothed (as 4) followed by further fine abrasion to remove ca. 50 μm.

The stress distributions in Figure 2.16(d) show very different magnitudes of near-surface stresses resulting from the applied processes. The recast EDM layer (1) produces tensile stresses within the first two calculation increments, which can subsequently be removed using a fine abrasive stone (2). The shot-peening process produces compressive stresses (3), which can be smoothed with little effect on the stress distribution (4). The removal of further material by fine abrasion removes the more intensely stressed material and shifts the as-peened stress distribution closer to the surface by an amount equal to the thickness of the material removed. Tests of this type are very useful for the development of surface preparation methods and parameters.

2.7.4 Coated Cylinder Bore – Adaptation of the Integral Method

For the measurement of residual stresses in components with coatings it is necessary to produce sets of Integral Method coefficients that match the configuration of the structure. In the example shown here, the bore of an aluminium cylinder has been coated with a 100–125 μm (0.004 to 0.005 inch) layer of nickel. The ratio of Young's moduli (coating:substrate) is 3:1. Figure 2.17(a) shows a detail from a typical 2D axisymmetric finite element model created for the evaluation of Integral Method coefficients. In this model, the coating (shown in red) has been assigned suitable material properties to represent the nickel layer; the mesh details are set so that the coating thickness coincides with a layer of elements. The gage application, drilling and strain recording processes proceed as for a conventional hole (Figure 2.17(b)). It is necessary to use a specially adapted Integral Method program to evaluate the residual stresses in which any smoothing of relaxed strains or residual stresses across the coating/substrate interface is suppressed. Different sets of coefficients are required for different coating thicknesses or ratios of elastic constants, which can make this type of analysis time consuming.

The example stress distribution (Figure 2.17(c)) shows how residual stresses in the coating are tensile but small in magnitude with no large depthwise gradients. There is a significant discontinuity in stresses at the interface. The first drilled part of the substrate contains a pattern of residual stresses that appear to be related to the bore machining process, while at depths further from the surface, compressive circumferential stresses are sustained to a depth of 1 mm.

2.8 Performance and Limitations of Methods

2.8.1 Practical Considerations

With meticulous preparation and closely controlled experimental procedures, center hole drilling can produce high quality residual stress results. The quality of results is affected by features of the specimen and target site:

- specimen bulk stress does not exceed 60% of the material yield stress. In practice, near-yield stresses close to the surface within the first few drilling increments does not appear to affect the integrity of results,
- the specimen surface is reasonably smooth, flat and readily accessible,
- the target strain gage rosette is installed on a thick section at a position remote from edges, steps or other features.

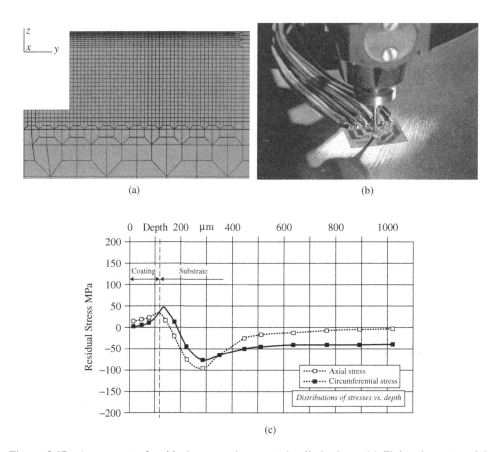

Figure 2.17 Assessment of residual stresses in a coated cylinder bore. (a) Finite element model incorporating coating layer, (b) Incremental drilling at the coated cylinder bore surface and (c) Distributions of residual stresses in the coating and substrate. Images courtesy of Stresscraft

Where the above factors lie within the ranges recommended in ASTM E837 [2] and Section 2.3, then the performance of the hole method is directly related to the quality of the rosette and its installation, leadwires and strain measuring instrument, the quality of the drilling machine and cutter and details of experimental procedure followed by the operator.

The hole drilling procedure usually produces a result, or set of results, in the form of a stress distribution. Any procedural problems can typically be identified by review of the measured data and the computed residual stresses, together with the operator's in-test observations and post-test examination of the drilled hole, gage and drilling cutter. In extreme cases, for example after gage bond failure, drill breakage or excessive drill wear, the test would need to be discarded and repeated.

2.8.2 Common Uncertainty Sources

The large number of activities produces a correspondingly large number of potential sources of uncertainties. Many of these have been identified in the NPL Good Practice

Guide [3] and, where possible, the likely contribution (major or minor) of each source estimated. Contributions to uncertainties include:

- Specimen and strain gage rosette installation
    - surface condition (roughness, flatness, etc.)
    - material properties (E, v and σ_y), isotropy and homogeneity
    - access to measurement location
    - presence of stress gradients (depthwise and in-plane)
    - geometry (near-edge or -hole), thickness and curvature
    - gage position on specimen surface
    - strength and thickness of the gage bond
- Hole drilling
    - hole misalignment (displacement and inclination with respect to gage axis)
    - definition of datum depth and depths of individual depth increments
    - hole profile (diameter, roundness, taper and corner radius)
    - drilling parameters (speed, feed, orbit eccentricity), condition (wear) and stress induced during drilling
    - temperature increases caused by drilling
- Strain recording
    - instrumentation bridge balance (zero strain)
    - strain gage/instrumentation output noise
    - instrumentation accuracy (resolution and linearity)
- Residual stress calculation
    - applicability of Integral Method coefficients for material, specimen and hole
    - data handling (treatment of raw strain data and computed stresses)
- Operator
    - operator training, skill and experience.

In the Good Practice Guide [3], the skill of the operator was identified as the most important single factor for achieving satisfactory residual stress measurements.

2.8.3 Typical Measurement Uncertainties

In the Code of Practice for evaluating uncertainties resulting from the hole drilling procedure, Oettel [6] has identified a number of uncertainty sources and evaluated the contributions made to these to the uncertainties in a worked example of computed uniform stresses. Scafidi et al. [43] have also identified uncertainties in uniform stress calculation and have presented a worked example that demonstrates that results can be produced with a maximum bias of about 10%.

Eccentricity of the drilled hole from the target gage can create a potentially large source of uncertainties in computed stresses. Beghini et al. [44] have proposed a stress evaluation correction method using influence coefficients from finite element models. In practical terms, misalignment during setting up and clearance/wear in locating elements of the drill head and optical head are likely to be the greatest sources of eccentricity. Sensitivity to the effect of hole eccentricity is increased by the use of single-sided strain gage rosettes such as the Type B pattern shown in Figure 2.3, and reduced by use of opposing pairs of gage elements such as the Type C pattern.

For a straightforward installation and drill set up on a thick, flat machined moderately stressed specimen, remote from edges and where good alignment is achieved between the drill and gage, the combined uncertainties (excluding strain and increment depth uncertainties) may be limited to 4% or 5% of the computed residual stress magnitude (over the entire depth range). For "uniform stress" measurements under ideal conditions, typical measurement uncertainties may be around 5%. For more challenging measurements, for example on curved or uneven surfaces, then measurement uncertainties may increase to the 5–10% range.

For stresses that vary with depth, determined by incremental hole drilling, the relationships between relaxed strains, increment depths and residual stresses calculated using the Integral Method, are more complex than for the uniform stress case. The overall effects of strain and depth uncertainties cannot be determined directly and another approach is required to establish reasonable uncertainty estimates for the computed stress values. Hole drilling tests (with additional instrumentation) can be used to establish the strain and depth uncertainties. Typical values may be:

- uncertainties in strain outputs of up to ± 2 $\mu\varepsilon$ from all sources involving the strain gage rosette, leadwires and the bridge/amplifier instrument,
- uncertainties in the depths of drilled hole increments of :
  - ± 2 μm for a stepper motor controlled drilling machine,
  - ± 6 μm for a manual micrometer controlled drilling machine.

The effect of these uncertainties on the output stress values can be calculated by repeated input of test strain and increment depth data to which have been added random uncertainties within the ranges of values listed above. A number of sets of strain data including the random strain and depth uncertainties are processed and the resulting stress distributions superimposed until there are no further changes in the uncertainty envelopes above and below the nominal stress distribution line.

Figure 2.18 shows an example stress distribution (solid line) to which have been added uncertainty envelopes (dashed lines) over the full depth range. Two distributions of stresses (plotted in grey) show typical results computed from sets of strains and hole depths that include random uncertainty components; these are seen to oscillate about the nominal measured stress line within the envelope shown. In the example shown, the hole was drilled in a steel specimen using a stepper motor controlled driller with hole depth increments varying from 32 μm at the surface to 128 μm at the full hole depth. The resultant stress uncertainties approach a maximum value of ± 40 MPa near to the surface because of the relatively large effect of input uncertainties on the low levels of strains measured in the small hole depth increments. Uncertainties reduce to a minimum of ± 12 around the mid-depth and then increase to ± 24 MPa at the full depth because of reducing sensitivity. The distribution and uncertainty example shown here are those for a stress direction

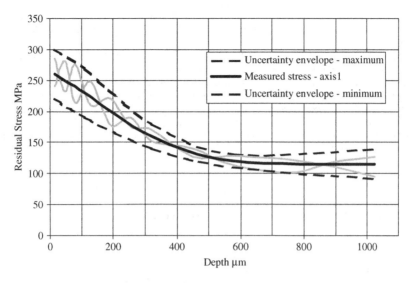

Figure 2.18 An example of uncertainties in residual stresses calculated from relaxed strain and drilling depth uncertainties. Image courtesy of Stresscraft

parallel to one of the gage elements. For principal stresses, the uncertainties increase as the principal stress directions rotate from the axes of gage elements 1 and 3.

The uncertainties in stresses described above result from strain and depth uncertainties only and are significantly greater than corresponding uncertainties for the uniform stress case. These uncertainties are dependent on the material Young's modulus and Poisson's ratio and increase as the increment size decreases and as the drilled hole diameter decreases. The uncertainty component resulting from strain noise alone is independent of the residual stress magnitude. The selection of the sizes of increment used in the Integral Method calculation can have a profound effect on the magnitudes of stress uncertainties resulting from strain and depth uncertainties. Residual stress distributions showing clear oscillations in stress magnitudes between successive increments indicate that the quality of strain data and/or drilling depth control are insufficient to provide a high level of stress distribution detail; recalculation of residual stresses using larger increments helps to restore some stability to the calculations.

It is recommended that this type of assessment is carried out for the materials and hole/calculation increments to be used in the test to demonstrate that uncertainties are within acceptable limits.

References
Hole Drilling

[1] Mathar, J. (1934) "Determination of Initial Stresses by Measuring the Deformation Around Drilled Holes." *Transactions ASME* 56(4):249–254.
[2] ASTM (2008) Determining Residual Stresses by the Hole-Drilling Strain-Gage Method. Standard Test Method E837-08. *American Society for Testing and Materials*, West Conshohocken, PA.

[3] Grant, P. V., Lord, J. D., Whitehead, P. S. (2002) The Measurement of Residual Stresses by the Incremental Hole Drilling Technique. *Measurement Good Practice Guide No.53*, National Physical Laboratory, Teddington, UK.
[4] Schajer, G. S. (1996) "Hole-Drilling and Ring Core Methods." Chapter 2 in *Handbook of Measurement of Residual Stresses*. Ed. J. Lu. Fairmont Press, Lilburn, GA.
[5] Vishay Measurements Group, Inc. (1993) Measurement of Residual Stresses by the Hole-Drilling Strain-Gage Method. Tech Note TN-503-6. Vishay Measurements Group, Inc., Raleigh, NC. 16pp.
[6] Oettel, R. (2000) The Determination of Uncertainties in Residual Stress Measurement (using the Hole Drilling Technique). Code of Practice 15, Issue 1, EU Project SMT4-CT97-2165.
[7] Soete, W., Vancrombrugge, R. (1950) "An Industrial Method for the Determination of Residual Stresses." *Proceedings SESA* 8(1):17–28.
[8] Kelsey, R. A. (1956) "Measuring Non-Uniform Residual Stresses by the Hole Drilling Method." *Proceedings SESA* 14(1):181–194.
[9] Rendler, N. J., Vigness, I. (1966) "Hole-drilling Strain-gage Method of Measuring Residual Stresses." *Experimental Mechanics* 6(12):577–586.
[10] Bijak-Zochowski, M. (1978) "A Semidestructive Method of Measuring Residual Stresses." *VDI-Berichte* 313:469–476.
[11] Schajer, G. S. (1981) "Application of Finite Element Calculations to Residual Stress Measurements." *Journal of Engineering Materials and Technology* 103(2):157–163.
[12] Flaman, M. T. (1982) "Brief Investigation of Induced Drilling Stresses in the Centre-hole Method of Residual Stress Measurement." *Experimental Mechanics* 22(1):26–30.
[13] Schajer, G. S. (2010) "Relaxation Methods for Measuring Residual Stresses: Techniques and Opportunities." *Experimental Mechanics* 50(8):1117–1127.

Ring Core

[14] Milbradt, K. P. (1951) "Ring Method Determination of Residual Stresses." *Proceedings SESA* 9(1):63–74.
[15] Keil, S. (1992) "Experimental Determination of Residual Stresses with the Ring Core Method and an On-Line Measuring System." *Experimental Techniques* 16(5):17–24.
[16] Ajovalasit, A., Petrucci, G., Zuccarello, B. (1996) "Determination of Non-uniform Residual Stresses Using the Ringcore Method." *Journal of Engineering Materials and Technology* 118(1):1–5.

Deep Hole Drilling

[17] Merrill, R. H. (1967) Three-Component Borehole Deformation Gage for Determining the Stress in Rock. *U.S. Bureau of Mines*, 7015, 38pp.
[18] Leggatt, R. H., Smith, D. J., Smith, S. D., Faure, F. (1996) "Development and Experimental Validation of the Deep Hole Method for Residual Stress Measurement." *Journal of Strain Analysis* 31(3):177–186.
[19] Procter, E., Beaney, E. M. (1987) "The Trepan or Ring Core Method, Centre-Hole Method, Sach's Method, Blind Hole Methods, Deep Hole Technique." *Advances in Surface Treatments* 4:166–198.

Moiré

[20] McDonach, A., McKelvie, J., MacKenzie, P., Walker, C. A. (1983) "Improved Moiré Interferometry and Applications in Fracture Mechanics, Residual Stress and Damaged Composites." *Experimental Techniques* 7(6):20–24.
[21] Nicoletto, G. (1991) "Moiré Interferometry Determination of Residual Stresses in the Presence of Gradients." *Experimental Mechanics* 31(3):252–256.
[22] Wu, Z., Lu, J., Han, B. (1998) "Study of Residual Stress Distribution by a Combined Method of Moiré Interferometry and Incremental Hole Drilling." *Journal of Applied Mechanics* 65(4) Part I:837–843, Part II:844–850.
[23] Schwarz, R. C., Kutt, L. M., Papazian, J. M. (2000) "Measurement of Residual Stress Using Interferometric Moiré: A New Insight." *Experimental Mechanics* 40(3):271–281.

ESPI

[24] Nelson, D. V., McCrickerd, J. T. (1986) "Residual-stress Determination Through Combined Use of Holographic Interferometry and Blind-Hole Drilling." *Experimental Mechanics* 26(4):371–378.
[25] Focht, G., Schiffner, K. (2003) "Determination of Residual Stresses by an Optical Correlative Hole Drilling Method." *Experimental Mechanics* 43(1):97–104.
[26] Steinzig, M., Ponslet, E. (2003) "Residual Stress Measurement Using the Hole Drilling Method and Laser Speckle Interferometry: Part I." *Experimental Techniques* 27(3):43–46.
[27] Suterio, R., Albertazzi, A., Amaral, F. K. (2006) "Residual Stress Measurement Using Indentation and a Radial Electronic Speckle Pattern Interferometer – Recent Progress." *Journal of Strain Analysis* 41(7):517–524.
[28] Wu, S. Y., Qin, Y. W. (1995) "Determination of Residual Stresses Using Large Shearing Speckle Interferometry and the Hole Drilling Method." *Optics and Lasers in Engineering* 23(4):233–244.

Digital Image Correlation

[29] McGinnis, M. J., Pessiki, S., Turker, H. (2005) "Application of Three-dimensional Digital Image Correlation to the Core-drilling Method." *Experimental Mechanics* 45(4):359–367.
[30] Nelson, D. V., Makino, A., Schmidt, T. (2006) "Residual Stress Determination Using Hole Drilling and 3D Image Correlation." *Experimental Mechanics* 46(1):31–38.
[31] Schajer, G. S., Winiarski, B., Withers, P. J. (2013) "Hole-Drilling Residual Stress Measurement With Artifact Correction Using Full-Field DIC." *Experimental Mechanics* 53(2):255–265.

Strain Gage Installation

[32] Vishay Precision Group, Inc. (2011) *Surface Preparation for Strain Gage Bonding*. Instruction Bulletin B-129-8.
[33] Vishay Precision Group, Inc. (2011) *Strain Gage Installations with M-Bond 200 Adhesive*. Instruction Bulletin B-127-14.
[34] Vishay Precision Group, Inc. (2010) *The Three-Wire Quarter-Bridge Circuit*. Application Note TT-612.
[35] Vishay Precision Group, Inc. (2010) *Strain Gage Soldering Techniques*. Application Note TT-609, 13 November.

Uniform Stress Calculation

[36] Schajer, G. S. (1991) "Strain Data Averaging for the Hole-Drilling Method." *Experimental Techniques* 15(2):25–28.
[37] Beghini, M., Bertini, L., Raffaelli, P. (1994) "Numerical Analysis of Plasticity Effect in the Hole-Drilling Residual Stress Measurement." *Journal of Testing and Evaluation* 22(6):522–529.
[38] Schajer, G. S., Yang, L. (1994) "Residual-stress Measurement in Orthotropic Materials using the Hole-Drilling Method." *Experimental Mechanics* 34(4):217–236.
[39] Pagliaro, P., Zuccarello, B. (2007) "Residual Stress Analysis of Orthotropic Materials by the Through-hole Drilling Method." *Experimental Mechanics* 47(2):217–236.

Profile Stress Calculation

[40] Schajer, G. S. (1988) "Measurement of Non-Uniform Residual Stresses Using the Hole-Drilling Method." *Journal of Engineering Materials and Technology* 110(4) Part I:338–343, Part II:344–349.
[41] Vangi, D. (1994) "Data Management for the Evaluation of Residual Stresses by the Incremental Hole-Drilling Method." *Journal of Engineering Materials and Technology* 116(4):561–566.
[42] Schajer, G. S. (2007) "Hole-Drilling Residual Stress Profiling with Automated Smoothing." *Journal of Engineering Materials and Technology* 129(3):440–445.

[43] Scafidi, M., Valentini, E., Zuccarello, B. (2011) "Error and Uncertainty Analysis of the Residual Stresses Computed by Using the Hole Drilling Method." *Strain* 47(4):301–312.
[44] Beghini, M., Bertini, L., Mori, L. F. (2010) "Evaluating Non-Uniform Stress by the Hole Drilling Method with Concentric and Eccentric Holes." Part I and Part II, *Strain* 46(4):324–336 and 46(4):337–346.

3

Deep Hole Drilling

David J. Smith
University of Bristol, Bristol, UK

3.1 Introduction and Background

The Deep Hole Drilling (DHD) method belongs to the class of mechanical strain relaxation techniques designed to measure stresses and residual stresses in materials. The technique described in this chapter is now used extensively to measure both applied and locked-in stresses in many manufactured engineering materials and components. Its origins as a technique rise from its use in rock mechanics [1,2]. The procedure involves drilling a pilot borehole into the rock to the depth required for stress measurement (Figure 3.1). A device, often called a strain cell, is inserted into the borehole and then pressurized to ensure good contact with the sides of the borehole or glued into the borehole. Then material surrounding the pilot borehole is then over-cored to relax the stresses in the surrounding rock, with the strain cell monitoring the relaxed strains. An elasticity analysis is then used to convert the strains to stresses. Rock mechanics practitioners also refer to using "soft or hard inclusion" analysis for determining the stresses from the strains measured, [2–4] with the analysis depending on the type of strain cell used. An example of its application in rock mechanics is provided by Martin and Christiansson [3]. They describe borehole and over-coring methods where the pilot hole was 38 mm diameter and the over-core diameter 96 mm.

The process of introducing a pilot hole, inserting a strain cell and then over-coring has also been developed and applied to civil engineering structures, such as bridges and concrete constructions. Ryall [5] explains a "hard inclusion" method based on over-coring. A solid steel bar, representing the "hard inclusion", is bonded into the borehole using high-modulus cement grout. The steel bar is strain gaged to permit measurement of strains during over-coring. Borehole and over-core diameters were about 42 mm and

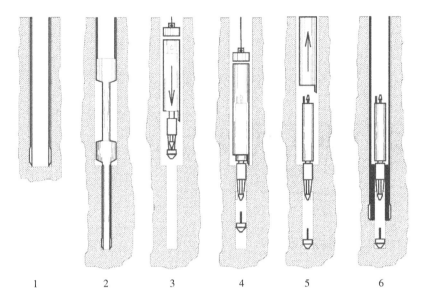

Figure 3.1 Steps in the overcoring measurements as applied for stress measurement in rock. (1) Advance a 76 mm diameter main borehole to measurement depth, (2) drill 36 mm diameter pilot hole, (3) lower installation tool, (4) install probe and gages bonded into pilot hole, (5) raise installation tool and (6) overcore the probe. Reproduced with permission from [4], Copyright 2003 Elsevier

150 mm respectively. The conversion of strains to stresses requires knowledge of the elastic properties of the structure, the cement grout and the steel bar.

The borehole and over-coring methods have been developed extensively for application in rock mechanics and in large man-made structures. Their main benefit is an ability to measure stress deep below the surface, either several hundreds of meters below the earth's surface or hundreds of millimeters deep within a civil engineering structure. The dimensions of the borehole and over-core used for these methods suggest that the sampling volume and low stress gradients are not too much of a concern. However, for many welded metal components, such as pressure vessels and pipes, large holes would not be very useful in measuring stress distributions.

Prior to the 1970s residual stress measurement in metallic components was conducted using fully destructive techniques and obtaining stresses deep within the component. Some of these are reviewed in Chapter 1. However, it was apparent that the over-core methods used in rock mechanics could also be adapted. For example, Beaney [6] and Proctor and Beaney [7] developed methods via the work of Ferrill et al. [8], with similar advances being made in Europe [9]. The term "Deep Hole Drilling" has been adopted to describe the method to distinguish it from near surface residual stress measurement methods such as center hole drilling. Initially, deep hole drilling used pilot or reference holes that were relatively large, about 8 mm, but certainly smaller than those used in rock mechanics, to ensure strain gages (or other diameter measurement devices) could be installed down the hole. However, there are only so many strain gages wires that are able to go down 8 mm holes. Consequently, only a very restricted number of measurements through the depth of the component could be made. The subsequent introduction of an air probe [10] was an important advance because it enabled measurement of the diameter of the reference hole at any location along the pilot hole.

Many developments have been made over the last 25 years to increase the accuracy and applicability of the DHD method to a range of engineering materials and components. Nonetheless, the basic procedure for the measurement process remains much the same. Figure 3.2 shows schematically a cross-section of a large metal component.

In step 1 reference bushes are installed at the entrance and exit faces of the line of measurement. A hole is created through the component and the reference bushes. In metals, the hole is often created using a gun-drill. Step 2 involves measurement of the diameter of the hole around its circumference and along its length. In step 3 a device is set up to measure distortions of the core that occur during the trepanning process (i.e. over-coring). Often electro-discharge machining is used to do the trepanning and when trepanning is completed, the core, containing the reference hole is retained in place by the rear reference bush. Step 4 is the final stage where the reference hole is re-measured, again along its length and around its diameter. The change is diameter between steps 2 and 4, together with the distortions of the core, are used to determine the relaxed stresses.

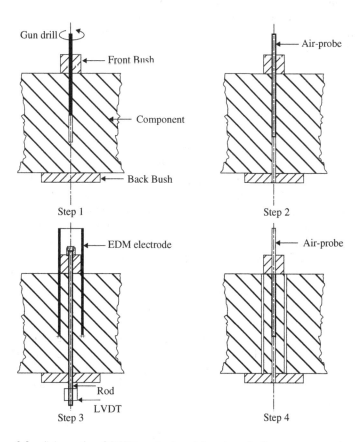

Figure 3.2 Schematic of DHD method and the steps in the measurement process

If there are no external forces applied to the component, these stresses are the locked-in or residual stresses. Alternatively, if there are external loads applied to the component, the measured stresses will be a combination of the applied and residual stresses.

The remainder of this chapter is in five sections. Section 3.2 describes the basic principles to convert measured distortions to stresses. Section 3.3 explains the experimental technique for the DHD method. Work to validate the method is summarized in Section 3.4. To illustrate the practical application of the method a selection of case studies is given in Section 3.5. The chapter closes, in Section 3.6 with a summary and a view to the future.

3.2 Basic Principles

The DHD method seeks to measure the distribution of stresses along the axis of the pilot or reference hole. The hole acts as a strain gage, so that when the stress is released, changes in diameter are measured together with changes in the through-thickness dimension of the trepanned core. Therefore a relationship is required between the original residual stresses acting at the measurement location and the measured changes in hole diameter and core height. In the initial developments of the DHD method it was assumed that measured distortions could be converted to stresses using an elastic analysis [10]. This approach, now called the conventional DHD method, is explained first in this section. Later, plasticity, which can occur during trepanning, is considered. A revised technique, called the incremental DHD method, is also explained.

3.2.1 Elastic Analysis

There are several approaches for converting measured distortions to stresses, including use of a finite element analysis [7,11], an eigenstrain method [12] and a relatively simplified scheme [10,13]. The development of the latter is explained here.

The extracted core, containing the reference hole, is divided into a number of blocklengths, each bounded by two parallel planes normal to the reference hole axis as shown in Figure 3.3. The co-ordinate scheme is also shown in Figure 3.3, with $x - y$ in the plane of the reference hole and z along its axis. Each blocklength is idealized as a plate containing a central hole. The diameter measurements, obtained at selected increments through the thickness of the specimen, are assumed to be the average diameter of the reference hole (at a given angle) within each blocklength.

It is also assumed that the through-thickness direction is a principal direction, the material properties of the specimen are isotropic, the state of stress is uniform within each blocklength before the reference hole is drilled, and the behavior of each blocklength is independent of other blocklengths.

Overcoring leads to the stresses within the core being completely relaxed elastically. The first step in the analysis is to calculate within a given blocklength the radial displacements at the edge of the reference hole due to the trepanning operation. The plate is assumed to be subjected to a uniform uniaxial stress. Assuming fully elastic behavior, the state of stress and deformation in a uniformly loaded plate containing a hole is the same regardless of whether the load is applied before or after the drilling of the hole. Hence the state of deformation at the reference hole before trepanning can be obtained from standard elastic solutions [14].

Deep Hole Drilling

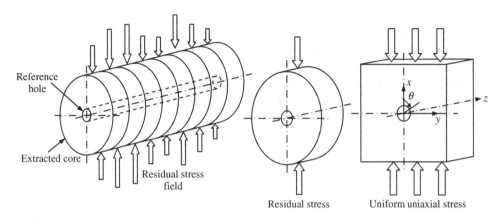

Figure 3.3 Extracted core, section of core and a simplification of the section to a plate containing a central hole

The radial displacement at a hole in a plate subject to uniform stress, σ_{xx} is:

$$u_r(r,\theta) = \frac{\sigma_{xx} a_r}{E}\left\{\left[(1+\nu)\frac{a_r}{2r}+(1-\nu)\frac{r}{2a_r}\right]\right.$$
$$\left. + \left[(1+\nu)\frac{r}{2a_r}\left(1-\frac{a_r^4}{r^4}\right)+\frac{2a_r}{r}\right]\cos 2\theta\right\}. \quad (3.1)$$

where a_r is the hole radius, E is the Young's modulus of the material, ν is Poisson's ratio, and θ is the angle measured from the axis of the applied stress, σ_{xx}.

The effect of trepanning out a core of material containing the reference hole is to restore the core to zero stresses. Hence the radial displacements at the edge of the reference hole caused by the trepanning operation are equal and opposite to the displacements that would occur due to the application of a uniform far-field stress to an infinite plate with a hole. The analysis accounts for the effect of the presence of the reference hole on the stress and strain field before trepanning. The analysis procedure enables the original stresses at the location of the reference hole to be calculated.

Displacements are measured at the hole edge at an angle (θ) and therefore, Equation (3.1) provides non-dimensional distortions given by

$$\tilde{\varepsilon}(\theta) = \frac{u_r(\theta)}{a_r} = \frac{\sigma}{E}(1+2\cos 2\theta). \quad (3.2)$$

The principle of superposition can then be used to combine the effects of individual uniaxial stresses $\sigma_{xx}, \sigma_{yy}, \sigma_{xy}$ and σ_{zz}. For the general case of uniform 2D plane stresses acting on a circular hole the non-dimensional distortions are:

$$\tilde{\varepsilon}(\theta) = \frac{u_r(\theta)}{a_r} = \frac{d(\theta)-d_0(\theta)}{d_0(\theta)}$$
$$= -\frac{1}{E}\{\sigma_{xx}(1+2\cos 2\theta)+\sigma_{yy}(1-2\cos 2\theta)+\sigma_{xy}(4\sin 2\theta)-\nu\sigma_{zz}\} \quad (3.3)$$

where $d_0(\theta)$ and $d(\theta)$ are the diameters of the reference hole before and after trepanning respectively. The x-direction coincides with the $\theta = 0$ direction.

At the entrance and exit of a circular hole the distortion of the hole is no longer accurately described by Equation (3.3) [15]. Garcia Granada et al. [13] suggested introducing two parameters, A and B, to account of the near entrance and exit face distortions and the measured distortions are a function of the through-thickness position, z.

$$\widetilde{\varepsilon}(\theta, z) = \frac{1}{E}\{f(\theta, z)\sigma_{xx} + g(\theta, z)\sigma_{yy} + h(\theta, z)\sigma_{xy} - \upsilon\sigma_{zz}\} \tag{3.4}$$

where

$$f(\theta, z) = A(z)[1 + B(z)2\cos(2\theta)] \tag{3.5a}$$

$$g(\theta, z) = A(z)[1 - B(z)2\cos(2\theta)] \tag{3.5b}$$

$$h(\theta, z) = 4A(z)B(z)\sin(2\theta) \tag{3.5c}$$

$A(z)$ and $B(z)$ represent measures of uniform expansion and eccentricity respectively of the hole. In earlier work $A(z)$ was selected to be 1 and $B(z)$ was determined from FE analysis for different plate thickness and varied from 0.98 at the surface to 0.85 at the mid-thickness of a plate [13]. To avoid the need to develop a numerical solution for matrix inversion (described later) it is assumed that $A(z) = B(z) = 1$. Earlier work [13] showed that the largest error was about 12% for a uniaxial residual stress and occurred at the surface, and reduces to about 5% for equibiaxial stresses.

Equation 3.4 provides the relationship between in the stresses and the in-plane distortions of the reference hole. Additionally the distortion of the core containing the reference hole can be measured and this is converted to strain using

$$\varepsilon_z = \frac{\Delta h_z}{\Delta h_{avg}} \tag{3.6}$$

where Δh_{avg} is the trepan depth increment and Δh_z is the change of height of the core during the trepan.

Using the theory described above a through-thickness residual stress distribution is calculated from measured distortions using a compliance matrix. Since the trepanned core is assumed to be composed of a stack of independent annular slices, stresses at a given depth are found independently from those at other depths. The reference hole distortions are measured at a set of n depths $z = \{z_1, z_2, \ldots, z_n\}$ and a set of m angles $\theta = \{\theta_1, \theta_2, \ldots, \theta_m\}$, where $m \geq 3$. Therefore, at each depth z_i, the measured distortions are assembled into a vector of m components.

$$\{\widetilde{\varepsilon}(z_i)\} = [\widetilde{\varepsilon}(\theta_1, z_i), \widetilde{\varepsilon}(\theta_2, z_i), \ldots, \widetilde{\varepsilon}(\theta_m, z_i), \widetilde{\varepsilon}_{zz}(z_i)]^T \tag{3.7}$$

The distortion vector is then related to the stresses

$$\{\widetilde{\varepsilon}(z_i)\} = -[M(z_i)]\{\sigma(z_i)\} \tag{3.8}$$

where the compliance matrix is

$$[M(z_i)] = \frac{1}{E} \begin{bmatrix} f(\theta_1, z_i) & g(\theta_1, z_i) & h(\theta_1, z_i) & -\nu \\ \vdots & \vdots & \vdots & \vdots \\ f(\theta_m, z_i) & g(\theta_m, z_i) & h(\theta_m, z_i) & -\nu \\ -\nu & -\nu & 0 & 1 \end{bmatrix} \quad (3.9)$$

and

$$\{\sigma(z_i)\} = [\sigma_{xx}(z_i), \sigma_{yy}(z_i), \sigma_{xy}(z_i)]^T \quad (3.10)$$

The functions f, g and h in the M matrix are given by Equation (3.5), with $A = B = 1$.

Finally, the unknown stress components $\{\sigma(z_i)\}$ are calculated from the measured distortions using least squares, so that

$$\{\sigma(z_i)\} = [\sigma_{xx}(z_i), \sigma_{yy}(z_i), \sigma_{xy}(z_i), \sigma_{zz}(z_i)]^T \quad (3.11)$$

This equation provides the optimized (or best fit) stresses from the measured distortions. As will be shown later, in some circumstances, the distortion of the core is not measured, in which case either the stress σ_{zz} is assumed to be zero (plane stress) or the distortion ε_{zz} is zero (plane strain). The compliance matrix, Equation (3.9), can be modified to reflect these assumptions.

3.2.2 Effects of Plasticity

The presence of high levels of residual stress near the yield strength in a metallic component can cause plasticity to occur during the material removal processes of the DHD method. However, there is no straightforward relationship between the measured displacement changes and the residual stress. Furthermore, it may not be evident that plasticity occurs. Plasticity introduces errors in the deep hole drilling measurements of residual stress for two reasons. The first is because a yielded region forms around the drilled hole due to stress concentration effects. These perturb the residual stress field. The second is because additional yielding takes place during the trepanning operation, invalidating the assumption of purely elastic unloading. To account for these effects it is proposed [16,17] that trepanning is interrupted at a given depth with distortions of the hole measured at this intermediate position. It has been found [16,17] that the changes in diameter at this position capture the elastic distortions representative of the recovery of the residual stresses. These distortions are then used in the standard elastic analysis described by equation 3.11. The incremental technique relies on obtaining measurements of hole distortions progressively for increments of trepanning depth giving diameters d_j where j is the increment of interrupted trepanning steps. The changes in diameter are normalized to give normalized distortions $\tilde{\varepsilon}_j = (d - d_{0j})/d_{0j}$ at the jth trepan increment, while changes in core length are normalized to give normalized through-thickness distortions $\varepsilon_{zj} = \Delta h_{zj}/\Delta h_{avgj}$.

In practice, only a limited set of measurements along the reference hole axis can be obtained corresponding to the trepan increments. At each trepan increment the

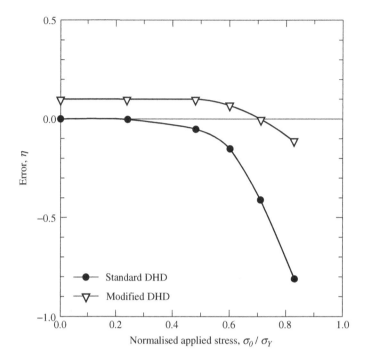

Figure 3.4 Error versus normalised stress for conventional and incremental DHD techniques, where error is defined as $\eta = \frac{\sigma_m}{\sigma_o} - 1$. Reproduced from [17], Copyright 2011 Elsevier

measured hole distortions are introduced into the standard DHD analysis procedure, Equation (3.11), to provide in-plane and out-of-plane stress components at each trepan increment, j, $\{\sigma_j(z_j)\}$ and are then compared and combined to create the finalized, discrete measurement results.

To illustrate the influence of plasticity on the DHD method, a finite element model was developed by Mahmoudi et al. [17] to simulate the reconstruction of stresses applied to a solid cylinder. Figure 3.4 shows their results, where the error in reconstruction is shown as a function of the applied stress normalized with respect to the yield stress. A negative error means that the measured residual stress is lower than the actual residual stress. The conventional DHD calculation is accurate at low magnitudes of applied stress but becomes unacceptably inaccurate for stresses greater than about $\frac{\sigma_0}{\sigma_Y} = 0.5$. The incremental DHD calculation introduces a new error for low magnitudes of applied stress (about 10%). However, the accuracy of the incremental DHD calculation is much better than the conventional one for higher magnitudes of applied stress.

3.3 Experimental Technique

Figure 3.2 schematically illustrates the overall procedure described in Section 3.1. A number of important components are required to employ the DHD technique [18], these include a combined reference frame and specimen table, a gun-drill system, a hole diameter measurement system, an electro-discharge machine and a computer for control and data logging. The reference frame and specimen table are used to ensure correct alignment of the sample and the devices for drilling (gun-drill), measurement of the reference hole (air-probe) and trepanning.

The reference frame and specimen table provide support for the measured component and to align the various components of the DHD process to the sample via the reference frame. Using a gun-drill creates a precision hole through the component. For the majority of the work reported later in this chapter the reference hole diameter was either 3.175 mm or 1.5 mm. Notably, using a gun-drill provides a highly repeatable surface finish, an axial deviation of about 0.1 μm per mm of depth, and a diameter tolerance of about ±10 μm for reference hole diameters less than 10 mm.

Prior to creating the reference hole, it is essential to install reference bushes [18,19]. Figure 3.5 illustrates this and shows the cross-section of a welded component with a front and rear reference bush glued to the component. Both bushes act as reference diameters and also permit estimates of experimental uncertainty to be determined [19]. The reference bushes do not change in shape since no residual stress is relaxed in them. The front bush also acts as a starting point for drilling. Figure 3.5 also shows an outer front bush. This provides support and alignment for gun-drilling, diameter measurement and trepanning of the core using an electro-discharge machine.

When drilling is completed, the diameter of the reference hole is measured as a function of angle, θ and position, z along the reference hole. The major development adopted by Leggatt et al. [10] was to use an air probe for diameter measurement. This system is widely used in the manufacturing industry to check the tolerance of drilled holes. Compressed air is passed through a tube and fed through diametrically opposite nozzles at the end of the tube. By restricting the flow of the air (as occurs when the probe is passed down a hole), the change in back-pressure can be monitored using a pressure transducer. Precise calibration of the probe is required to relate the back-pressure to diameter. This technique has been developed further [18,19] so that hole diameters can be measured accurately to better than 0.5 μm.

Typically, diameters are measured at every 10° around the circumference and at every 0.2 mm along the reference hole. This requires computer-controlled servo-motors that move the measurement device along and around the reference hole and also needs devices to record the position of the probe and the diameter measurement.

When the measurement of the reference hole diameter is finished equipment to trepan the over-core replaces the measurement system. There are many methods for machining the over-core and include electro-chemical machining [7] and mechanical cutting using diamond tipped hole saws [20]. An effective process for metal components is electro-discharge machining. This process removes material by means of repetitive spark discharges between the electrode tool and the component. Provided the correct operating parameters are selected, the electro-discharge machining provides a relatively stress free machining process. Servo-controlled mechanisms are required to rotate the electrode around and co-axial to the axis the reference hole. It is essential that the electro-discharge machining system is accurately positioned along the axis of the reference hole.

Figure 3.5(b) shows schematically the arrangement for the electro-discharge machining electrode, usually a copper tube, co-axial to the reference hole. When 3.175 mm and 1.5 mm diameter reference holes are used typical electrode inner diameters are 10 mm

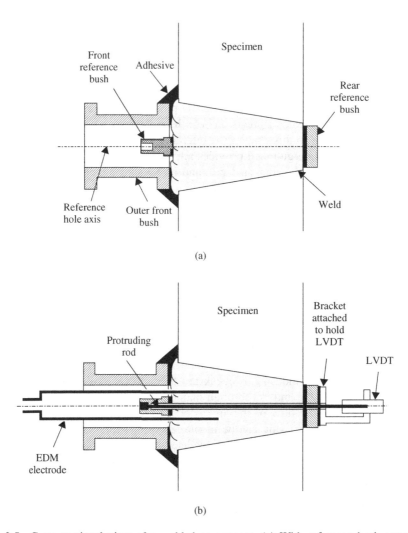

Figure 3.5 Cross-sectional view of a welded component. (a) With reference bush arrangement and (b) subsequent set-up for trepanning using an electro-discharge machining tube electrode and measurement of the axial distortion of the core. Reproduced from [18]

and 5 mm diameter respectively [18,19]. During electro-discharge machining a dielectric is passed down the center of the electrode and exits down the outside. Continuous flushing removes material away from the tip of the electrode with the electrode moving slowly forward and eventually finishing at the rear-bush. The presence of the glue holding the rear-bush acts as barrier to further trepanning.

Changes in the length of the trepanned core are monitored simultaneously during the trepanning process. The distortion of the core is transmitted via a rod connected to the front bush. This rod passes through the reference hole and protrudes out of the rear face, as shown in Figure 3.5(b), connected to a linear variable differential transformer that records the distortion.

The final step in the DHD method is the re-measurement of the reference hole in the same way as described earlier. If it is required to implement the incremental DHD method it is essential to interrupt the electro-discharge machining trepan. This is done at various (pre-selected) positions through the depth, z, of the component. The diameter measurement system is re-introduced and realigned to obtain diameter measurements at the preselected positions.

The principal outcomes of the experimental procedure are records of the diameter measurements before and after trepanning. These are obtained as a function of angle around the circumference of the reference hole and positions along the reference hole. These records are then used to determine the distortions corresponding to Equation (3.7). Knowledge of the material's Young's modulus and Poisson's ratio, together with the angular positions of the diameter measurements are required to form the compliance matrix M, Equation (3.9). Stresses are then computed from the measured distortions using Equation (3.11).

3.4 Validation of DHD Methods

Several studies have been conducted to validate the DHD method using samples subjected to external loading in the elastic range to create a "known" stress or by applying a load history in the plastic range to create a "known" internal stress state. The "known" stresses are then compared with measured stresses to determine the overall uncertainty in the stresses. This procedure often relies on using analytical methods to establish the "known" stress. Examples include rectangular bars subjected to elastic deformation by tensile or torsional loads [21], or prior elastic–plastic deformation [10,22,23], shrink fitted assemblies, [18,16,24], or tubes subjected to hydrostatic pressure [21] or overstressed (autofrettaged) tubes [25]. If each sample is seen as a standard (i.e., the measurand) then the measurement is a means of calibrating the measurement method. Identification of a significant difference between the standard and the measurements indicates whether there is any systematic bias in the measurement methods.

In this section a number of examples are described where the DHD method has been validated experimentally. Also included is a numerical validation of the DHD method using a finite element model [26], to simulate the measurement process and compare with the originally determined residual stress.

3.4.1 Tensile Loading

Experiments [21] using aluminium rectangular bars, containing initial reference holes, permitted an assessment of the accuracy of the measurement of hole distortion using the air-probe system. Figure 3.6 shows a typical example of the measured hole distortion or strain for a 20 mm thick aluminium rectangular bar subjected to an applied stress of 57 MPa, where 18 angular distortions were measured. Also shown is the fitted curve using Equation (3.11) and assuming $E = 70$ GPa.

Figure 3.7 summarizes the results of a series of tensile tests on various aluminium rectangular bars for two reference hole sizes, 3.175 and 1.5 mm. The measured stress, averaged through the thickness of each bar, is compared with the applied stress. Error bars for each data value correspond to one standard deviation obtained through the thickness for

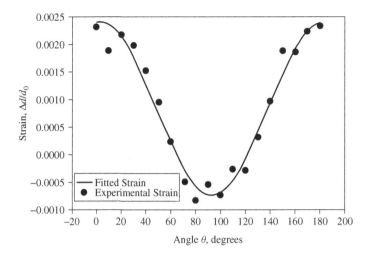

Figure 3.6 Typical example of a least squares fit to measured diametral strain

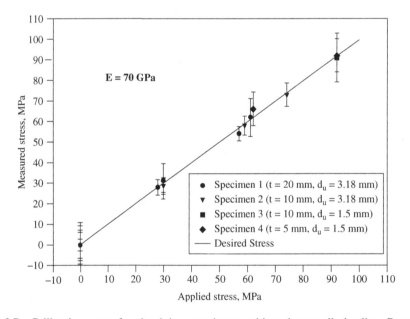

Figure 3.7 Calibration curve for aluminium specimens subjected to tensile loading. Reproduced with permission from [21], Copyright 2002 Sage

each applied stress. Irrespective of the applied load, up to a stress of 100 MPa, the standard deviation was about ±10 MPa, corresponding to a resolved strain of ±150 με. For a 3.175 mm diameter air-probe this equates to a measurement resolution of less than 0.5 μm.

By adopting test samples suggested by Proctor and Beaney [7], George et al. [21] undertook additional validation studies, measuring the distortion of holes in steel cylinders subjected to external pressure and beams in elastic bending. In steel the estimated error was found to be ±30 MPa. These studies were confined to stresses below about one third of the yield stress of the materials.

3.4.2 Shrink Fitted Assembly

A second validation study [16], was concerned with residual stresses created by shrink fitting. When a ring is shrunk onto a shaft, the shaft is subjected to uniform pressure resulting in a compressive residual stress in the shaft. When the ring has a smaller length than the shaft the compressive stress along the center line of the shaft is distributed so that the compressive stress is zero at the ends of the shaft. Figure 3.8(a) shows the dimensions of an aluminium alloy shrink fitted ring onto a shaft. Residual stresses were measured along the axis of the shaft as shown in Figure 3.8 using the conventional DHD technique [16]. Along this line, the components of residual stress σ_{xx} and σ_{yy} are both equal to the radial residual stress, σ_{rr}. The reference hole diameter used for these measurements was 3.175 mm and the diameter of the trepanned core was 10 mm. No measurements were made of the axial strain since the axial residual stresses, σ_{zz} were assumed to be negligible. The measurement used Equation (3.11) to convert measured distortions to stresses assuming that $\sigma_{zz} = 0$.

Figure 3.8(b) shows the measured radial residual stress distribution as a function of position, z, along the axis of the shaft. The peak compressive stress, 103 MPa, occurred midway along the length of the shaft. The radial residual stresses at the ends of the shaft were approximately zero. The error bars were derived by Goudar et al. [27] and used the uncertainties in material properties, uncertainty related to the calibration and use of an air probe and uncertainty in the location of the air-probe in the reference hole.

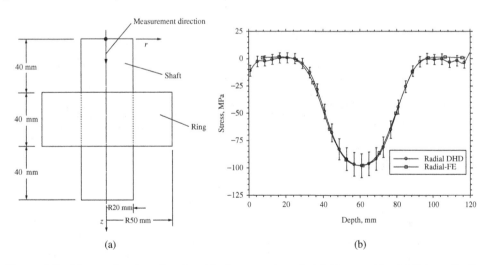

Figure 3.8 Measured and predicted residual stresses in a shrink fit assembly. (a) Schematic diagram of the shrink fit assembly and DHD measurement direction. Reproduced with permission from [16], Copyright 2009 Springer. (b) Comparison between DHD measurement and FE prediction, including uncertainty on the measurement. Reproduced from [27], Copyright 2011 John Wiley & Sons

Also shown in Figure 3.8 is a prediction from a finite element analysis described by Mahmoudi et al. [16]. There is excellent agreement between FE and experiments. Since the stresses were about one quarter of the yield strength of the aluminium alloy, the results confirmed that using the analysis described in Section 3.2 for the conventional DHD analysis was sufficiently accurate and that there was no evidence of plasticity during the DHD process.

3.4.3 Prior Elastic–plastic Bending

The third validation example relates to the formation of "known" internal stress resulting from prior elastic–plastic loading. Residual stresses are introduced in metal rectangular beams when they are subjected to 4-point bending to induce elastic and plastic deformation and unloaded. Residual stresses are then created through strain incompatibility when unloading the beams. This method was used by Leggatt et al. [10] and Goudar et al. [22] to generate a uniaxial residual stress distribution that varies through the depth of the beam. A bi-axial residual stress distribution can also be generated using thick discs subjected to ring loading [19,23].

Figure 3.9 illustrates an example of recent findings [22] for a residual stress generated in a 50 mm square section and 250 mm long steel beam. DHD measurements are shown together with the "known" stress field determined from a FE simulation of the creation of the residual stresses in the beam. The conventional DHD method was applied because FE

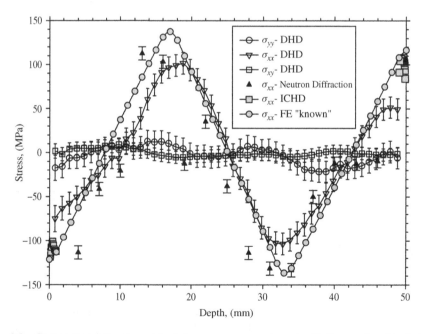

Figure 3.9 Comparison between residual stress measurements and with "known" residual stresses in an unloaded elastic–plastic beam, measurement through the complete depth of the beam. Reproduced with permission from [22], Copyright 2013 Springer

simulations of the DHD method revealed that no plastic relaxation of the residual stresses during trepanning is expected.

There is not exact agreement with the "known" stress distribution and it is found that the measurement method does not capture the expected peak tensile and compressive stresses at depth of about 17 and 33 mm. This has also been found by others for uniaxial [10] and biaxial [23] residual stress distributions.

Also measurements using neutron diffraction and incremental surface hole drilling methods were made in the same beam. These results are also shown in Figure 3.9. The application of simple bounds to the combined data indicates an uncertainty of about ±50 MPa. This is greater than the measurement accuracies of the individual measurement techniques. This is also greater than the uncertainty associated with measurement of simple "known" stresses created by external loading.

3.4.4 Quenched Solid Cylinder

The final example in this section is the measurement of residual stresses created in a quenched cylinder, and demonstrates how plasticity impacts on the DHD method. Hossain et al. [28] confirmed that quenching of a solid stainless steel cylinder generates residual stresses that are equal to the yield strength near to the surface of the cylinder and in excess of the yield strength at the interior due to imposed triaxial constraint. Figure 3.10 shows the distribution of radial residual stresses through the depth of a cylinder.

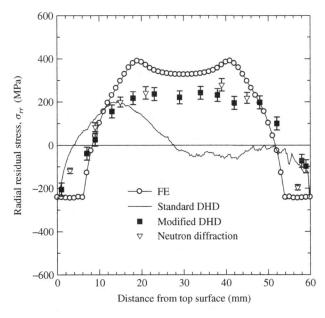

Figure 3.10 Comparison of predicted and measured radial residual stresses versus distance from the top surface of a quenched cylinder. Reproduced with permission from [16], Copyright 2009 Springer

An initial experimental application of the conventional DHD method [29] revealed that the residual stresses could not be fully reconstructed across the diameter of the quenched cylinder. This is evident in Figure 3.10 for depths greater than about 15 mm. However, the development and experimental application of the incremental DHD method provided reconstructed stresses closer to the predicted values. Additional neutron diffraction measurements on the cylinder confirmed the incremental DHD measurements. These results suggest that the larger predicted stresses from the FE analysis arise from a number of simplifications made in the model and include the material hardening and absence of time dependency in the material model.

3.5 Case Studies

A wide variety of examples of the practical application of the DHD method to engineering components can be drawn upon, some of which were summarized earlier by Kingston et al. [30]. These include a large forged steel roll with a diameter of 435 mm, a steel rail section, a section of a welded submarine hull and a friction stir welded titanium tube. Here, it is intended to provide a number of examples, particularly instances where measurements were either made using other techniques or the measurements were used to confirm predictions from finite element simulations. Examples include welded nuclear components, equipment used in the steel rolling industry and finally, fibre composite materials. In all the case studies the DHD method was applied because it was the only practical method that could be employed for large components with measurement depths in excess of 50 mm. The alternative would be a fully destructive measurement.

3.5.1 Welded Nuclear Components

The first case study is based on major investigations in Japan, [31–33] and in the USA, [34,35]. In each case a series of full-scale mock-ups of safe-end nozzles for pressurized water reactors (PWR) were manufactured. The purpose of the work was to determine the contribution of residual stresses to stress corrosion cracking (SCC). This cracking is one of the largest and most common problems faced by nuclear power station operators the world over, notably in nickel-based alloys used for cladding, buttering or weld filler materials in PWR and boiling water reactors (BWR).

In the Japanese investigation full-scale dissimilar-metal weld (DMW) mock-ups of 735 mm internal diameter, and 73 mm wall thickness at the weld were manufactured, [31–33]. In the USA research [34,35] full-scale dissimilar-metal welds were examined with additional structural weld overlays introduced. Weld overlay is used extensively as a repair and mitigation technique for SCC in pressurizer nozzle dissimilar-metal welds. Two DMW mock-ups were manufactured, one nozzle (mock-up A) in the "virgin" state, and a second mock-up B, nominally identical to A but with the addition of a full structural weld overlay. Figure 3.11 shows a cross-section through the DMW nozzle, with the overlay illustrated as the hatched area over the main weld. The outer diameter of the nozzle at the location of the DMW was about 203 mm with a wall thickness 32 mm. The installation of the weld overlay increased the wall thickness to 50.5 mm. Residual stress

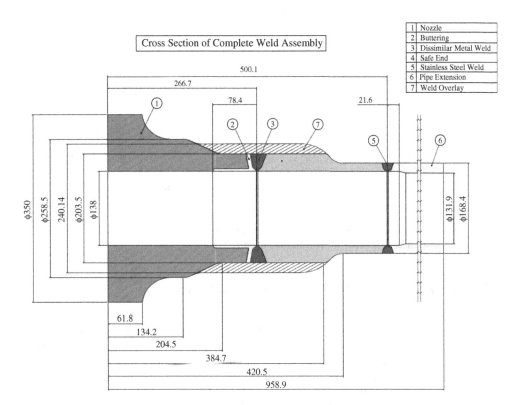

Figure 3.11 Schematic cross-section of a PWR pressurizer nozzle, including a full structural weld overlay. All dimensions in mm

measurements were made through the centerline of the DMW at different locations around the circumference of mock-ups A and B, using both conventional and incremental DHD methods. The reference hole and core diameters in the DHD methods were 1.5 mm and 5 mm respectively. Measurements of the axial distortion of the core during trepanning were not made.

Predictions of residual stresses were made also using a thermo-mechanical finite element analysis assuming an axisymmetric model of the nozzle. Further details can be found in [35]. A comparison between predictions and measurements of the axial and hoop stresses in the nozzle are shown in Figure 3.12. The through-wall depth has been normalized with respect to the wall thickness with the overlay absent. The FE predictions are for a range of heat input times, and the measurements correspond to different angular positions around the circumference.

There is very good agreement between measurement and prediction in the hoop direction except at 0° location. In both mock-ups the axial measured stresses were slightly more tensile than the predicted stresses. This is thought [35] to be due to through-wall bending that was not accounted for in the FE analysis. Conspicuously, the addition of the weld overlay did not make the inner wall more compressive but did make the zone of compression expand deeper from the inner surface towards the outer surface.

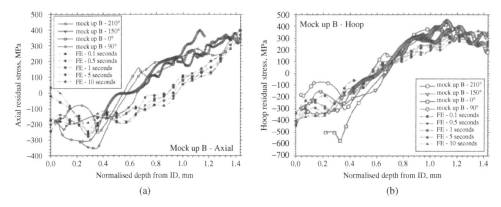

Figure 3.12 Comparison of measured and predicted residual stresses in a PWR nozzle with a full structural overlay. Reproduced with permission from [35]

3.5.2 Components for the Steel Rolling Industry

The second case study is related to residual stresses in large components used in steel rolling. Much earlier work on the application of residual stress techniques to steel rolls has been confined to near surface measurements [36]. The developments in the DHD method to measure large components led to its application to forged and quenched steel rolls [30] and various cast iron sleeves [36]. One example is a bimetallic sleeve. The sleeve was spun-cast, with an outer layer of HiCr (2.5%C. 13%Cr) cast onto an inner layer of a GSC150 steel. A residual stress profile was measured using the conventional DHD method, along a radial line in the center of the sleeve.

Since the depth of measurement was large (up to 200 mm) a 5 mm diameter reference hole with a 15 mm trepan diameter was used and distortion measurements confined to the reference hole, with axial distortions of the trepanned core not measured. The resulting profile of the hoop stress is illustrated in Figure 3.13, demonstrating that the HiCr steel was in compression, with the inner cast material in tension. Hardness profiles and optical micrographs revealed that the transition from compressive to tensile residual stresses coincided with a transition between the two materials [36].

3.5.3 Fibre Composites

The final practical case study is concerned with residual stresses in carbon fiber composites. The application of the DHD technique was explored by Bateman et al. [37] on a 22 mm thick carbon fiber-epoxy composite laminate. This work required the isotropic elastic analysis, given in Section 3.2, to be revised to permit the formation of a compliance matrix for orthotropic materials. Experimental validation of the DHD method for a unidirectional composite using the revised compliance matrix was also undertaken. Figure 3.14 shows the through-thickness arrangement of the 22 mm thick composite laminate. Overcoring was completed using a diamond tipped hole saw with a 17 mm outside diameter leaving an intact 8 mm diameter core.

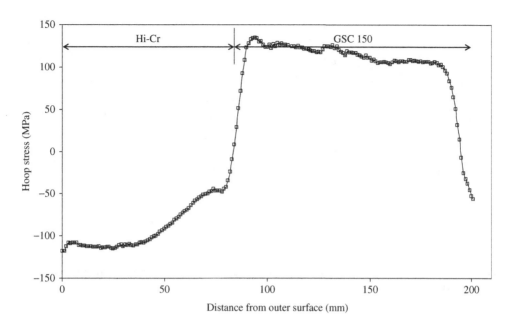

Figure 3.13 Measured residual stresses in a spun-cast bimetallic Hi-Cr, G-Cr-Ni-Mo 65 (GSC 150) sleeve intended for a hot-rolling sleeve, measured hoop residual stress distribution. Reproduced with permission from [36], Copyright 2009 Maney

A typical measured residual stress distribution of the in-plane stresses are shown in Figure 3.14. The highest stresses are of the order of 40 MPa in the fiber direction for the set of 0° plies nearest the surface of the laminate. The highest stress transverse to the fiber was less than 10 MPa. It is pleasing to see that the residual stresses were approximately symmetric about the center section of the laminate.

3.6 Summary and Future Developments

Usually, metallic structures and components formed by welding, casting, forging and machining contain highly varying residual stress fields through a component's thickness. The DHD method is one of several mechanical strain relaxation methods suitable for obtaining these stress distributions. Although initially developed for application in rock mechanics and large civil engineering structures, the case studies show that the DHD method has now been widely used to measure residual stresses in components manufactured and fabricated from metals and composites. It has advantages over other through-thickness methods. It is semi-invasive, but this means it only has the ability to obtain a distribution along a line corresponding to the reference hole created as part of the DHD process. However, this permits it to be applied to components of complex shape such as nozzle-to-nozzle intersections [38,39], where other methods could not be applied. The DHD method can also be applied to in-situ welded components, usually full-scale mock-ups [40], without recourse to using a specialized laboratory. The analysis

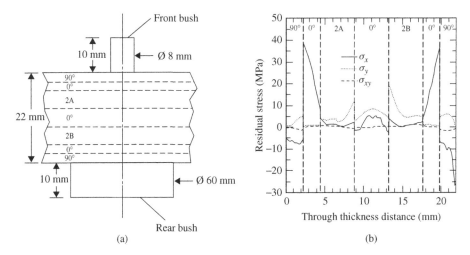

Figure 3.14 Measured residual stresses in a thick section composite laminate. Reproduced from [37], Copyright 2005 Elsevier

for reconstruction of the stresses from measured distortions also made a number of simplifications (Section 3.2) which are shown, from a variety of validation studies (Section 3.4), to be relatively robust.

As with the fully destructive methods, there is no practical limit in size to which the DHD method can be applied, but, as the case studies presented in Section 3.5 demonstrate, it does require, as explained in Section 3.3, highly specialized tools, with special care required for co-axial alignment over large distances. Measurement of in-situ residual stresses is of particular interest, especially at different stages in the life of a component and in highly aggressive environments, and this is leading to the creation of remotely controlled DHD devices [41].

Other mechanical strain relaxation methods such as the contour method (see Chapter 5) have the advantage of being able to obtain maps of the stress distribution in components. While results from the DHD method are confined to a line distribution, full 3D distributions of the residual stresses can be derived by introducing the line distributions into finite element models. The limited experimental results, combined with the necessary boundary conditions, can determine the residual stress distribution in the remaining sections of the component [36]. This and the development of remote controlled DHD devices is the subject of future research.

Acknowledgments

Most of the work reported in this chapter is the result of many hours of research and development by the team at the University of Bristol with financial support provided by UK government in the form of research grants and from direct industrial funding. In particular I would like to thank my Bristol academic colleagues: Martyn Pavier and Chris Truman. Bristol DHD researchers have been invaluable and include Neil Bonner, Andres Granada-Garcia, Dan George, Ed Kingston, Danut Stefanescu, Xavier Ficquet, Sayeed

Hossain, Ali Sisan, Amir Mahmoudi, Foroogh Hosseinzadeh, Nova Simandjuntak, Soheil Nakhodchi, Dev Goudar and Gang Zheng. I am grateful for the support of a Royal Academy of Engineering Professorship which is co-funded by EDF-Energy, Rolls-Royce and the Royal Academy of Engineering.

References
Introduction and Background

[1] Leeman, E. R., "The Measurement of Stress in Rock, Parts I–III." *J So Afr Inst Min Met* Part I, September 1964, pp. 45–81; Part II, September 1964, pp. 82–114; Part III, November 1964, pp. 254–284; February 1965, pp. 408–423, July 1965 pp. 656–664; October 1965, pp. 109–119.

[2] Fairhurst, C. (2003) "Stress Estimation in Rock: A brief history and review," *Int. J. of Rock Mechanics & Mining Sciences*, 40: 957–973.

[3] Martin, C. D. and Christiansson, R. C. (1991) "Overcoring in Highly Stressed Granite: Comparison of USBM and Modified CSIR Devices," *Rock Mechanics and Rock Engineering* 24: 207–235.

[4] Ljunggren, C., Chang, Y., Janson, T. and Christiansson, R. (2003) "An Overview of Rock Stress Measurement Methods," *Int. J. Rock Mechanics & Mining Sciences*, 40: 975–989.

[5] Ryall, M. J. (1999) "Theory and Application of the Hard Inclusion Techniques for the Measurement of in Situ Stresses in Prestressed Concrete Box-girder Bridges," *Proc. Instn Civ. Engrs Structs & Bldgs*, 134: 57–65.

[6] Beaney, E. M. "Measurement of sub-surface stress." Report RD/B/N4325, Central Electricity Generating Board, July 1978.

[7] Proctor, E. and Beaney, E. M. (1987) "The Trepan or Ring Core Method, Centre-hole Method, Sach's Method, Blind Hole Methods Deep Hole Technique" in *Advances in Surface Treatments; Technology-Application-Effects*, edited by A. Niku-Lari, 4: 165–199.

[8] Ferrill, D. A., Juhl, P. B. and Miller, D. R. "Measurement of Residual Stresses in Heat Weldments," *Weld. Res.*, Nov 1966, 540s.

[9] Zhdanov, I. M. and Gonchar, A. K. "Determining the Residual Welding Stresses at a Depth in Metal," *Autom. Weld.*, September 1978, 31(9): 22–24 (translation of *Avtomaticheskaya Suarka*).

[10] Leggatt, R. H., Smith, D. J., Smith, S. and Faure, F. (1996) "Development and Experimental Validation of the Deep Hole Method for Residual Stress Measurement," *Journal of Strain Analysis for Engineering Design*, 31(3): 177–186.

Basic Principles

[11] Walker, J. (1987) "Development of the Elasticity Matrix Used in Deep Hole Residual Stress Measurement," CEGB Report No. TPRD/B/0895/R87.

[12] A. T. DeWald and M. R. Hill (2003) "Improved Data Reduction for the Deep-hole Method of Residual Stress Measurement," *J. Strain Analysis* 38(1): 65–78.

[13] Granada-Garcia, A. A., George, D., and Smith, D. J., "Assessment of Distortions in the Deep Hole Technique for Measuring Residual Stresses," in Proceedings of the 11th Exp Mech Conf, Oxford, Aug, 1998, pp. 1301–1306.

[14] Timoshenko, S. and Goodier, J. N. (1951) *"Theory of Elasticity,"* Mcgraw-Hill, 2nd Edition.

[15] Sternberg, E. and Sadowsky, M. A. (1949) "Three Dimensional Solution for the Stress Concentration Around a Circular Hole in a Plate of Arbitrary Thickness," Trans. ASME, *J. Appl. Mech*, 16: 27–38.

[16] Mahmoudi, A. H., Hossain, S., Truman, C. E., Smith, D. J., Pavier, M. J. (2009) "A New Procedure to Measure Near Yield Residual Stresses Using the Deep Hole Drilling Technique." *Experimental Mechanics*, 49(4): 595–604.

[17] Mahmoudi, A. H., Truman, C. E., Smith, D. J., Pavier, M. J. (2011) The Effect of Plasticity on the Ability of the Deep Hole Drilling Technique to Measure Residual Stress, *International Journal of Mechanical Sciences*, 53: 978–988.

Experimental Technique

[18] Kingston, E. J. (2003) *"Advances in the Deep-hole Drilling Technique for Residual Stress Measurement,"* PhD Thesis, University of Bristol, UK.

[19] Goudar, D. M. (2011) *"Quantifying Uncertainty in Residual Stress Measurements Using Hole Drilling Techniques,"* PhD Thesis, University of Bristol, UK.

[20] Bateman, M. G., Miller, O. H., Palmer, T. J., Breen, C. E. P., Kingston, E. J., Smith, D. J. and Pavier, M. J. (2005) "Measurement of Residual Stress in Thick Composite Laminates Using the Deep-hole Method," *International Journal of Mechanical Sciences*, 47(11): 1718–1739.

Validation of DHD Methods

[21] George, D., Kingston, E. and Smith, D. J. (2002) "Measurement of Through-Thickness Stresses Using Small Holes," *Jnl. Strain Analysis*, 37(2): 125–139.

[22] Goudar, D. M. and Smith, D. J. (2013) "Validation of Mechanical Strain Relaxation Methods for Stress Measurement," to appear in *Experimental Mechanics*.

[23] Mabe, W. R., Koller, W. J., Holloway, A. M. and Stukenborg, P. R. (2006) "Deep Hole Drill Residual Stress Measurement Technique Experimental Validation," *Mat Sci Forum, Vols* 524–525: 549–554.

[24] Hosseinzadeh, F., Mahmoudi, A. H., Truman, C. E. and Smith, D. J. (2011) "Application of Deep Hole Drilling to the Measurement and Analysis of Residual Stresses in Steel Shrink-Fitted Assemblies," *Strain* (2011) 47 (Suppl. 2), 412–426.

[25] George, D. and Smith, D. J. (2000) "Application of the Deep Hole Technique for Measuring Residual Stresses in Autofrettaged Tubes," in ASME, *Pressure Vessels and Piping*, 406: 25–31.

[26] Hossain, S., Truman, C. E. and Smith, D. J. (2012) "Finite Element Validation of the Deep Hole Drilling Method for Measuring Residual Stresses," *International Journal of Pressure Vessels and Piping* 93–94.

[27] Goudar, D. M., Truman C. E., and Smith, D. J. (2011) "Evaluating Uncertainty in Residual Stress Measured Using the Deep-Hole Drilling Technique," *Strain*, 47: 62–74.

[28] Hossain, S., Truman, C. E., Smith, D. J., and Daymond, M. R. (2006) "Application of Quenching to Create Highly Triaxial Residual Stresses in Type 316 H Stainless Steels," *International Journal of Mechanical Sciences* 48: 235–243.

[29] Hossain, S. (2005) "Residual Stresses Under Conditions of High Triaxiality," PhD Thesis, University of Bristol, UK.

Case Studies

[30] Kingston, E. K., Stefanescu, D., Mahmoudi, A. H., Truman, C. E., and Smith, D. J. (2006) "Novel Applications of the Deep-hole Drilling Technique for Measuring Through-thickness Residual Stress Distributions," *Journal of ASTM International*, 3(4), Paper ID JAI12568.

[31] Muroya, I., Ogawa, N., Ogawa, K., Iwamoto, Y., and Hojo, K., *"Residual Stress Evaluation of Dissimilar Weld Joint Using Reactor Vessel Outlet Nozzle Mock-up Model"* (report-1), Proceedings of PVP2008, ASME Pressure Vessels and Piping Division Conference July 27–31, 2008, Chicago, USA, Paper PVP2008-61829.

[32] Ogawa, K., Muroya, I., Iwamoto, Y., Chidwick, L. O., Kingston, E. J., and Smith, D. J., *"Measurement of residual stresses in the dissimilar metal weld joint of a safe-end nozzle component,"* Proceedings of PVP2009 2009 ASME Pressure Vessels and Piping Division Conference July 26–30, 2009, Prague, Czech Republic, Paper PVP2009-77830.

[33] Ogawa, K., Kingston, E., and Smith, D. J. (2007) *"Measurement of Residual Stresses in Full-scale Welded Components Typically Found in Japanese Nuclear Power Plants,"* Proceedings of the 19th Structural Mechanics in Reactor Technology, Toronto, Canada, CD.

[34] Goudar, D. M., Kingston, E. J., Smith, M., Goodfellow, A., Marlette, S., Feyer, P., *"Through Thickness Residual Stress Measurement in a Full Structural Weld Overlay on PWR Pressuriser Nozzle,"* Trans. SMIRT 21, 2011, paper 153.

[35] Smith, M., Muransky, O., Goodfellow, A., Kingston, E., Freyer, P., Marlette, S., Wilkouski, G. M., Brust, B., Shim, D., *"The Impact of Key Simulation Variables on Predicted Residual Stresses in Pressuriser*

Nozzle Dissimilar Metal Weld Mock-ups. Part 2 Comparison of simulation and measurements", Proceedings of ASME 2010 Pressure Vessels & Piping Division, Bellevue, Washington, USA, July, 2010.

[36] Hosseinzadeh, F., Smith, D. J. and Truman, C. E. (2009) "Through Thickness Residual Stresses in Large Rolls and Sleeves for the Metal Working Industry," *Material Science and Technology*, 25(7): 862–873.

[37] Bateman, M. G., Miller, O. H., Palmer, T. J., Breen, C. E. P., Kingston, E. J., Smith, D. J. and Pavier, M. J. (2005) "Measurement of Residual Stress in Thick Composite Laminates Using the Deep-hole Method," *International Journal of Mechanical Sciences*, 47 (11): 1718–1739.

Summary and Future Developments

[38] Smith, D. J., Kingston, E., Thomas, A. and R. Ehrlich (2002) "Measurement of Residual Stresses in Steel Nozzle Intersections Containing Repair Welds," *ASME, PVP* 434: 67–72.

[39] Kingston, E., Smith, D. J. and Watson, C., *"Measurement of Residual Stress in Tube Penetration Welds for Ferritic Steel Hemispherical Pressure Vessel Heads,"* Proceedings of the 19th Structural Mechanics in Reactor Technology, Toronto, Canada, 2007, CD.

[40] Ogawa, K., Kingston, E. J., Smith, D. J., Saito, T., Sumiya, R. and Okuda, Y., *"The Measurement and Modelling of Residual Stresses in a Full-scale BWR Shroud-support Mock-up,"* Proceedings of 2008 ASME Pressure Vessels and Piping Conference, Chicago, USA, Paper 6158, CD.

[41] Ficquet, X., Smith, D. J., and Truman, C. E., *"Residual Stress Measurement Using a Miniaturised Deep Hole Drilling Method,"* Proceedings of the 13th International Conference on Experimental Mechanics, Alexandroupolis, Greece, 2007, paper 386, CD.

4

The Slitting Method

Michael R. Hill
Department of Mechanical and Aerospace Engineering, University of California, Davis, California, USA

4.1 Measurement Principle

The slitting method is a mechanical relaxation technique capable of measuring a single in-plane normal component of residual stress through the thickness of a material. The method originally developed by Finnie and co-workers [1–5], has been reviewed in detail by Prime [6], and is the subject of a recent monograph by Cheng and Finnie [7]. The method has been called by various names in earlier work, including "crack compliance method," or simply the "compliance method," but ASTM Task Group E28.13.02 suggested using "slitting method" going forward to highlight a useful parallel between slitting and the "hole drilling method." Slitting has great utility for practical laboratory residual stress measurements because it is relatively simple to perform, can be done quickly, and offers excellent repeatability [8]. This chapter is intended to complement the extensive technical background existing presently in the literature, much of it summarized in [6] and [7], with a practical treatment useful to those interested in making measurements. Another source for useful practical information on slitting is the internet resource maintained by Prime [9].

In slitting, a planar slit is introduced by incrementally cutting into a material containing residual stresses in steps of increasing depth. Deformation near the slit, arising from release of residual stress on the (newly traction-free) slit faces, is measured as a function of increasing slit depth and used to determine residual stress existing normal to the slit plane prior to slitting. Much of this chapter considers a common implementation of slitting, where through thickness residual stress in a metallic flat block is determined from strain measured at the back face of the block by a strain gage, see Figure 4.1. However, the method has sufficient flexibility to enable measurements of residual stresses in a variety

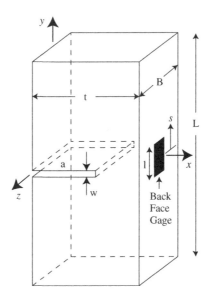

Figure 4.1 Block geometry for slitting. t = thickness of the part in the slitting direction, L = part length, B = part depth, a = slit depth, w = slit width, l = gage length, and s = distance between the centers of the gage and slitting plane. Copyright 2013, Hill Engineering, LLC

of materials such as metals, glass, crystal and plastic, and geometries such as blocks, beams, plates, rods, tubes, and rings [10–18].

In typical practice for metallic materials, the slit is cut using a wire electric discharge machine "wire EDM" [19], released strain is measured using a metallic foil strain gage and a bridge-type strain indicator. The residual stress is computed from the measured strains through a linear system called the compliance matrix, which is determined from finite element analysis, for example, [13,14,16,20]. Figure 4.2 shows the results of a typical measurement on a block coupon to evaluate the residual stresses in a quenched aluminum plate [21]. Strain versus slit depth data collected during the experiment are shown as symbols in Figure 4.2(a), while Figure 4.2(b) shows the computed residual stresses corresponding to the best-fit line in Figure 4.2(a). Error bars reported for stress represent two-standard-deviation uncertainties, which are computed based on the error propagation analysis described by Prime and Hill [22].

4.2 Residual Stress Profile Calculation

The procedure for calculation of residual stress given strain versus slit depth data is similar to the calculation procedures for other mechanical relaxation methods with incremental cutting, for example, hole drilling. The determination of residual stress from measured strain is an inverse problem, which for slitting is most commonly solved by using a series expansion for residual stress. Much of the published work assumes a polynomial series for stress, which converts the problem of finding residual stress versus depth into

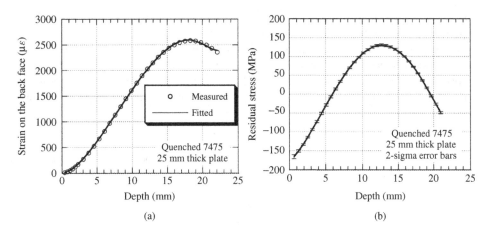

Figure 4.2 Results from a typical slitting measurement: (a) measured strain versus slit depth and (b) computed residual stress with two-sigma error bars. The fitted strain in (a) corresponds to residual stress in (b). Copyright 2013, Hill Engineering, LLC

a better-posed problem of finding a set of unknown coefficients of the polynomial series from measured strain data.

The stress calculation relies on a geometry-specific elasticity solution relating residual stress to measured strain, where the solution is embodied in the compliance matrix. The compliance matrix is most commonly determined using geometry-specific finite element analysis. Earlier work, based on finite element stress analysis, provides a scheme to compute analytically the compliance matrix for flat-block slitting of a long block ($L/t \geq 4$) having a small strain gage ($0.005 < l/t < 0.1$) mounted on the back face of the slit plane ($s = 0$), and these ready-to-use compliances simplify slitting for the novice practitioner [23]. In general application and in most published work, a set of finite element analysis results is used to provide the compliance matrix for stress calculation, where the analysis models the geometry of the experiment, including slit dimensions and gage locations. While this can present a significant burden, formulation of an accurate compliance matrix is essential for getting useful results.

In a flat block such as in Figure 4.1, slitting will determine the component of the residual stress tensor that acts normal to the slit plane. Measurement of the relieved strain vs. slit depth $\varepsilon_{yy}(a)$ enables subsequent calculation of the stress profile with depth, $\sigma_{yy}(x)$. The Legendre polynomial basis is convenient for describing through-thickness residual stress because, when the constant and linear terms are omitted, each term of the series automatically satisfies force and moment equilibrium conditions. Residual stress is therefore expressed as (dropping the yy subscript for simplicity)

$$\sigma(x) = \sum_{j=2}^{m} A_j P_j(x) \qquad (4.1)$$

where $P_j(x)$ are known Legendre basis functions and A_j are a set of multipliers to be determined to fit the measured strain data. The highest order Legendre term, m, appearing in the series is chosen during data reduction, for example, by error minimization [16,22], and generally has a value within the range $m = 4$ to 12.

Assuming linear elastic material behavior, the compliance matrix is a linear system relating unknown basis function coefficients to measured strain. A single element of the compliance matrix, C_{ij}, is the strain ε that would be occur at a stated strain gage location for a slit of depth a_i with residual stresses σ_{inp} exactly equal to a given Legendre polynomial term P_j

$$C_{ij} \equiv \varepsilon(a_i)|_{\sigma_{inp}=P_j(x)} \tag{4.2}$$

Using the principle of superposition, the strain that would occur as a function of slit depth for the residual stress in Equation (4.1) is

$$\varepsilon(a_i) = \sum_{j=2}^{m} C_{ij} A_j \tag{4.3}$$

or, adopting matrix notation

$$\underline{\varepsilon} = \underline{\underline{C}} \underline{A} \tag{4.4}$$

where a single underscore denotes a vector and a double underscore denotes a matrix. Typical experiments employ more than 30 strain vs. depth data pairs and require fewer than 12 basis function coefficients, so that Equation (4.4) is over-determined. Use of fewer than 25 data pairs can lead to significant increases in uncertainty [24]. The vector of basis function coefficients may be determined by inverting Equation (4.4) in a least squares sense, using the pseudoinverse and a vector of measured strain versus slit depth data, ε_{meas}

$$\underline{A} = [(\underline{\underline{C}}^T \underline{\underline{C}})^{-1} \underline{\underline{C}}^T] \underline{\varepsilon}_{meas} \tag{4.5}$$

The general approach to developing the compliance matrix is to use a well-refined finite element mesh with geometric details that match the experimental geometry. For a simple block geometry, a typical finite element mesh for the compliance matrix is symmetric about the slit plane, has high refinement close to the symmetry plane, and has small, uniform element size near the slit. The analysis can be carried out in three-dimensions, but a two-dimensional model can often provide sufficient accuracy. Recent work by Ayinder and Prime [25] enables compliances computed using a two-dimensional analysis, in plane stress or plane strain, to be adapted to three-dimensional situations that are between the two planar extremes. The finite element analysis is designed to proceed in a series of steps, where each step represents a given slit depth. Practically speaking, a zero-width slit can be progressed through a symmetric mesh with uniform elements by releasing the boundary conditions on a number of nodes along the symmetry plane that corresponded to a given slit depth, while a finite-width slit can be progressed by removing elements adjacent to the slit plane. With most commercial finite element codes, either of those options can be automated by using script-based editing of a text finite element input file, for example, in a programming language like python, perl, or awk, and a command-line interface to the operating system, for example, the bash shell available in unix or linux. A useful series of steps might represent slit depths $a_i/t = 0.005, 0.01, 0.02, \ldots, 0.96$. Each step is then subdivided into a set of increments, where each increment represents loading of slit face elements by a distribution of normal pressure corresponding to Legendre polynomials $P_2(x), P_3(x), \ldots, P_m(x)$. The result of this multi-step (and increment) finite element analysis is a state of deformation for the defined range of slit depths and polynomial orders of interest.

If the mesh is designed to have nodes located at the boundaries of a strain gage location, the compliance matrix can be determined for a given gage from displacement results provided by the multi-step finite element analysis. For a strain gage mounted at $s = 0$ on the back face of the block of Figure 4.1, at a step and increment in the analysis representing a_i/t and P_j, the compliance matrix element C_{ij} is computed by dividing the y-direction displacement at the node located at the gage boundary ($x/t = 1$ and $y = l/(2t)$) by $l/(2t)$ (that is, strain is computed as the change of length divided by initial length). Using a single node is specific for a two-dimensional mesh symmetric about the slit and symmetric gage location ($s/t = 0$), but using differences of nodal displacements is a general approach that can be applied for any gage location (and in a three-dimensional mesh) [26]. Repeating the analysis over all steps and increments provides a gage-specific compliance matrix having a number of rows equal to the number of slit depths and a number of columns equal to the number of polynomial terms ($m - 1$). The analysis can be repeated to obtain compliance matrices for other strain gage sizes and/or strain gage locations. Example columns of the polynomial-basis compliance matrix for a flat block, from the analysis described in [23], are shown in Figure 4.3 as input crack face tractions and resulting strain versus slit depth.

Measurements can be made with more than a single strain gage, and the use of multiple gages, read simultaneously at each slit depth, can improve the stability of the slitting method [16]. For example, using two strain gages on a flat block, one gage on the back face and a second gage near the start of the slit (e.g., at $x/t = 0$ and $y/t \approx +0.1$), has been found to reduce the uncertainty of near-surface residual stress [16]. Combinations of gages can be analyzed by assembling individual compliance matrices into a partitioned

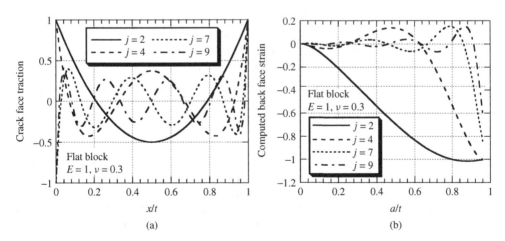

Figure 4.3 Results from finite element analysis of a flat block having elastic properties $E = 1$ and $v = 0.3$: (a) Legendre polynomial crack face tractions, of various orders j (b) resulting strain versus slit depth. Copyright 2013, Hill Engineering, LLC

system [13,16] of the form

$$\underline{\underline{C}}_{1,2,3} = \begin{bmatrix} \underline{\underline{C}}_1 \\ \underline{\underline{C}}_2 \\ \underline{\underline{C}}_3 \end{bmatrix} \quad (4.6)$$

where $\underline{\underline{C}}_{1,2,3}$ represents a three gage installation comprised of gages at locations *1, 2* and *3*. The basis function coefficient vector would be determined from this system using Equation (4.5) and a partitioned vector of measured strain

$$\underline{\varepsilon}^{meas}_{1,2,3} = \begin{bmatrix} \varepsilon^{meas}_1 \\ \varepsilon^{meas}_2 \\ \varepsilon^{meas}_3 \end{bmatrix}. \quad (4.7)$$

It should be noted that the order of the individual entries in the partitioned system does not affect the stress calculation because $\underline{\underline{C}}_{1,2,3}$ is a row-wise permutation of $\underline{\underline{C}}_{2,1,3}$ (and so forth for other combinations) and such permutations do not affect Equation (4.5), provided that the order of partitioning is consistent among Equations (4.6) and (4.7).

When considering different gage placements, analysis of the compliance matrix provides a means to select one placement over another. One gage placement is considered better than another if it provides lower uncertainty in residual stress for a given uncertainty in measured strain. The uncertainty in the residual stress results from uncertainty in the basis function coefficient vector $\underline{A}$ of Equation (4.1). Uncertainty in the coefficient vector may arise from several sources, with a primary source being uncertainty in strain measurement. Since the coefficient vector $\underline{A}$ depends on measured strain through the pseudoinverse of $\underline{\underline{C}}$ (term in square brackets in Equation (4.5)), the degree to which uncertainty in measured strain is amplified or attenuated depends on the stability of $\underline{\underline{C}}$. For a given uncertainty $\partial \varepsilon^{meas}$ in the vector of measured strain, the uncertainty $\partial \underline{A}$ in the residual stress coefficient vector is bounded by [27]

$$\frac{|\partial \underline{A}|}{|\underline{A}|} \leq \sqrt{\frac{\lambda_{max}}{\lambda_{min}}} \frac{|\partial \varepsilon^{meas}|}{|\varepsilon^{meas}|} \quad (4.8)$$

where $|\bullet|$ is the vector norm, and λ_{max} and λ_{min} are the largest and smallest eigenvalues of $\underline{\underline{C}}^T \underline{\underline{C}}$ (which is inverted in Equation (4.5)). Therefore, the amplification of uncertainty in measured strain depends on the ratio of the largest to smallest eigenvalues of $\underline{\underline{C}}^T \underline{\underline{C}}$, the square root of which is a quantity known as the condition number κ of the matrix $\underline{\underline{C}}$

$$\kappa = \sqrt{\frac{\lambda_{max}}{\lambda_{min}}} \quad (4.9)$$

A better gage placement therefore has a smaller condition number, and provides lower sensitivity to uncertainty in measured strain. Optimal strain gage placements for slitting may be developed by systematic analysis of the condition number [13]. As a general rule-of-thumb, application of St. Venant's Principle suggests that uncertainty is reduced by placing one or more strain gages close to the region within which the stress values are required.

While a polynomial basis analysis for stress is useful, more recent work has used the regularized unit pulse analysis described by Schajer and Prime [29], which assumes constant stress over each cut-depth increment (and therefore piecewise constant stress over the slit depth). While assuming piecewise constant stress generally leads to an unstable stress calculation, Schajer and Prime showed that Tikhonov regularization provides a stable calculation that is robust to the influences of measurement noise. The unit-pulse analysis is superior to a polynomial-basis analysis when measuring stress distributions that are expected to have very high gradients with depth, or to be discontinuous, because it avoids the implicit assumption of a smoothly varying stress profile inherent in a polynomial basis.

In general, the stress calculation for the unit-pulse analysis uses a compliance matrix that is mathematically similar to that used for the polynomial-basis analysis, but with different numerical values. Where the polynomial-basis compliance matrix typically has many more rows (slit depths) than columns (basis functions), the unit-pulse analysis compliance matrix is square (equal number of slit depths (rows) and depth increments having unit stress (columns)); where the polynomial-basis compliance matrix is full (non-zero strain for all slit depths and each basis function), the unit-pulse compliance matrix is lower triangular (strain at a given depth arises from stress at shallower depths, but not deeper ones). For the polynomial basis, C_{52} is the strain for a slit of depth a_5 resulting from a pressure distribution corresponding to the second polynomial acting on the entire slit face. For the unit pulse analysis, C_{52} is the strain for a slit of depth a_5 resulting from uniform pressure acting on the slit face over the interval $a_1 \leq x \leq a_2$. The unit-pulse compliance matrix also may be developed using finite element analysis, by following the procedure laid out above for the polynomial basis, but with unit pulse functions replacing the polynomials (a trivial change).

Given the compliance matrix, calculation of residual stress for the unit-pulse analysis uses Tikhonov regularization to provide a smooth stress versus depth profile robust to uncertainties in strain [27]. In slitting, the unknown residual stress profile is generally considered to be a smooth function of depth from the surface, except in special cases that include material inhomogeneity or discontinuous processing. The polynomial-basis analysis provides smoothness on account of the basis functions it employs. For the unit-pulse analysis, Tikhonov regularization provides a smooth result by including a penalty function in the stress calculation that scales the non-smoothness of the stress result, estimated as a vector norm of point-wise derivatives of stress with respect to depth (commonly the second derivative, computed as a divided difference), by a multiplicative weighting factor, β, called the regularization parameter. For a zero weighting factor β, the square compliance matrix for the unit-pulse method gives a stress solution that exactly corresponds to the measured strain data. However, the stress solution contains substantial noise that is amplified from the small experimental errors in the strain data. A positive weighting factor β has the effect of smoothing the stress solution, but it also slightly deviates the mathematically corresponding strains from the actually measured strains. The difference between the mathematical and actual strains is called the "misfit." Following the Morozov criterion, optimum regularization is achieved when the norm of the misfit equals the standard error in the strain measurements (see [28]). For smooth residual stress fields, the polynomial-basis and unit-pulse analyses generally provide similar results, but the unit-pulse analysis is very useful when stresses have abrupt changes, as can arise in coated, carburized, or discontinuously processed components.

4.3 Stress Intensity Factor Determination

While slitting is often used to determine residual stress, the method can also be applied to determine the residual stress intensity factor as a function of slit depth, which has found application in fracture mechanics. Considering the slit as a crack, Schindler et al. [30] have shown that the residual stress intensity factor at crack length a, $K_{Irs}(a)$, can be determined from an influence function $Z(a)$ and the change in strain $\varepsilon(a)$ (at a specified gage location) due to an infinitesimal change in slit depth da

$$K_{Irs}(a) = \frac{E'}{Z(a)} \frac{d\varepsilon(a)}{da} \quad (4.10)$$

where E' is equal to E in plane stress and $E/(1-v^2)$ in plane strain. The influence $Z(a)$ function is specific to a given geometry, strain gage location, and set of boundary conditions, but is independent of loading. Influence functions for beams, rectangular plates, and cylinders are available, most having been published by Schindler [29,31]. Having $Z(a)$, determination of $K_{Irs}(a)$ is straightforward, involving differentiation of the strain versus depth data and algebraic computations. Equation (4.10) has been applied to determine $K_{Irs}(a)$ in bodies containing unknown residual stresses [32,33], and has been useful in correlating the fracture behavior of bodies containing various levels of residual stress [32,34].

4.4 Practical Measurement Procedures

As with any experimental method, attention to detail in executing a slitting experiment is critical to a good outcome. The following discussion is divided into five parts: establishing a measurement location and reference frame; strain gage application; instrumentation and cutting; post-cutting metrology; and data analysis.

While it can be straightforward to establish a measurement location and coordinate reference frame on a given part, most real components exhibit departures from ideal shape (and those subjected to residual stress treatments can be significantly distorted). These departures from expected shape can cause significant errors in slitting, and following good practice typical of precision machining helps to control these errors. A first essential element is establishing the location and orientation of the slitting plane, which may be done with a granite flat, height gage, parallel, and perpendicular typical of precision machining. On the flat block of Figure 4.1, a reference plane orientation having a normal along $-x$ can be established by locating three points on the front face of the coupon, one each at (x, y, z) locations of $(0, +L/2, +B/2)$, $(0, -L/2, +B/2)$, $(0, +L/2, -B/2)$. Given this reference plane, two lines representing the intersections of the slit plane with the front $(-x)$ and back $(+x)$ faces of the component can be located at a distance $L/2$ along y from the block edge having normal along $+y$; light scribe lines may be used to mark the intersection of the slit plane with the front and back faces of the component.

Given the slit plane location and orientation, strain gages are applied at desired locations using typical procedures, but with care for protection from EDM hazards. Strain gages are typically centered at the mid-length ($z = 0$) of the slit plane intersection with the back face (i.e., at $s = 0$), and sometimes adjacent to the slit plane intersection with the front face, for example, at $y/t \approx +0.1$. Strain gage manufacturers provide useful guides for gage installation, for example, [35], and cyanoacrylate adhesive is generally suitable for use in slitting. Waterproofing of the gage installation is necessary when cutting with EDM (deionized water typically provides the electrolyte and flushing of cutting debris), and can be accomplished by using readily available coatings, for example, room-temperature volatizing silicone such as 3145 RTV or nitrile rubber. The adhesive should be masked to ensure that the coating overlaps the gage and the adhesive by at least 0.5 mm, so that water does not migrate under the gage by wicking along the bond line; for gages near start of the slit, the coating should be masked so it will not interfere with the EDM cutting. Smoothing of coating edges prevents coating lift-off due to high-pressure flushing jets used in EDM. EDM also introduces electrical interference, and bridge-based strain measurement devices should be isolated by grounding and shielding of lead wires; the best results are obtained when strain data are read with the EDM wire powered off between increments of cut depth. Figures 4.4 and 4.5 show example strain gage installations, with the latter illustrating the precision needed when using a front-face gage close to the slit.

EDM cutting offers good dimensional control and, since it does not rely on mechanical force, does not impose significant residual stress when cutting, if used at low material removal rates [19]. A typical wire electrode has 0.25 mm diameter, which produces a slit approximately 0.27 mm wide. While the slit depth increments can be made very fine (e.g., $t/200$ or smaller with half the wire diameter being a practical lower limit), it is typical to use larger increments, for example, $t/40$, over most of the slit depth with finer increments near expected gradients in stress, for example, near the start of the slit, material boundaries, or processing discontinuities.

Careful planning is required when mounting the specimen in the EDM cutting tool. Minimal clamping is generally desirable so that the sample can deform in response to stress release; any constraints imposed in clamping must be included in the model used to create the compliance matrix. For the block of Figure 4.1, it is usual to clamp the part at one end only, far from the slitting plane, for example, near $y = L/2$. It is then necessary to register the machine coordinate frame with the slitting plane, which is best done with reference to the points used to establish the slit plane markings on the front and back

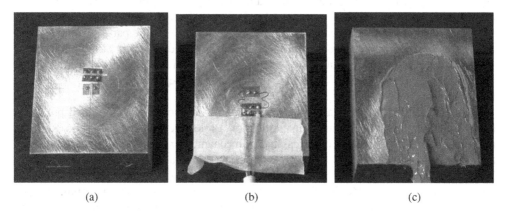

(a)   (b)   (c)

Figure 4.4 Strain gage application on back face of a coupon (= 50.8 mm in vertical direction). (a) Strain gage application (b) lead wire attachment (c) waterproof coating (3145 RTV). Copyright 2013, Hill Engineering, LLC

Figure 4.5 Post-experiment, microscopic image of a front face strain gage next to the EDM slit, after removal of waterproof coating. The gage grid length = 0.79 mm and slit width = 0.25 mm). Note the precision achieved by microscopic gage placement: the trimmed gage backing (darker area under the gage grid), masked gage-bonding adhesive (vertical line just left of the gage backing), and the masked waterproof coating (transition from shiny (coated) to dull (uncoated) surface just right of the slit). Copyright 2013, Hill Engineering, LLC

face, as described above. Again, typical practice in precision machining is suitable for registration of the slitting plane in the machine frame. The EDM wire is typically located relative to the workpiece by touch, using electrical continuity between the wire and the workpiece, and good planning should be made for keeping useful reference locations clear of leadwires and/or coatings that would impede location of the EDM wire relative to the workpiece. Figure 4.6 shows an example of an irregular welded sample, where slitting was performed at the site indicated. Two orthogonal features on the part provided a coordinate frame having a *baseline* along a machined feature and an origin at the intersection of the baseline with a second machined *perpendicular* feature. This coordinate frame was used to locate the slitting site, strain gage placements, and registration of the part and the EDM frames.

Following slitting, gage locations and slit dimensions should be determined precisely using digital photogrammetry under magnification. Protective coatings should be carefully removed from the gages using chemical and/or mechanical processes. Digital pictures can then be captured under suitable (e.g., 50X optical) magnification, with a length scale visible in the frame or a reference calibration available; measurements can then be made with imaging software. Measurements should be made for the final slit depth, the slit width, and the location of the center of each strain gage relative to the edge, or center, of the slit and suitable coupon features. The measured gage locations, and gage sizes (available in gage manufacturer specifications), are then used to construct the finite element mesh used to develop the compliance matrix, so that nodal locations coincide with

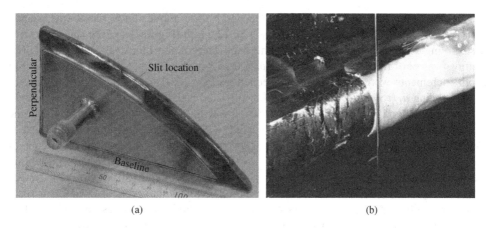

Figure 4.6 Example of irregular sample: (a) shows part shape, with highlights showing two machined features, one declared a baseline and the other perpendicular to the baseline, that are used to locate the slit and angle of cutting and (b) part set up in EDM with wire electrode near the start of the slit, strain gage coated with nitrile rubber is to right of the electrode. Copyright 2013, Hill Engineering, LLC

measured gage locations. The measured final slit depth is used to determine the depth increments in the experiment by repetitively subtracting slitting increments from the final slit depth, therefore accounting for any constant offset between the planned depths and those actually cut. The compliance matrix can be interpolated in cut depth to account for offset between planned and actual slit depths. Strain versus slit depth data are then used with the compliance matrix, following the data reduction procedure described above, to compute residual stress.

Prior to taking on experimental work, the practitioner should review a summary of additional experimental details for slitting assembled by Prime [36].

4.5 Example Applications

Significant efforts to develop laser shock peening (LSP) have used slitting to characterize the compressive residual stress layer introduced by the process. LSP process capability may be quickly established by measuring residual stress in flat blocks that have been uniformly treated with LSP over one face. In the present example, residual stress was determined in a series of blocks made from beta solution-treated and over-aged (BSTOA) Ti-6Al-4V, which is a coarse grained alloy (0.5 mm typical grain size) used in airframe and rotorcraft structures [37]. Background on LSP is available elsewhere, for example, [38,39], and the present series of blocks was used to determine the effects of two LSP process parameters on residual stress: the laser irradiance (power per area), and the number of treatment layers, where each layer represents full coverage of the processed surface by a raster of individual, neighboring spots. A third process parameter, the pulse duration (time), was held fixed at 20 ns. Earlier work shows that an increase in either irradiance or number of treatment layers will increase the depth of the near-surface compressive residual stress produced by LSP.

The slitting implementation followed the description above. Each block was square, with $L = B = 50.8$ mm, and had thickness $t = 12.7$ mm (Figure 4.1). The measurements used a single strain gage, mounted on the back face of the block, as in Figure 4.4(a). Cutting was performed by wire EDM with a 0.25 mm diameter ($\approx 0.02t$) electrode. Slit depth increments were small near the peened surface ($0.01t$) and larger in the interior ($0.04t$), being $a_i/t = 0.01, 0.02, \ldots, 0.08, 0.10, \ldots, 0.28, 0.32, \ldots, 0.84$.

Figure 4.7(a) and Figure 4.7(b) respectively show the strain vs. slit depth data, and the corresponding residual stresses computed using a polynomial-basis compliance matrix. The results in Figure 4.7(b) show that increased irradiance produces greater depth of compressive residual stress, with an increase from 2 to 10 GW/cm^2 roughly doubling the depth of compression. The results also show that the depth of compression can be increased by processing the surface multiple times, with eight layers at 2 GW/cm^2 producing a similar stress as two layers at 6 GW/cm^2. Published studies have shown that the large depth of compression generated by LSP, relative to the depth provided by conventional processes like glass bead peening (Figure 4.7(b)), improves resistance to fatigue cracking, and that the degree of improvement correlates to the level of measured residual stress, for example, [40]. The ability to alter LSP process parameters to achieve the significantly different stress versus depth profiles shown in Figure 4.7(b) enables trades between potential benefits from deeper near surface compression and potential detriments from higher subsurface tension, required to equilibrate the compression.

A different program of work carried out tests to evaluate whether comparable values of fracture toughness may be determined from coupons that contain a range of residual stress [32]. Coupons for fracture toughness tests were fabricated in high strength aluminum, and residual stresses were induced in coupon subsets by LSP. Each coupon subset had a unique LSP treatment design, such that the stress intensity factor due to residual stress varied from positive to neutral to negative over a range of five different coupon subsets. Residual stresses on the plane of fracture were measured using slitting.

The coupon design is a standard compact tension (C(T)) geometry, with a localized area of LSP applied on the coupon faces. Figure 4.8(a) shows the coupon dimensions and the

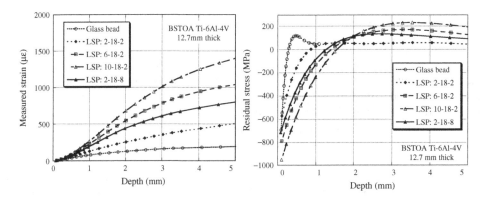

Figure 4.7 Results of slitting measurements on peened blocks that used a single strain gage on the back face each coupon: (a) strain versus depth data, and (b) computed residual stresses; the legend key shows peening process, and for LSP, process parameters as a dash sequence of irradiance (GW/cm^2) − pulse duration (ns) − number of layers. Copyright 2013, Hill Engineering, LLC

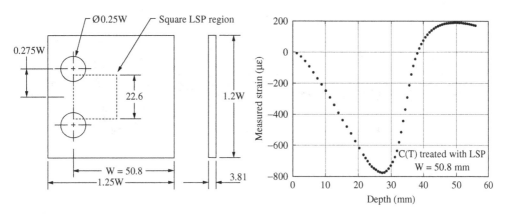

Figure 4.8 (a) Compact tension sample used to determine the ability of measured residual stress to correlate fracture behavior (dimensions in mm) with the LSP treatment area shown for condition LSP-3N; and (b) strain versus depth data for one of the coupon conditions. Copyright 2013, Hill Engineering, LLC

LSP processed area for one of the coupon conditions, LSP-3N. Strain measured during slitting across the fracture plane (left to right in Figure 4.8(a)) of a coupon in condition LSP-3N is shown in Figure 4.8(b). Residual stress was computed using a polynomial-basis analysis and also using a unit pulse analysis. The difference between measured and fitted strain, called the strain misfit, for each analysis is shown in Figure 4.9(a). Misfit for the polynomial-basis analysis reaches 40 $\mu\varepsilon$, or 5% of peak measured strain, while misfit for the unit-pulse analysis is on the order of measurement precision afforded by metallic foil strain gage instrumentation ($\pm 3\ \mu\varepsilon$). The large misfit for the polynomial-basis analysis indicates that the analysis results are not likely to be useful. Computed residual stress from the two analysis methods differ significantly (Figure 4.9(b)), with the unit-pulse analysis results having small uncertainties (barely visible in the plot) and showing a steep stress gradient from 32 to 37 mm, which spans the edge of the LSP treated area at 35.3 mm. The polynomial-basis analysis results have large uncertainties, reflecting their large strain misfit, and show a much shallower gradient at the edge of the LSP treated area. A comparison of the results of the two methods suggests that the polynomial-basis analysis is incapable of fitting the underlying stress field, with its high stress gradient arising from the sharp processed area boundary, while the unit-pulse analysis does not suffer from the same problem.

4.6 Performance and Limitations of Method

The influences of plasticity should be considered in applications of slitting, as they should be for any of the mechanical release techniques. The stress computation for slitting depends on superposition and therefore elastic behavior of the coupon during slitting. Because a slit is similar to a crack, concepts of plastic zone size from fracture mechanics are helpful in considering errors due to plasticity. The most comprehensive study of plasticity effects in slitting was published by Prime [41], who used elastic–plastic finite element simulation to show that plasticity-induced errors generally come from plasticity

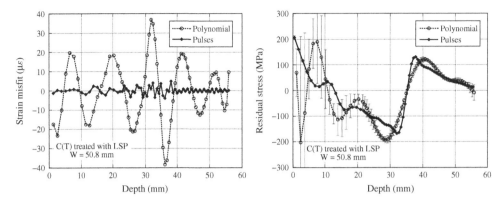

Figure 4.9 Data from measurements of residual stress in C(T) coupons where a polynomial-basis analysis was insufficient and a unit-pulse analysis gave a useful result: (a) strain misfit gave a poor fit for the polynomial basis, $m = 12$, (b) computed residual stress showed a high stress gradient with the unit-pulse analysis, and large uncertainties from the polynomial-basis analysis. Copyright 2013, Hill Engineering, LLC

at the bottom of the slit (cut tip). Prime provided a scheme to bound plasticity-induced errors in residual stress by correlating stress error with an *apparent* residual stress intensity factor, computed from the slitting data using Equation (4.10). This enables a simple check to determine whether the influence of plasticity may introduce significant error in residual stress. In the author's experience, while plasticity errors are a concern, only seldom have they been significant. The simple check provided by Prime allows a useful warning of potential problems.

Performance of residual stress measurement methods can be understood through a number of different approaches, each of which can provide complementary information. One approach is *intra-laboratory repeatability*, which measures the dispersion of a set of measurements given identical measurement inputs, where all measurements are performed under identical conditions, inclusive of the operator, methods, equipment, and facility. This is separate from *inter-laboratory repeatability*, which is also useful but includes the influences of different operators, equipment, and facilities. A second approach is *cross-method validation*, which determines the degree of correlation between the results of a given measurement technique and of other measurement techniques. A third approach is *phenomenological correlation*, which assesses the ability of the measurement data to correlate a directly observable, and useful, phenomenon known to depend on the measurand. This third approach is necessary for residual stress because the measurand cannot be known directly, and so *truth data* for another useful phenomenon are used to indicate measurement quality indirectly. For mechanical performance of materials, two examples of useful phenomena affected by residual stress are fatigue crack growth and low-energy fracture; for manufacturing considerations, cutting induced distortion is a useful phenomenon.

In an example study of intra-laboratory repeatability, data for slitting were developed using a set of five 17.8 mm thick blocks cut from a single plate of 316L stainless steel that had uniform LSP applied to one face to induce a deep residual stress field [8]. Typical slitting method techniques were employed with a single metallic foil strain gage

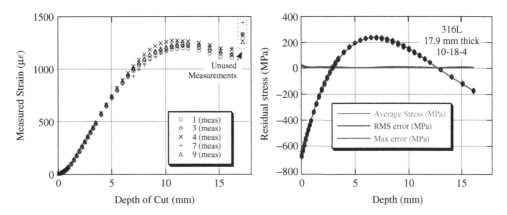

Figure 4.10 Repeatability data from a set of flat-block coupons cut from a plate treated uniformly with LSP: (a) strain versus depth data, and (b) residual stress data. Copyright 2013, Hill Engineering, LLC

on the back face of the coupon and incremental cutting by wire EDM. Measured strain vs. depth data were analyzed to determine the stress versus depth profile for each block. A statistical analysis of residual stress for all blocks provided an average residual stress vs. depth profile, and the variability of residual stress about the average among the set of measurements. The average depth profile had a maximum value of −668 MPa at the peened surface, as shown in Figure 4.10(b). The maximum variability about the average occurred at the surface and had a standard deviation of 15 MPa and an absolute maximum deviation of 26 MPa, or 2% and 4% of the peak stress level, respectively. This work shows slitting to have a very high level of measurement repeatability, which to the author's knowledge is unsurpassed by other residual stress measurement techniques. However, repeatability data support conclusions related to precision, and engineering requirements for accuracy, in addition to precision, are not supported by repeatability data alone.

Cross-method validation for slitting was performed using a similar set of flat block coupons, which were cut from a single plate that had uniform LSP on one face [42]. The material selected was mill annealed Ti-6Al-4V, which has microstructure of fine, equiaxed grains that provides good compatibility with X-ray diffraction (in contrast to the large-grained BSTOA Ti-6Al-4V microstructure discussed earlier). The laser shock peened plate was 8.7 mm thick, 50 mm wide and 50 mm long, and was cut into four blocks, each 25 mm square. Residual stresses were measured in three of the coupons, one coupon with slitting, one with X-ray diffraction and layer removal (XRD) (see Chapter 6), and one coupon with the contour method (see Chapter 5). Slitting used a single strain gage on the back face and a polynomial basis. The slitting and XRD results reflect a spatial average of stress at each depth, where slitting reflects an average along the z-direction of Figure 4.1 and XRD reflects an average in the $y - z$ plane over the X-ray spot and in-plane oscillation area, along with difficult to quantify contributions (that grow with depth) due to stress gradients in the etch pit area (here, the word "average" is useful but imprecise, as the measurement results are a convolution (or, weighted average) of the stress existing in the domain of the sampled and/or removed material). The design of the present coupon, having a single face peened uniformly with LSP, mitigates issues

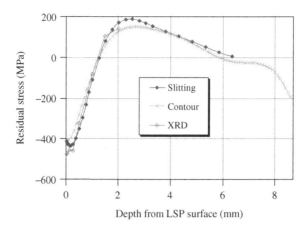

Figure 4.11 Results of cross-method validation study on a set of flat-block coupons cut from a plate treated uniformly with LSP. Copyright 2013, Hill Engineering, LLC

related to stress gradients in the $y - z$ plane. Figure 4.11 shows results from the three measurement methods. Overall, the results of all methods are consistent, showing a similar stress distribution. There are some significant differences near the surface that are limited to depths below 0.2 mm; at the surface there is about 100 MPa difference between XRD and the other two methods, a 20 to 25% discrepancy. Near-surface discrepancies are not uncommon between XRD and other methods, and may be due to a number of issues including uncertainties in slitting and contour results that are generally highest near the surface, in comparison with the uncertainties in XRD results that are generally lowest near the surface (in fine-grained, non-textured materials).

Phenomenological correlation for residual stress measurement has been performed in the context of fatigue and fracture performance of metallic materials. Work in the high-strength steel 300 M was conducted to assess the ability to residual stress treatments to improve its high-cycle fatigue (HCF) performance [39]. The HCF performance of materials under constant amplitude loading is affected by the cyclic stress amplitude primarily, with the cyclic mean stress having an important secondary effect. The amplitude and mean cyclic stresses are often combined into an equivalent stress, expressed in terms of the amplitude and mean according to a simple equation, with a recent study by Dowling [43] finding the Walker equivalent stress model to have good accuracy over a wide range of materials including 300 M. Residual stress can be included in an equivalent stress model for HCF using linear superposition, so that residual stress appears in the mean stress because it does not fluctuate with time. Figure 4.12 shows that high cycle fatigue performance of 300 M samples treated with LSP [39] is consistent with the fatigue performance of untreated material when residual stresses measured by slitting are included in the equivalent stress. The same study [39] includes a comparison of slitting and XRD residual stress measurements, as well as a correlation of the behavior of fatigue strength in sharply notched coupons using a fracture mechanics based fatigue analysis that includes residual stress; a review of that earlier work is left to the interested reader.

Residual stress effects on fracture toughness in high strength aluminum C(T) coupons also shows consistent phenomenological correlation. An example of residual stress

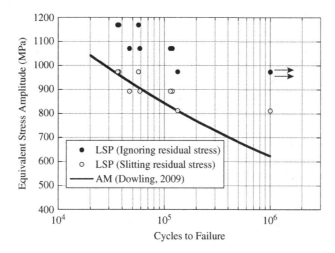

Figure 4.12 Results of phenomenological correlation of high-cycle fatigue in high strength steel: data plotted in terms of applied stress are inconsistent with trend for untreated, as-machined (AM) coupons; data plotted in terms of an equivalent stress, that incorporates residual stress measured by slitting, are consistent with the AM trend [41], which indirectly validates the measured residual stress. Copyright 2013, Hill Engineering, LLC

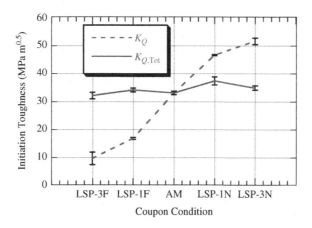

Figure 4.13 Results of phenomenological correlation of fracture in high aluminum C(T) coupons: data plotted in terms of applied stress intensity factor K_Q are inconsistent; data plotted in terms of the total stress intensity factor $K_{Q,Tot}$, which includes residual stress, are consistent. This result indirectly validates the measured residual stress. Copyright 2013, Hill Engineering, LLC

measurement in these coupons was described above in Figure 4.8 and Figure 4.9. Fracture tests described elsewhere [32] show large differences in toughness from LSP processing across a range of coupon subsets when residual stresses are ignored, see Figure 4.13. However, slitting measurements of residual stress on the crack plane, and of the residual stress intensity factor, enable correction of the toughness data through superposition and linear elastic fracture mechanics. The resulting corrected fracture toughness values shown in Figure 4.13 are in good agreement across the range of coupon subsets because they include contributions from both applied and residual stress that together act to drive fracture. The fracture toughness data may be used to make conclusions about the *accuracy* of residual stress measured by slitting, but accuracy conclusions are occluded through the fracture phenomena and the models used to account for the contributions of residual stress on the observed fracture behavior (i.e., fracture mechanics and linear superposition).

The collection of data described in this section demonstrate the following attributes of the slitting method: a high level of repeatability, with intra-laboratory repeatability standard deviation better than 2% of peak stress magnitude; agreement with other techniques to a level better than 10%, except very near the surface, at depths shallower than 0.2 mm, where agreement is within 20%; and, an ability to correlate fatigue life to a factor of 2 and low-energy fracture toughness to 10%.

4.7 Summary

The slitting method is a mechanical relaxation technique that can be performed in the laboratory and provides useful results. It is well suited in making measurements of residual stress through the entire thickness of parts having a prismatic cross section, such as beams, plates, disks, cylinders, or less simplistic sections. The method is best suited to measuring stress fields that vary through the depth of the slit plane, and do not vary across the width of the slit plane; with reference to Figure 4.1, when residual stress varies with z, slitting will return a convolution of the stress variation along z. Slitting is robust to variations of microstructure and texture because these generally do not affect elastic behavior, and can be extended to parts with graded or composite elastic behavior by including appropriate property variations in the stress analysis used to develop the compliance matrix. In summary, slitting has proven useful, with a high degree of repeatability and a broad range of demonstrated capability.

References

[1] Vaidyanathan, S., Finnie, I. (1971) "Determination of Residual Stress from Stress Intensity Factor Measurements," *Applied Mechanics Reviews* 52(2):75–96.
[2] Finnie, I., Cheng, W. (2002) "A Summary of Past Contributions on Residual Stresses," *Materials Science Forum* 404–407:509–514.
[3] Finnie, I., Cheng, W. (1996) Residual stress measurement by the introduction of slots or cracks, Localized Damage IV Computer Aided Assessment and Control of Localized Damage – Proceedings of the International Conference 1996. Computational Mechanics Inc., Billerica, MA, USA, 37–51.
[4] Cheng, W., Finnie, I. (1985) "A Method for Measurement of Axisymmetric Axial Residual Stress in Circumferentially Welded Thin Walled Cylinders," *Journal of Engineering Materials and Technology* 107(3):181–185.
[5] Cheng, W., Finnie, I., Vardar, O. (1991) "Measurement of Residual Stresses Near the Surface Using the Crack Compliance Method," *Journal of Engineering Materials and Technology* 113(2):199–204.
[6] Prime, M. B. (1999) "Residual Stress Measurement by Successive Extension of a Slot: The Crack Compliance Method," *Applied Mechanics Reviews* 52(2):75.
[7] Cheng, W., Finnie, I. (2007) *Residual Stress Measurement and the Slitting Method*, Springer: New York.
[8] Lee, M. J., Hill, M. R. (2007) "Intra-laboratory Repeatability of Residual Stress Determined by the Slitting Method," *Experimental Mechanics* 47(6):745–752.

[9] Prime, M. B. (2012) The Slitting (Crack Compliance) Method for Measuring Residual Stress http://www.lanl.gov/residual/
[10] Aydiner, C. C., Ustundag, E., Prime, M. B., Peker, A. (2003) "Modeling and Measurement of Residual Stresses in a Bulk Metallic Glass Plate," *Journal of Non-Crystalline Solids*, 316(1):82–95.
[11] Midha, P. S., Modlen, G. F. (1976) "Residual Stress Relief in Cold-extruded Rod," *Metals Technology* 3(11):529–533.
[12] Cook, C. S. (1987) Straightening operations and residual stresses in tubing, 10th ASM Conference on Advances in the Production of Tubes, Bars, and Shapes, Orlando, FL, ASM International, Metals Park, OH, USA, 61–68.
[13] Rankin, J. E., Hill, M. R. (2003) "Measurement of Thickness-average Residual Stress Near the Edge of a Thin Laser Peened Strip," *Journal of Engineering Materials and Technology* 125(3):283–293.
[14] Prime, M. B., Hill, M. R. (2002) "Residual Stress, Stress Relief, and Inhomogeneity in Aluminum Plate," *Scripta Materialia* 46(1):77–82.
[15] Prime, M. B., Prantil, V. C., Rangaswamy, P., Garcia, F. P. (2000) "Residual Stress Measurement and Prediction in a Hardened Steel Ring," *Materials Science Forum* 347:223–228.
[16] Hill, M. R., Lin, W. Y. (2002) "Residual Stress Measurement in a Ceramic-metallic Graded Material," *Journal of Engineering Materials and Technology* 124(2):185–191.
[17] DeWald, A. T., Rankin, J. E., Hill, M. R., Schaffers, K. I. (2004) "An Improved Cutting Plan for Removing Laser Amplifier Slabs from Yb:S-FAP Single Crystals Using Residual Stress Measurement and Finite Element Modeling," *Journal of Crystal Growth* 265(3–4):627–641.
[18] Taylor, D. J., Watkins, T. R., Hubbard, C. R., Hill, M. R., Meith, W. A. (2012) "Residual Stress Measurements of Explosively Clad Cylindrical Pressure Vessels," *Journal of Pressure Vessel Technology* 134(2):011501.
[19] Cheng, W., Finnie, I., Gremaud, M., Prime, M. B. (1994) "Measurement of Near Surface Residual Stresses Using Electric Discharge Wire Machining," *Journal of Engineering Materials and Technology* 116(1):1–7.
[20] Rankin, J. E., Hill, M. R., Hackel, L. A. (2003) "The Effects of Process Variations on Residual Stress in Laser Peened 7049 T73 Aluminum Alloy," *Materials Science Engineering* 349(1–2):279–291.
[21] Makino, A., Nelson, D. V., Hill, M. R. (2011) "Hole-within-a-hole Method for Determining Residual Stresses," *Journal of Engineering Materials and Technology*, 133:021020.
[22] Prime, M. B., Hill, M. R. (2006) "Uncertainty, Model Error, and Order Selection for Series-Expanded, Residual-Stress Inverse Solutions," *Journal of Engineering Materials and Technology* 128(2):175–185.
[23] Lee, M. J., Hill, M. R. (2007) "Effect of Strain Gage Length When Determining Residual Stress by Slitting," *Journal of Engineering Materials and Technology* 129(1):143–150. See also an important erratum (2009) 131(4):047001.
[24] Prime, M. B., Pagliaro, P. (2006) Uncertainty, Model Error, and Improving the Accuracy of Residual Stress Inverse Solutions, Proceedings of the 2006 SEM Annual Conference and Exposition on Experimental and Applied Mechanics, June 4–7, 2006, St Louis, MO, USA, paper number 176.
[25] Aydiner, C. C., Prime, M. B. (2013) "Three-dimensional Constraint Effects on the Slitting Method for Measuring Residual Stress," *Journal of Engineering Materials and Technology*, accepted for publication.
[26] Schajer, G. S. (1993) "Use of Displacement Data to Calculate Strain Gauge Response in Non-Uniform Strain Fields," *Strain* 29(1):9–13.
[27] Strang, G. (1986) *Introduction to Applied Mathematics*, Wellesley-Cambridge Press, Wellesley, MA, 460–461.
[28] "Hole-Drilling Residual Stress Profiling with Automated Smoothing," *JEMT* 129:440–445.
[29] Schajer, G. S., Prime, M. B. (2006) "Use of Inverse Solutions for Residual Stress Measurement," *Journal of Engineering Materials and Technology* 128(3):375–382.
[30] Schindler, H. J., Cheng, W., Finnie, I. (1997) "Experimental Determination of Stress Intensity Factors due to Residual Stresses," *Experimental Mechanics* 37(3): 272–279.
[31] Schindler, H. J., Bertschinger, P. (1997) Some Steps Towards Automation of the Crack Compliance Method to Measure Residual Stress Distributions, Proceedings of the Fifth Int Conference on Residual Stresses, 682–687.
[32] Ghidini, T., Dalle Donne, C. (2007) "Fatigue Crack Propagation Assessment Based on Residual Stresses Obtained Through Cut-Compliance Technique," *Journal of Fatigue and Fracture of Engineering Materials and Structures* 30(3):214–222.
[33] Hill, M. R., VanDalen, J. E. (2008) "Evaluation of Residual Stress Corrections to Fracture Toughness Values," *Journal of ASTM International*, 5(8):11.

[34] Newman, J. C., Ruschau, J. J., Hill, M. R. (2010) "Improved Test Method for Very Low Fatigue-Crack-Growth-Rate Data," *Fatigue and Fracture of Engineering Materials and Structures*, 34(4):270–279.
[35] Vishay Measurements Group (1992) Student Manual for Strain Gage Technology: A Brief Introduction and Guide to Selection, Installation, Instrumentation (Bulletin 309E), Vishay Measurements Group, Raleigh, NC.
[36] Prime, M. B. (2000) Experimental procedure for crack compliance (slitting) measurements of residual stress, LA-UR-03-8629, Los Alamos, NM, http://www.lanl.gov/residual/testing.pdf
[37] Cotton, J. D., Clark, L. P., Phelps, H. R. (2002) "Titanium Alloys on the F-22 Fighter Airframe," *Advanced Materials and Processes*, 160(5):25–28.
[38] Fabbro, R., Peyre, P., Berthe, L., Scherpereel, X. (1998) "Physics and Applications of Laser-shock Processing," *Journal of Laser Applications* 10(6):265–279.
[39] Montross, C. S., Wei, T., Ye, L., Clark, G., Mai, Y-W. (2002) "Laser Shock Processing and its Effects on Microstructure and Properties of Metal Alloys: A Review," *International Journal of Fatigue* 24(10):1021–1036.
[40] Pistochini, T. E., Hill, M. R. (2011) "Effect of Laser Peening on Fatigue Performance in 300 M Steel," *Fatigue and Fracture of Engineering Materials and Structures* 34(7):521–533.
[41] Prime, M. B. (2010) "Plasticity Effects in Incremental Slitting Measurement of Residual Stresses," *Engineering Fracture Mechanics* 77(10):1552–1566.
[42] DeWald, A. T. (2007) *Benchmarking of Residual Stress Measurement Methods on Thin Titanium Coupons*, Hill Engineering, LLC, Rancho Cordova, CA.
[43] Dowling, N. E., Calhoun, C. A., Arcari, A. (2009) "Mean Stress Effects in Stress-life Fatigue and the Walker Equation," *Fatigue and Fracture of Engineering Materials and Structures*, 32(3):163–179.

5

The Contour Method

Michael B. Prime[1] and Adrian T. DeWald[2]
[1]*Los Alamos National Laboratory, New Mexico, USA*
[2]*Hill Engineering, LLC, California, USA*

5.1 Introduction

5.1.1 Contour Method Overview

The contour method, which is based upon solid mechanics, determines residual stress through an experiment that involves carefully cutting a specimen into two pieces and measuring the resulting deformation due to residual stress redistribution. The measured displacement data are used to compute residual stresses through an analysis that involves a finite element model of the specimen. As part of the analysis, the measured deformation is imposed as a set of displacement boundary conditions on the model. The finite element model accounts for the stiffness of the material and part geometry to provide a unique result. The output is a two-dimensional map of residual stress normal to the measurement plane. The contour method is particularly useful for complex, spatially varying residual stress fields that are difficult (or slow) to map using conventional point wise measurement techniques. For example, the complex spatial variations of residual stress typical of welds are well-characterized using the contour method.

Contour method measurements are typically performed on metallic parts, which can be cut using a wire electric discharge machine (EDM). There are no specific size restrictions, but the measurement signal (displacement) scales with specimen size and performing measurements on parts smaller than 5 mm by 5 mm in cross-section requires extreme precision. There are no restrictions on the shape of the specimen due to the fact that complex geometry is accounted for using a finite element model of the part.

The contour method is the youngest method covered in this book, having been first presented at a conference in 2000 [1] and then in a journal in 2001 [2]. Therefore, consensus best practices are not as well established as for other methods. A basic measurement procedure is provided along with comments about potential alternate approaches, with references for further reading.

5.1.2 Bueckner's Principle

The contour method theory is a variation on Bueckner's superposition principle. Bueckner presented the relevant theory in 1958 [3] and discussed it further in later publications [4,5]. However, Buckner's papers present no figure like those presented in this chapter. The apparent first use of such a figure for Bueckner's principle was by Barenblatt in 1962 [6]. A very similar principle and figure was presented independently in Paris' landmark 1961 paper on fatigue crack growth [7], and he credits the principle to a 1957 report he wrote for the Boeing Company. It is conceivable that other work predates that of Bueckner and Paris. Bueckner's principle is also quite similar to the better known inclusion problem presented by Eshelby in 1957 [8]. In any case, Bueckner's principle is indispensable in fracture mechanics work [9] and has proven invaluable when used appropriately.

5.2 Measurement Principle

5.2.1 Ideal Theoretical Implementation

The ideal theoretical implementation of the contour method is presented here first before discussing the assumptions and approximations that will be required for a practical implementation. Figure 5.1 presents a 3D illustration for a thick plate in which the longitudinal stress varies parabolically through the thickness of a plate. Figure 1.12 in Chapter 1 gives a complementary 2D illustration of the principle. Step A in Figure 5.1 is the undisturbed part and the residual stresses that one wishes to determine. In B, the part has been cut in two on the plane $x = 0$ and has deformed because of the residual stresses released by the cut. In C, the deformed cut surface is forced back to its original shape and the resulting change in stress is determined. Superimposing the stress state in B with the change in stress from C gives the original residual stresses throughout the part:

$$\sigma^A(x, y, z) = \sigma^B(x, y, z) + \sigma^C(x, y, z) \quad (5.1)$$

where σ refers to the entire stress tensor and the superscripts refer to the various steps of Figure 5.1. Because σ_x, τ_{xy}, and τ_{xz} are zero on the free surface in B, the described superposition principle *uniquely* determines the original distribution of those residual stresses on the plane of the cut, that is, at $x = 0$ in A [2].

If one could measure in-plane displacements on the cut surface, this theory would be complete. However, measurement of the transverse displacements is not experimentally possible; instead, some reasonable assumptions and approximations are required.

5.2.2 Practical Implementation

Proper application of the superposition principle combined with a few assumptions allows one to determine the normal residual stresses experimentally along the plane of the

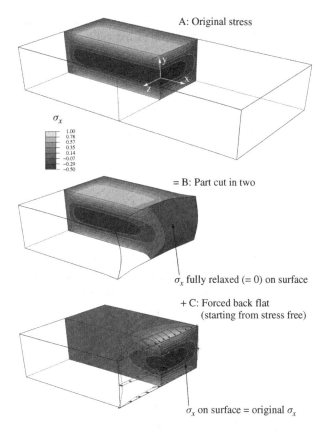

Figure 5.1 Superposition principle for the contour method. Stresses are plotted on one quarter of the original body

cut: $\sigma_x^A(0, y, z)$. Experimentally, the contour (surface height map) of the free surface is measured after the cut (in B). Measurement of the surface contour provides information about the displacements in the normal (x) direction only. Therefore, the analytical approximation of Step C will elastically force the surface back to its original configuration in the x-direction only, leaving the transverse displacements unconstrained. (In a finite element (FE) model, leaving the transverse displacements unconstrained on the free surface results in automatic enforcement of the free surface conditions $\tau_{xy} = 0$ and $\tau_{xz} = 0$.) Thus, the contour method can identify the normal stresses σ_x only, and not the shear stresses τ_{xy} and τ_{xz}. In spite of the presence of any shear stresses and transverse displacements, one need only average the contours measured on the two halves of the part to determine the normal stress, σ_x. The shear stresses released on the plane of the cut affect the surface displacements anti-symmetrically. Thereby, when the average is computed, the effects of transverse displacements and shear stresses are cancelled out and the result is the surface displacements due to release of the residual stress normal to the surface [2].

Since the stresses normal to the free surfaces in B must be zero in Equation 5.1, Step C by itself gives the correct stresses on the plane of the cut:

$$\sigma_x^B(0, y, z) = 0$$
$$\Rightarrow \qquad , \qquad (5.2)$$
$$\sigma_x^A(0, y, z) = \sigma_x^C(0, y, z)$$

which is the standard implementation of the contour method.

5.2.3 Assumptions and Approximations

5.2.3.1 Elastic Stress Release and Stress Free Cutting Process

The superposition principle assumes that the material behaves elastically during the relaxation of residual stress and that the material removal process does not introduce stresses of sufficient magnitude to affect the measured displacements. These assumptions are common to relaxation methods and have been studied extensively as described in Chapters 2 and 4 on hole drilling and incremental slitting. Plasticity errors will be discussed in more detail for the contour method later in this chapter.

5.2.3.2 Starting with Flat Surface in Analysis

One approximation to the theory is made purely for convenience in the analysis: the deformed shape of the body is not modeled before analytically performing Step C in Figure 5.1. Because the deformations are quite small for engineering materials, and the analysis is linear, the starting point for this step can be a flat surface and the displacement boundary conditions will then force the surface into the opposite of the measured shape. The results are the same, and the analysis is simpler.

5.2.3.3 Part is Symmetric About Cut Plane

Averaging the contours on the two halves to remove shear stress effects requires another assumption: that the stiffness is the same on the two sides of the cut. For homogeneous materials, this assumption is certainly satisfied when a symmetric part is cut precisely in half. In practice, the part only needs to be symmetric within the region where the stiffness has a significant effect on the deformations of the cut surface, which can be estimated as extending from the cut surface by no more than 1.5 times the Saint Venant's characteristic distance. The characteristic distance is often the part thickness, but is more conservatively taken as the maximum cross-sectional dimension. If the part is asymmetric, an FE analysis can be used to estimate possible errors, which tend to be small until the part is very asymmetric.

5.2.3.4 Anti-symmetric Cutting Errors Average Away

Figure 5.2 shows that averaging the two contours removes any errors caused by anti-symmetric cutting effects – those that cause a low spot on one side and a mating high

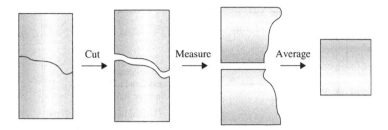

Figure 5.2 The effect of crooked cut averages away

spot on the other side [10]. The two main causes of anti-symmetric errors are the cut itself wandering, as illustrated in Figure 5.2, or the part moving during cutting as stresses are relaxed and the part deforms.

5.2.3.5 Symmetric Errors: Cutting Irregularities

There are other errors that cause symmetric or asymmetric effects that do not average away. Most such error sources are relatively straightforward and can be avoided with good experimental practice. Local cutting irregularities, such as wire breakage or overburning at some foreign particle, are usually small length scale (order of wire diameter) and are removed by the data smoothing process or manually from the raw data. A change in cut width can occur in heterogeneous materials since the EDM cut width varies for different materials. A change in the part thickness (in the wire direction) can also cause this. A "bowed" cut [2,11] can usually be avoided by using good settings on the wire EDM [11].

5.2.3.6 Bulge Error

It must also be assumed that the cut removes a constant width of material when measured relative to the state of the body prior to any cutting. From a theoretical point of view, the relevant assumption for the superposition principle in Figure 5.1 is that the material points on the cut surface are returned in Step C to their original locations. (The averaging of the contours on the two surfaces takes care of the issue with not returning material points to their original location in the transverse direction.)

Figure 5.3 illustrates the "bulge" error, a symmetric error that can causes bias in the contour method results [10]. The cutting process makes a cut of constant width w in the laboratory reference frame. As the machining proceeds, stresses relax and the material at the tip of the cut deforms; however, the physical cut will still be only w wide. This means that the width of material removed has been reduced when measured relative to the original state of the body. Therefore, forcing the cut surface back flat as in Step C of Figure 5.1 will not return the material to its original location, which causes an error in the stress calculation. The bulge effect is symmetric and will not be averaged away. The effect occurs when the stress state at the cut tip changes relative to the original stress state, which is caused by specimen deformation as the cut progresses. Therefore, the error can be minimized by securely clamping the part. In addition, it is worth noting that this effect scales with the width of the cut.

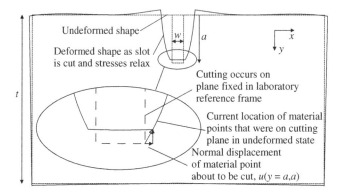

Figure 5.3 The "bulge" error occurs when the material at the cut tip deforms prior to the cut, changing the effective cut width. For simplicity, the typical round-bottomed EDM slot is illustrated as flat. Reproduced with permission from [10], Copyright SEM 2011

5.2.3.7 No Assumptions of Isotropic or Homogeneous Elasticity Required

Bueckner's principle and the contour method do not require any assumption that the material be elastically isotropic or homogeneous, only that the linear elastic behavior be accurately reflected in the FE model used to calculate stress. Most contour method measurements assume that the material being measured is isotropic and homogeneous, which is a good assumption for many materials. However, it is possible to include anisotropic elastic constants in the material definition file and to have these constants vary spatially throughout the part. The nuclear power plant weld presented later in this chapter is an example of heterogeneous elasticity.

5.3 Practical Measurement Procedures

5.3.1 Planning the Measurement

It is useful to spend time planning the measurement at the outset to avoid potential issues that may arise. Carefully look through the list of experimental steps and analytical steps and consider how each task will be executed. Recognize that fixturing the part and making the cut are the most important experimental aspects of the contour method, and poor technique will lead to increased errors. Measuring the surface to sufficient precision is relatively easy.

The contour method is a specialized technique that is only appropriate for particular measurements. If near-surface stresses are of primary interest consider using a different technique. Also, very small parts and/or small magnitude and localized stress fields may be difficult to resolve using the contour method. In cases of concern, it is useful to simulate the experiment ahead of time.

5.3.2 Fixturing

As discussed in Section 5.2.3, the original plane of the cut should be constrained from moving as the stresses are relaxed during cutting. Such constraint requires clamping both

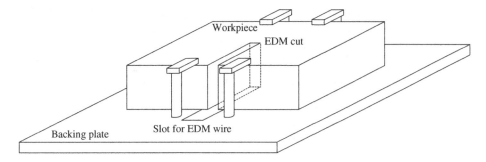

Figure 5.4 Illustration of double sided clamping arrangement used for the contour method

sides of the cut to a rigid fixture. Figure 5.4 shows an example clamping arrangement. In general, more clamping constraint is better. Care should be taken not to clamp in a manner that will induce stress into the part. Some novel approaches have been used to try to obtain maximum constraint [12,13] including self-restraint by leaving a ligament of uncut material [14,15]. Some results have been reported when the specimen was only clamped on one side [16,17], often because the same cut was used for a slitting measurement [10,18,19], which generally leads to very different contours measured on the two halves after cutting. After averaging the contours, the results are often still good, but sometimes an increased bulge error is evident [18]. In general, the use of novel fixturing arrangements should be used with great caution.

5.3.3 Cutting the Part

For the contour method, the ideal machining process for cutting the part has the following characteristics: a straight (planar) and cut with a smooth surface, minimal cut width (kerf), not removing any further material from already cut surfaces, and not causing any plastic deformation or inducing any residual stress. Wire electric discharge machining (wire EDM) is currently the method of choice. In wire EDM, a wire is electrically charged with respect to the workpiece, and spark erosion causes material removal. The cutting is non-contact, whereas conventional machining causes localized plastic deformation from the large contact forces. The part is submerged in temperature-controlled deionized water during cutting, which minimizes thermal effects. The wire control mechanisms can achieve positional precision of a fraction of a micrometer, especially for a straight cut.

Since the bulge error increases with cut width, a smaller wire diameter is recommended when possible. Too small a wire diameter, however, will sometimes result in undesirable wire breakage during the experiment or can lead to unreasonably long cutting times. Table 5.1 shows some general guidance on minimum wire diameter relative to cut thickness. That thickness depends on the part orientation during cutting, and the orientation that minimizes thickness is generally preferred. The smallest robust wire diameter will depend on the material being cut and the EDM machine, so the table should only be used as a starting point for selecting wire size. Sometimes, a larger wire size is chosen to ensure a more robust cut [11].

Table 5.1 Rough guide to suggested wire sizes. The ranges overlap because the choice will also depend on the sample material and the EDM machine

Specimen thickness	EDM wire diameter
< 15 mm	100 μm
10 mm – 100 mm	150 – 200 μm
> 50 mm	250 μm

The best results have been obtained using EDM wires made of brass. Although no systematic study has been reported, the use of other wires such as tungsten or zinc-coated brass have seemed to resulted in lower quality cuts [2,20,21].

The cut quality is a primary factor in determining the quality of contour method results. In addition to selecting the proper wire size and type, it is important to select cutting conditions that produce a cut that represents, as close as possible, the conditions described above for the ideal cutting process. It is generally advisable to use a "skim" or "finish" cut setting for the machine. Skim cut settings are lower power than conventional rough cut settings (which are optimized for speed) and are intended to provide a better surface finish and minimal recast [22]. Specific cut parameters and settings are machine specific and there is typically a library of cut settings for different configurations included in the machine's control unit. Typical settings for cutting a specific material with a specific wire will include settings for a single rough cut and then for three or four sequential skim cuts. The setting for the first or second skim cut are often the best choice for the contour method, because the settings for the final skim cut often result in wire breakage or extremely slow cuts.

To cut the part, set it up on the EDM and secure it with clamps after the part and the clamps have come to thermal equilibrium with the water in the EDM tank. Align the cut path to the part and program the EDM cut to cut through the entire cross section in a single pass. Upon completion of the cut, the parts should be removed from the EDM, taking care to preserve the integrity of the cut surfaces, and rinsed to remove any loose debris that may have adhered to the surface.

5.3.4 *Measuring the Surfaces*

The surfaces created by the cut should be measured. In general, the form of the surface contours will have a peak-to-valley magnitude on the order of 10 μm to 100 μm. Accurate measurement of surface height fields on this level requires precision metrology equipment. A coordinate measuring machine (CMM) is a useful and widely available device for this purpose.

The two halves created by the EDM cut should be placed on the CMM with their "cut" surfaces exposed (the term "cut" is used here to describe the surface where residual stress measurement is performed). The metrology device should be programmed to acquire points over the entire surface with point spacing sufficient to resolve the form of the displacement field. If nothing is known ahead of time that could guide measurement density, it is possible first to measure the parts with a coarse spacing to estimate the form of the displacement field and then to measure again with a fine spacing that is sufficient to capture the necessary detail. A simple uniform grid of 50 by 50 points is a useful starting point for a CMM with a 2 mm ruby stylus. A relatively large stylus such as this is desirable because it will smooth out some of the features on the "rough" EDM surface. When possible, both surfaces should be measured using the same measurement point locations, bearing in mind that one coordinate direction will be reversed when comparing the two surfaces. Since CMM measurements occur at about one-per-second frequency, the measurements can take several hours. Therefore, temperature stability is important and the CMM should be isolated from thermal fluctuations. In addition, it is helpful for later alignment of the two surfaces to collect a series of points tracing the perimeter of each "cut" surface by placing the CMM tip slightly below the surface and touching the sides of the part. More details of CMM measurements for the contour method are reported elsewhere [23].

Other methods can be used to measure the surface, but measuring the surface contour is relatively easy and has never been the limiting factor for the contour method measurements. Non-contact optical scanners have been used widely [12,24,25] and demonstrated to give nearly identical final stress results to a CMM [21,26]. The optical scanners generally provide noisier results because they capture the roughness of the EDM cut. Therefore, significantly denser measurement points are required. However, optical scanners can measure points more quickly, which also might reduce thermal fluctuation issues. The optical scanners generally cannot measure in the transverse direction, which means one cannot directly measure the part perimeter.

The PhD dissertation by Johnson [23] gives further detailed information on measurements and analysis relating to the contour method.

5.4 Residual Stress Evaluation

In general, several steps are required to process contour data and calculate stresses. Practitioners should use care to make sure that the processed data remains as true as possible to the original data. It is suggested that intermediate results be examined carefully after each step.

5.4.1 Basic Data Processing

5.4.1.1 Align the Coordinate Frames

The two data surfaces, one from each side of the "cut," should be aligned to the same coordinate frame such that the material points prior to cutting are coincident on the two surfaces. The two cut surfaces appear as mirror images, so one of the Cartesian coordinate directions needs to be reversed so as to connect corresponding points on the two surfaces. This coordinate reversal can be seen in the third panel of Figure 5.2, when the lower section has been reversed to register with the surface of the upper section. If further alignment is required then it is necessary to perform rigid body translations and rotation in the plane of the cut surface to set both surfaces in the same coordinate frame. For example, if the measured surface is approximately oriented in the yz-plane then it is necessary to translate one surface in y and z and also rotate it about the x-axis until it sits on top of the other surface (when viewed along the x-axis). The perimeter trace is very useful for this alignment. The other rigid body translation (x-direction) and rotations (about y and z) will not affect the results and can be ignored. It is generally convenient, however, to fit each surface to a best fit yz-plane and to subtract this from the data (which will bring each surface close to the yz-plane).

Following surface alignment, the perimeter trace should be decoupled from the data sets. The perimeter trace may be used to support FE model construction and after that it is no longer needed.

5.4.1.2 Construction of FE Model

A finite element model representing half of the original part, for example, the shape of one of the two pieces after it has been cut in half, should be constructed based upon measurements of the part. If available, the perimeter trace of the cross-section from a CMM represents a useful starting point for the model. If the cross-section is relatively simple, then measurements using a linear measurement tool such as calipers provide a useful alternative. This cross section can typically be "extruded" in the third dimension based on simple dimensional measurements. The finite element model should represent the cross-section of the part at the measurement plane and should have a similar stiffness relative to displacements being applied on the measurement surface. Features in the part "far" from the measurement plane are unlikely to influence this stiffness and can typically be ignored.

A finite element mesh should be generated on the model. It is useful to bias the mesh for higher refinement near high gradients in the displacement surface and near edges of the measurement cross section. This can help to produce a converged solution in an efficient manner. First order hexahedral (brick) elements or second-order reduced-integration hexahedra are preferred. A useful starting point for the mesh density is 50 by 50 elements over the cross section. The element size can be relaxed to grow large away from the measurement plane without affecting the stress results.

Once complete, a list of nodes on the cut surface where displacement boundary conditions will be applied should be generated along with their coordinates. This list will be used to generate prescribed displacement boundary conditions.

5.4.1.3 Average the Two Sides

Once the two surfaces are aligned, the data from both surfaces should be averaged. This can be accomplished by taking the matching points from each surface and computing the average value.

5.4.1.4 Filter the Noise

The surface measurement data will contain some "noise" that is the result of measurement error and roughness on the EDM cut surface. The random noise and roughness are not caused by bulk residual stresses, however, they will significantly affect the calculated stress because the stress depends on the curvature of the displacement field and this high frequency content has a high curvature. For this reason, it is important to remove the noise from the data while preserving the overall form of the surface (which is the result of bulk residual stress).

A two-step process can be used to prepare the displacement data for stress computation. First, obvious outliers should be deleted from the data set. Outliers can result from unintended particles such as dust settling on the surface during measurement, artifacts from measurements near the edges of the perimeter, and gross measurement errors. Outliers can be identified by plotting the data surface and visually looking for points that are significantly away from the overall form of the surface.

Second, a method should be employed to extract the form of the surface while eliminating the roughness and noise. This is typically accomplished by fitting the data to a smooth surface (e.g., bivariate splines). There are commercial software packages available for straightforward implementation of this including MATLAB® which has a Spline Toolbox®. Spline smoothing has been most widely used [26] but in most formulations requires that the data points be on a regular, rectangular grid, which can require extra processing to grid the data. Alternately, the data can be fit to a continuously defined smooth surface, such as a bivariate Fourier series, without gridding the data [27,28]. A continuous surface sometimes cannot fit the data as well as local splines. A few alternative approaches for smoothing the data have also been reported in the literature [18,29,30]. Any method that filters the surface roughness while accurately capturing the overall form of the contour should be acceptable.

There is not yet a robust, objective method for selecting the optimal amount of smoothing. Examining plots of the misfit, the difference between the experimental displacements and the fit, is very helpful [23]. Often, the fit is selected using a linear or semi-log plot of the root mean square (RMS) misfit versus increasing spline knot density or order of series fit. The fit where the RMS misfit begins to flatten out is selected because it often represents the transition point between over-fitting and under-fitting the important features of the data [24,26,31]. Improved selection of smoothing might be possible by estimating the uncertainty in the stresses and picking the fit that minimizes the uncertainty [26].

5.4.1.5 Transfer to FEM

The final displacements, after averaging and filtering, should be inverted about the surface normal and interpolated/extrapolated to the node locations on the finite element surface as displacement boundary conditions. Only the displacement in the direction normal to the surface of the cut should be specified, and it must be specified for all nodes on the cut surface. Three additional point boundary conditions should be applied to the model to restrain rigid-body motion (but nothing more). Such a minimal constraint arrangement ensures that the calculated residual stress map satisfies equilibrium, and is the reason that rigid body motions of the measured contour do not affect the results. Figure 5.6 shows an example with the following rigid-body motions being constrained: translation in y and z and rotation about the x-axis. Once the boundary conditions are in place, the model should be allowed to reach equilibrium. The resulting stresses on the plane of the cut in the normal direction are the "results" from the contour measurement. These represent the original residual stresses in the part prior to sectioning.

5.4.1.6 Reporting Results

Practitioners should report sufficient details so that an expert reader can independently interpret the results. Descriptions of the part's material should be detailed and include the heat treatment state and the yield strength. The arrangement used to clamp the part during cutting should be illustrated or described. Description of cutting should include the EDM wire diameter and material, the cut settings, the rate of cut, and any wire breakages or other issues. Description of the surface contouring should include the instrument details, the measurement density, and the thermal conditions throughout the measurement duration. The resulting surface contour maps should be plotted or described including the peak-to-valley range of each side. The sequence of steps used to process the data should be described in detail. Ideally, the smoothed surface contour should be plotted as should the misfit between the data and the smooth surface. The misfit should be quantified as a root-mean-square average. The description of the FE calculation should include mesh and element details. Stresses near the perimeter of the measurement surface may exhibit higher measurement uncertainty than points in the interior, see Section 5.6.1. Typically, some data near the perimeter are discarded before plotting.

5.4.2 Additional Issues

5.4.2.1 Order of Data Processing Steps

The outline of experimental steps presented above is considered to be a straightforward approach to performing contour method measurements. In practice, the order of data processing steps in going from raw displacement measurements to transferring the displacements to the FEM can vary significantly while still achieving satisfactory results. For instance, the data averaging described above assumes that both sets of data contain points at the same location. This is not always the case. If the sets of surface points do not match, then it is necessary to interpolate them onto a common set of points before averaging. The set of common points could be one of the data surfaces, a regular grid, or the finite element node locations. As an alternative to interpolating onto a grid for averaging, one could smooth each surface independently and then average the smooth surfaces.

5.4.2.2 Extrapolation

Displacements must be defined on all nodes on the cut surface in the finite element model. Because of finite distance between measurement points, part alignment and other experimental issues, the data generally will not extend all the way to the perimeter of the surface. In some way, the displacements must be extrapolated to the perimeter, but the method will depend on other data processing choices. If gridded or otherwise regular, the data can be linearly extrapolated out to the edges [23,32]. Alternatively, the smoothed surface fitted to the data can be used to extrapolate to the perimeter, but then the choice of fit surface can have a strong effect on the extrapolation [18]. Any region where the data were extrapolated should be considered unreliable and the stresses there not reported in the final results.

5.4.2.3 No Filtering

With careful planning, it can be possible to perform a measurement without any post-measurement filtering of the displacement data [12,33]. If the finite element node points are known prior to performing the surface displacement measurements then the CMM can be programmed to take displacement measurements directly at these locations. For this to work effectively, the raw displacement data should be as smooth as possible, for example, using a large measurement stylus and averaging multiple measurement values. These displacement data can then be directly averaged and applied to the finite element model. If carefully implemented, this approach can result in stresses that are reasonably smooth. If necessary, the computed stresses can be smoothed at the end of the analysis.

5.4.2.4 Stress-free Test Cut

It is good practice to verify the cutting assumptions by performing a similar cut on a part with the same cross-section in the absence of residual stress [10,11]. Since stress-free material can be difficult to find, such a test is often performed by cutting a slice off of the end of the part. That region is nearly stress-free because of the adjacent free surface. The thin slice will often have unrepresentative deformations, but the cut surface on the larger piece can be examined. Displacements measured on the test cut surface should not have significant form, that is, the surface should be flat. In some reported cases, the experimental displacement data have been "corrected" based on the form observed in the stress-free cutting condition [2,20,34]. A brief example of this is discussed later in the experimental application to the stainless steel indented disk.

5.5 Example Applications

5.5.1 Experimental Validation and Verification

This section presents experimental applications of the contour method where there are independent residual stress measurements for comparison. The first example compares with neutron diffraction and is from a class of specimens that tends to provide good agreement with neutron diffraction: non-welds. The second example is a linear friction weld with very high stress magnitudes and gradients, also compared with neutron diffraction. The third example is a laser peened plate with near surface stress gradients, compared with incremental slitting and X-ray diffraction with layer removal.

There are currently approximately two dozen published comparisons of contour method measurements with other measurements, primarily neutron and synchrotron diffraction, but also some relaxation methods. Such independent validations commonly show very good agreement on non-welded specimens [2,20,35–39] and on friction welded specimens [23,40,41], but not always [21,30]. Fusion welds sometimes showed good to very good agreement [18,19,26,28,29,42–45] and sometimes noticeable disagreement in some regions [46–50]. It is difficult to assess which results are more accurate when the two methods disagree. Welds are challenging for diffraction methods because of spatial variation of the unstressed lattice spacing and because of intergranular effects (microstresses and strains), as discussed in Chapters 7 and 8. Welds can also be challenging for contour measurements because the stresses can be quite high and the local yield strength may be lowered by the thermal process, both effects increasing plasticity errors. In addition, some of the measurements reported in the literature (both neutron and contour) were not done using the best practices.

5.5.1.1 Indented Stainless Steel Disk Compared With Neutron Diffraction

This example is presented in Chapter 8 on Neutron Diffraction with additional details. 60 mm diameter, 10 mm thick disks of 316L stainless steel were plastically compressed through the thickness with opposing 15 mm diameter, flat-end, hard steel indenters in the center of the disk [20], as illustrated in Figure 8.7. The residual hoop stresses on a diametrical plane of two disks indented under the same experimental conditions were measured with the contour method, see Figure 8.9. Disk A was cut in half using wire EDM with a 50 μm diameter tungsten wire. To avoid repetition of errors attributed to that wire, disk B was cut using a 100 μm diameter brass wire. As controls, two stress-free unindented disks were also cut using the 50 μm and 100 μm diameter wires, respectively.

The contours of both cut surfaces of each disk were measured using a laser scanner [26]. For disk B, Figure 5.5 shows the average measured contour. The peak-to-valley amplitude of the contour is about 40 μm. The surface roughness level in the measured contour is typical of laser scanners and much larger than what is measured using a CMM with a spherical tip. The cut surfaces of the two stress-free, unindented disks were also measured. The contour of the disk that was cut using the 100 μm brass wire was flat to within the measurement resolution. The contour of the disk that was cut using the 50 μm tungsten wire was bowed: higher by about 6 μm on the top and bottom edges of the 10 mm thickness than in the mid-plane. In order to correct this effect, the contour on the unindented disk was subtracted from the contour of the indented disk A, which was cut with the same wire.

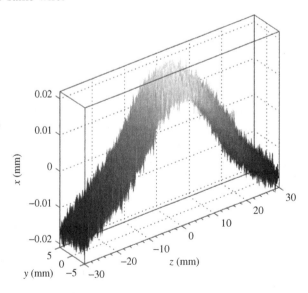

Figure 5.5 The average measured surface contour on one of the disks. Reproduced with permission from [32], Copyright 2010 Springer

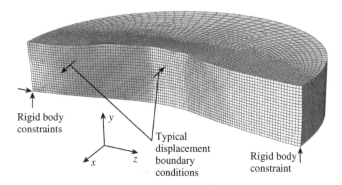

Figure 5.6 Finite element mesh of half-disk after displacement boundary conditions have deformed the cut surface into the opposite of the measured contour. Deformations exaggerated by a factor of 200

Figure 5.6 shows the finite element model used to calculate the residual stresses. The mesh for the half-disk used 51,920 linear hexahedral 8-node elements with reduced integration (C3D8R). The material behavior was considered elastically isotropic with an elastic modulus of 193 GPa and a Poisson's ratio of 0.3. To avoid clutter in the figure, only two typical displacement boundary conditions are illustrated, but all nodes on the cut surface had x-direction conditions applied.

Figure 5.7 shows the contour-method maps of residual hoop stress on the cross sections of the disks. In spite of the correction required on the Disk A data, the results agree to within about 20 MPa over most of the cross section. The contour results are compared to extensive neutron diffraction measurements on the same disk in Figure 8.10 in Chapter 8. The agreement between the contour method and neutron diffraction is excellent. Note that the contour results have a mild left-right asymmetry even through the specimens were prepared to be as axisymmetric as possible. The asymmetry probably reflects a slight bulge error as discussed in Section 5.2.3.

5.5.1.2 Linear Friction Weld Compared With Neutron Diffraction [51]

Sample blocks 38.1 mm tall by 50.8 mm wide by 12.7 mm thick machined from Ti-6Al-4V alloy bar stock were joined (at the 50.8 mm by 12.7 mm face) using linear friction welding (Figure 5.8). The resulting specimen was nominally 76.2 mm tall by 50.8 mm wide by 12.7 mm thick. The LFW process produces a narrow bond region and heat affected zone where the microstructure is altered from its original condition. Residual stresses were measured in the LFW test specimen using the contour method and neutron diffraction. Prior to the residual stress measurements the specimen edges were cut square and polished to reveal the LFW bond line. The final specimen dimensions are shown in Figure 5.8. The measurements were performed on the same test specimen, in sequence. First, neutron diffraction was used to measure σ_x, σ_y, and σ_z along the line through the center of the specimen shown in Figure 5.8. Following completion of the neutron diffraction measurement, the contour method was used to measure σ_y over the plane shown in Figure 5.8.

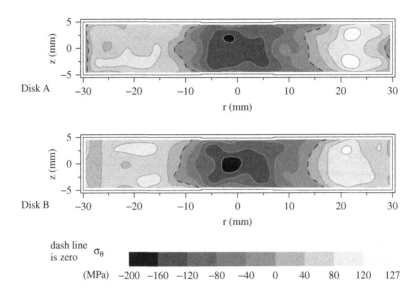

Figure 5.7 Hoop stresses measured on cross section of indented steel disks. The results agree very well with neutron measurements as shown in Figure 8.10. Reproduced with permission from [20], Copyright 2009 ASME

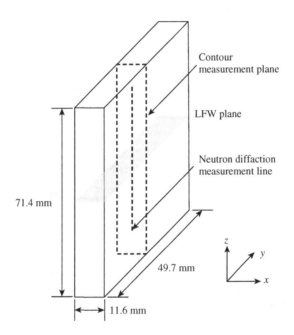

Figure 5.8 Illustration of LFW test specimen showing dimensions, reference coordinate frame, and measurement locations. Reproduced with permission from [51], Copyright 2013 SEM

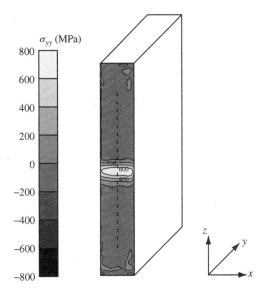

Figure 5.9 Two-dimensional map of the measured residual stress over the contour measurement plane. Reproduced with permission from [51], Copyright 2013 SEM

Figure 5.9 shows a plot of the two dimensional residual stress measured using the contour method. There is a concentrated region of high-magnitude tensile residual stress near the LFW joint. The peak stress magnitude is around 750 MPa. The tensile stress quickly diminishes to slight compression away from the LFW joint.

Data were extracted from the 2D contour surface along the same line where the neutron diffraction measurements were performed, and the comparison is plotted Figure 5.10. Overall, there is very good correlation between the two measurement techniques. The neutron diffraction data show slightly higher magnitude peak stress at the center of the weld (nominally 800 MPa versus 750 MPa, which is a 6% difference). The width of the tensile stress region is very similar for both sets of measurements. Since the results in Figure 5.10 differ by an almost constant shift of about 50 MPa, the most likely explanation for the modest differences between the two techniques would be an error in the unstressed lattice spacing used for the neutron stress determination.

The peak stress magnitudes are large, which makes plasticity a potential issue for the contour method measurement, but the agreement with neutron results indicate that plasticity was not a significant problem. The yield strength of the Ti-6Al-4V prior to welding was about 915 MPa. The occurrence of yielding during a contour measurement is a complicated phenomenon that depends on the full multiaxial stress state, the prior thermal and plastic history of the material, and the effectiveness of the part clamping. This LFW example should demonstrate that the contour method can at least sometimes get accurate results even with very high stresses.

5.5.1.3 Laser Peened Plate Compared with Slitting and X-ray Diffraction with Layer Removal

A plate of Ti-6Al-4V with original dimensions of 50 mm × 50 mm × 8.7 mm, shown in Figure 5.11, was processed using laser shock peening (LSP) over the top surface to induce

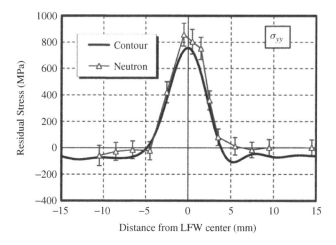

Figure 5.10 Line plot comparing the measured residual stress from the contour method and neutron diffraction experiments. Reproduced with permission from [51], Copyright 2013 SEM

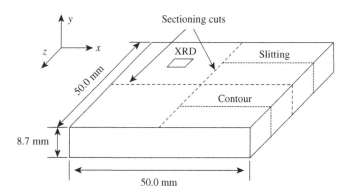

Figure 5.11 Illustration of laser shock processed test specimen used for residual stress measurement

a uniform layer of compressive residual stress. The plate was then cut into four equal size blocks, each nominally 25 mm × 25 mm × 8.7 mm, which are expected to contain similar amounts of residual stress. Residual stress measurements were performed on three of the four blocks using different techniques: the contour method, the slitting method, and X-ray diffraction with layer removal.

Figure 5.12 shows a two-dimensional map of the residual stress in the block measured using the contour method. As expected, the block has compressive residual stress near the laser peened face, which transitions to tensile stress near the interior, and then back to compressive residual stress on the back face. The tensile residual stress in the interior and the compressive stress on the back face are the results of the plate reacting to the strain induced by laser shock peening to achieve an equilibrium stress state. A line plot of residual stress versus depth at the center of the block is shown for the contour method, the slitting method, and X-ray diffraction with layer removal in Figure 5.13.

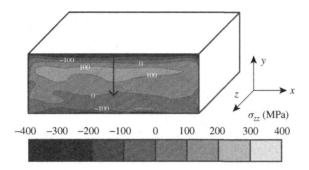

Figure 5.12 Two-dimensional map of the measured residual stress using the contour method

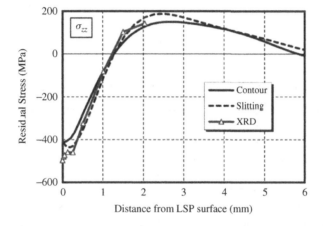

Figure 5.13 Line plot of residual stress vs. distance from the surface in a laser shock peened Ti-6Al-4V block measured using contour, slitting, and X-ray diffraction with layer removal

Overall there is excellent agreement between each of the three residual stress measurement techniques. The most significant differences are in the near surface region at depths less than 0.5 mm, but even at these locations the differences are relatively small, generally 10% and up to 20% in the extreme case.

5.5.2 Unique Measurements

This section presents experimental applications of the contour method that exemplify some of the unique capabilities of the technique. The first example is a very large and complicated weld joint taken from a nuclear power plant. The second example is railroad rails that show a very informative 2D distribution of stresses. In both of these examples, largely because of their size, there is essentially no other way to measure the 2D stress maps that contour provides.

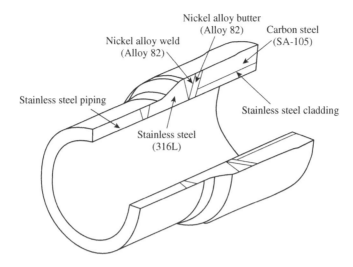

Figure 5.14 Illustration of nozzle specimens used for residual stress measurement

5.5.2.1 Large Dissimilar Metal Weld From a Nuclear Power Plant [52]

A contour method measurement was performed on a relief nozzle from the pressurizer of a canceled nuclear power plant. The nozzle was nominally 711 mm long with a 201 mm outer diameter and an inner diameter of 113 mm and contained a nickel based dissimilar metal weld used to join a carbon steel component to stainless steel piping, see Figure 5.14. Dissimilar metal welds are particularly susceptible to primary water stress corrosion cracking and residual stresses can have a significant influence on this failure mode.

As part of this study, the contour method was used to measure residual stress at six locations in two similar nozzles. Only a single measurement is included here due to space limitations, a two-dimensional map of the hoop residual stress in the nozzle. To facilitate measurement, the nozzle was first instrumented with strain gages and cut open to relieve the bending moment in the hoop direction. The residual stress released from this process was accounted for in the results using a finite element based approximation. Next, the contour method measurement was cut opposite the initial cut. Cutting was performed using 0.25 mm diameter brass wire.

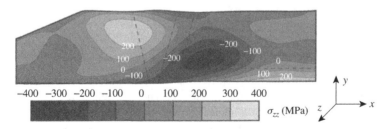

Figure 5.15 Two-dimensional map of the hoop residual stress for Nozzle #3 160-deg location. Reproduced from [52]

To account for the different materials, regions of the finite element model were assigned elastic properties unique to each material. Figure 5.15 shows a contour plot of the measured hoop residual stress in the nozzle. Significant compressive hoop residual stress exists on the inner diameter near the dissimilar metal weld. The region of compressive hoop residual stress grows larger through the Alloy 82 "butter" and into the carbon steel region. Compressive stresses by the ID are not likely to cause stress corrosion cracking, so are an encouraging result. However, weld repairs are not present in this nozzle, and have been shown by others to produce high residual stress on the ID, which could be problematic. Tensile hoop residual stress exists near the outer diameter, in a region that is shifted towards the stainless steel end of the nozzle.

5.5.2.2 Railroad Rail

For a study motivated by the 2000 fatal rail accident at Hatfield, UK, longitudinal residual stresses were mapped in two specimens of UK rail: a new roller-straightened rail and one that had undergone 23 years of service [53]. Both rails were BS 11 normal grade pearlitic steel with the standard 113A profile, and 76 cm long sections were measured. The sections were clamped securely and cut in two using EDM and a 0.25 mm diameter brass wire. The surfaces were then contoured by scanning with a Keyence LT-8105 confocal ranging probe, similar to the procedure described in [26]. The surface was scanned in the horizontal direction (x in Figure 5.16) at a sampling rate of 16 points/mm, and the rows were spaced vertically to give four rows per mm. This produced about 460,000 points on each surface. The peak-to-valley range of the contours was about 75 μm. The data points were interpolated onto a common grid mimicking the original data grid. A rectangular region that covered the entire cross-section was defined and then regions with

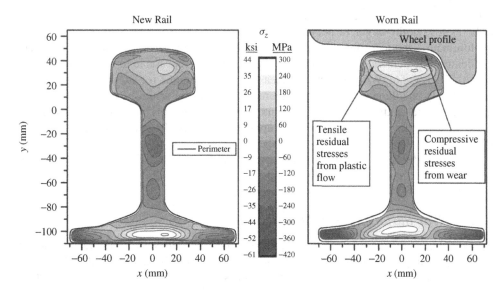

Figure 5.16 Contour method maps of longitudinal residual stresses in UK rails show the effects of roller straightening on the stresses in a new rail and the extensive effects of wear and plastic deformation on the stresses in a worn rail

missing data were filled in by extrapolation. At each grid location over the rectangle, the two data points were then averaged. This average point cloud was then fit to a smooth surface using bivariate (tensor product) cubic smoothing splines [26]. The stresses were calculated by forcing the cut surface, taken as initially flat, into the opposite shape of the measured contour in a 3D, elastic, FE analysis.

The resulting stress maps shown in Figure 5.16 were especially informative. The new rail shows a complicated stress pattern associated with plastic deformation from the roller straightening process [54]. The residual stresses are tensile in the head and foot of the rail, with the peak stresses located subsurface. There are balancing compressive stresses in the web and in the lateral regions of the foot. The worn rail results show that the stresses changed significantly. The stresses have become significantly compressive under the region of contact with the wheel, as had been observed previously with other techniques. The contour maps, though, show other changes that have never been experimentally observed before. The tensile stresses have increased in the subsurface region of the head in a matter consistent with plastic flow driven by the angled wheel contact. Subsurface initiated cracks in this region cause failures and a significant portion of train derailments occur because subsurface cracks are hard to detect. The contour results also show an increase in the magnitude of compressive stresses in the lateral region of the foot, which may be caused by plastic deformation since the rails are known to reduce in height over time.

5.6 Performance and Limitations of Methods

The contour method is nearly unique in its ability to measure a 2D cross-sectional map of residual stresses in even large parts. When used correctly and on appropriate specimens, the results are reasonably accurate and reliable. Based on all of the validations in the published literature, accuracies and uncertainties in the best test conditions can be estimated to be as low as the larger of about 10% or $\sigma/E \approx 0.00015$ (30 MPa in steel or 10 MPa in aluminum – but these numbers really depend on part size as discussed below). Several publications demonstrate that the contour method is also repeatable to these levels or better [20,27,33,55,56]. When selecting a measurement technique for stress distributions that are primarily one-dimensional, better accuracy can likely be achieved with other methods, like incremental slitting. To date, the contour method has only been applied using wire EDM to make the cut, which limits the application to metals and a few other materials that can be cut with EDM.

The remainder of this section discusses the conditions under which one can or cannot achieve the best results.

5.6.1 Near Surface (Edge) Uncertainties

Stresses near the perimeter of the measurement surface may exhibit higher measurement uncertainty than points in the interior depending on the nature of the stress field and the details of the surface displacement measurement and data processing. Furthermore, the assumption of constant cut width may be less accurate near the edges of the cut surface. For example, the EDM cut width may flare out a little bit at the top and bottom of the cut (entry and exit of the wire) or at the beginning or end of the cut [10,11]. Also the surface height map can be uncertain because, especially with non-contact scanners, it can be difficult to know exactly where the edge of the surface is and accurately determine the surface height. Contour method results can therefore be more uncertain near the edges of the cut. Depending on cut quality and measurement details, the uncertain region is typically about 0.5 mm. With special care, good results have been achieved closer to the edges [23,35,56]. Results should not be reported in the near-surface region unless such special care has been taken. Recently, an improvement in cut quality at the exit edge of the EDM wire has been achieved by using a sacrificial layer attached to the part surface [11].

5.6.2 Size Dependence

The contour method generally works better on larger parts. Other relaxation methods for measuring residual stress, for example, hole drilling and slitting, tend to be relatively size independent because, for a given stress magnitude, measured strains do not change when the part size is scaled up or down. Rather than strain, the contour method measures the surface shape, to infer displacements. For a given stress distribution, those displacements scale linearly with the part size. The EDM surface roughness and other cutting artifacts tend to remain relatively fixed in magnitude. Therefore, larger parts give more easily measured contours than smaller parts when all else is equal.

With current technology, a minimum peak-to-valley surface contour of $10-20\,\mu m$ is suggested in order to get reasonable results. If one has an idea of stresses, the expected contour can be estimated prior to the experiment with an elastic FE model. The smallest parts measured with the contour method have been about 2–6 mm thick [16,17,24,25,35,38,43,55,57–61], and in some of those cases the contour is averaged over the thin dimension resulting in a 1-D profile of the stress averaged through the thickness. The main limitation to achieving good results with smaller parts [62] and smaller contours than about $10\,\mu m$ is the cut and surface roughness quality currently achievable with EDM, which degrades the signal-to-noise ratio of the measured displacements. Measuring the surface contour is relatively easy and has never been the limiting factor for the contour method measurements.

5.6.3 Systematic Errors

5.6.3.1 Bulge

The bulge error (illustrated in Figure 5.3) results from elastic deformation and can be estimated using an FE model and even corrected for [10]. However, the error varies along the length of the cut tip, requiring a 3D model in the general case, depends on the circular shape of the cut tip, and must be estimated at incremental cut depths. Therefore, an FE estimate can be tedious. Figure 5.17 shows a 2D FE-based correction of the bulge error for a plastically bent beam [2,10]. The bulge effect estimated in Figure 5.17 is qualitatively typical of what can be expected: a reduction in the peak stress and a slight shift of the peaks. The error depends on the cut direction. Some results in the literature [28,48], especially those where the specimen was clamped on only one side [10,16,19], show indications of possible bulge errors.

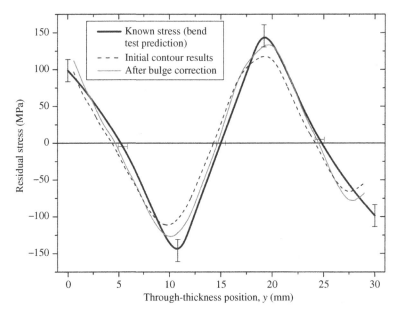

Figure 5.17 The bulge error in a plastically bent beam had a modest effect on the residual stress profile. Reproduced with permission from [10], Copyright SEM 2011

5.6.3.2 Plasticity

Like all relaxation methods, the contour method requires the assumption that as residual stresses are released, the material unloads elastically. The stress concentration at the cut tip may cause local yielding, which can affect the measured contour and therefore cause errors. The plasticity error is even more difficult to simulate than the bulge error, because in addition to needing a 3D model of many cut increments, it is also necessary to simulate reverse yielding behavior accurately. Only two studies of the plasticity effect are presently published in the literature. A simplistic 2D FE parameter study indicated relatively low plasticity errors for the contour method even with residual stresses at 70% or more of the yield strength [63]. A more sophisticated model indicated large errors for a particular specimen but started with residual stresses well above the yield strength [64]. In general, plasticity errors are mitigated by the round EDM cut tip and by strain hardening in the material. It is difficult to make any general statements about the sensitivity of the contour method to plasticity errors. Some contour method measurements have returned very high residual stresses but agreed well with diffraction measurements, indicating minimal plasticity errors [18–20,26,65]. There are also many examples where the agreement is not as good [30,48–50], and some of those likely have plasticity errors, but they are usually welds and it is sometimes difficult to assess the accuracy of the diffraction measurements used for validation.

For three reasons, it can be very misleading to compare measured residual stress magnitudes to the yield strength when assessing the potential for plasticity. First, the yield strength values usually reported are those of the as-received material. Often the same processes that produce the residual stress also change the yield strength, increasing it by strain hardening or decreasing it by thermal processes, which can significantly change the propensity for yielding and errors. Second, the measured residual stress values are generally for a single stress component, but a von Mises effective stress is what should be compared with the yield strength. Peak residual stresses are often in a region of high triaxialty, which makes the effective stress lower than the peak individual stress component. Third, cut tip yielding is driven by the integrated effect of all the released residual stresses, which can be characterized by the intensity factor, K_{Irs}, at the cut tip from the accumulated effect of releasing residual stress. As studied for the slitting method, the total effect will depend on the distribution of stresses rather than just the peak stress and can be low since residual stresses must satisfy force equilibrium and therefore compressive regions tend to mitigate tensile regions in the integrated effect [66].

5.7 Further Reading On Advanced Contour Method Topics

5.7.1 Superposition For Additional Stresses

The conventional contour method measures one stress component. There are two more advanced implementations where the contour method can be used to measure multiple stress components by using superposition of multiple measurements, each validated by comparison with neutron diffraction [32,37]. To explain the superposition methods, observe that Equation (5.2) in this chapter came from applying Equation (5.1) to a location, specifically the cut surface $x = 0$, where $\sigma_x^B = 0$. For other stress components and/or other locations, the σ^B stresses can be measured. In the "multiple cuts" method [32] a second cut is made to measure stresses on a new cut plane with the contour method. Such stresses would have been affected by the first cut, but the contour measurement and calculation for the first cut also determines the change in stress on the location of the second cut (and elsewhere). For the example of a second cut to measure σ_z on the plane at $z = 0$ in Figure 5.1, the original σ_z stresses are given by

$$\sigma_z^A(x, y, 0) = \sigma_z^B(x, y, 0) + \sigma_z^C(x, y, 0) \tag{5.3}$$

where the σ_z^B term is the contour method result from the second cut and σ_z^C over the plane of the second cut is just σ_z on that plane extracted directly from the same FEM calculation used to determine σ_x for the first contour cut. Because this method is very new, there are few examples in the literature [11,67], including one using the slitting method instead of contour [68]. In previous work with multiple cuts, the correction was not made to account for the effects of previous cuts [69].

The second superposition method involves using multiple methods instead of multiple cuts [37]. For example, the in-plane stresses *on the cut surface* ($x = 0$) in Figure 5.1 could be measured using X-ray diffraction or hole drilling once the EDM-affected layer is removed by electropolishing. For the example of σ_z, the original residual stresses on the cut plane are given by

$$\sigma_z^A(0, y, z) = \sigma_z^B(0, y, z) + \sigma_z^C(0, y, z) \tag{5.4}$$

where now the σ_z^B stresses are the surface stresses measured by another method and the σ_z^C term is extracted directly from the same FEM calculation used to determine σ_x

for the first contour cut. Because this method is very new, there are only a handful of examples in the literature [14,70].

5.7.2 Cylindrical Parts

Measuring hoop stresses in a cylindrical geometry requires special attention with the contour method [71]. In a cylinder, the residual hoop stresses can have a net bending moment through the thickness of a ring. For a contour method measurement of hoop stress, a radial cut would result in excessive stresses built up at the cut tip because of the bending and moment. Such high stresses could cause plasticity errors, which have been observed for such a cut with the contour method [23] and similarly with the slitting method [72]. Different approaches have been used to deal with this issue when measuring hoop stresses in cylinders [14,50,71], including the measurements presented previously on the nuclear reactor nozzle.

5.7.3 Miscellaneous

A contour method measurement was performed on a specimen with only a partial-penetration weld, which resulted in a discontinuous surface contour across the unbonded interface [50,71]. A special treatment was used in the surface smoothing and FE stress calculation to handle the discontinuity. Axial stresses in cylinders and rods have been measured only rarely [34], maybe because of the difficulty in clamping such parts for a cross-sectional cut. In some work, multiaxial stress states were determined using multiple cuts and an eigenstrain analysis to reconstruct the full stress tensor [24,28].

5.7.4 Patent

The residual stress measurement technique described herein was invented at Los Alamos National Laboratory and is protected by United States Patent Rights (Patent Number: 6,470,756 filed February 2001, granted October 2002) until 2021. The patent is administered by the Technology Transfer Division at Los Alamos National Laboratory. Patent rights are protected in the United States only. In some circumstances, there is an "experimental use" exemption for non-commercial research. This paragraph does not constitute legal advice.

Acknowledgments

The authors would like to thank their longtime collaborator Michael R. Hill of the University of Califonia, Davis, who has been as instrumental in the development of the contour method as anyone and should have co-authored this chapter, but he was too busy writing the slitting chapter. We would like to thank Gary Schajer not only for the difficult task of editing this book but also (Prime) for years of stimulating and fruitful research interactions. We would also like to thank Lyndon Edwards, John Bouchard, Tom Holden, Phil Withers, Don Brown, Bjorn Clausen, Pierluigi Pagliaro, David Smith, C. Can Aydiner and others for valuable interactions on the contour method or residual stress in general.

Los Alamos National Laboratory is operated by the Los Alamos National Security, LLC for the National Nuclear Security Administration of the U.S. Department of Energy under contract DE-AC52-06NA25396. The U.S. Government retains a non-exclusive, royalty-free license to publish or reproduce the published form of this contribution, or to allow others to do so, for U.S. Government purposes.

References

[1] Prime, M. B., Gonzales, A. R. (2000) The Contour Method: Simple 2-D Mapping of Residual Stresses, The 6th International Conference on Residual Stresses, IOM Communications, London, U.K., Oxford, UK 617–624.
[2] Prime, M. B. (2001) "Cross-sectional Mapping of Residual Stresses by Measuring the Surface Contour After a Cut," *Journal of Engineering Materials and Technology* 123(2):162–168.
[3] Bueckner, H. (1958) "The Propagation of Cracks and the Energy of Elastic Deformation," *Transactions of the American Society of Mechanical Engineers* 80:1225–1230.
[4] Bueckner, H. (1970) "Novel Principle for the Computation of Stress Intensity Factors," *Zeitschrift fuer Angewandte Mathematik & Mechanik* 50(9).
[5] Bueckner, H. F. (1973) "Field Singularities and Related Integral Representations," *Mechanics of Fracture*, GC Sih, ed. 239–314.
[6] Barenblatt, G. I. (1962) "The Mathematical Theory of Equilibrium Cracks in Brittle Fracture, Advances" in *Applied Mechanics*, H. L. Dryden, T. v. Kármán, G. Kuerti, F. H. v d Dungen, L. Howarth, eds., Elsevier 55–129.
[7] Paris, P. C., Gomez, M. P., Anderson, W. E. (1961) "A Rational Analytic Theory of Fatigue," *Trends in Engineering*, University of Washington, 13:9–14
[8] Eshelby, J. D. (1957) *The determination of the elastic field of an ellipsoidal inclusion, and related problems*, Proceedings of the Royal Society of London. Series A. Mathematical and Physical Sciences 241(1226):376–396.
[9] Tada, H., Paris, P. C., Irwin, G. R. (2000) *The Stress Analysis of Cracks Handbook* 3rd ed, The American Society of Mechanical Engineers, New York, NY.
[10] Prime, M. B., Kastengren, A. L. (2011) *The Contour Method Cutting Assumption: Error Minimization and Correction, Experimental and Applied Mechanics*, Volume 6, T. Proulx, ed., Springer: New York, 233–250. Currently available at http://www.lanl.gov/contour/. DOI 210.1007/1978-1001-4419-9792-1000_1040.
[11] Hosseinzadeh, F., Ledgard, P., Bouchard, P. (2013) "Controlling the Cut in Contour Residual Stress Measurements of Electron Beam Welded Ti-6Al-4V Alloy Plates," *Experimental Mechanics* 53(5):829–839.
[12] Hacini, L., Van Lê, N., Bocher, P. (2009) "Evaluation of Residual Stresses Induced by Robotized Hammer Peening by the Contour Method," *Experimental Mechanics* 49(6):775–783.
[13] Frankel, P. G., Withers, P. J., Preuss, M., Wang, H. T., Tong, J., Rugg, D. (2012) "Residual Stress Fields After FOD Impact on Flat and Aerofoil-shaped Leading Edges," *Mechanics of Materials* 55:130–145.
[14] Hosseinzadeh, F., Bouchard, P. (2013) "Mapping Multiple Components of the Residual Stress Tensor in a Large P91 Steel Pipe Girth Weld Using a Single Contour Cut," *Experimental Mechanics* 53(2): 171–181.
[15] Traore, Y., Bouchard, P., Francis, J., Hosseinzadeh, F. (2011) A novel cutting strategy for reducing plasticity induced errors in residual stress measurements made with the contour method, ASME Pressure Vessels and Piping Conference 2011, Baltimore USA, 17–21 July 2011.
[16] Murugan, N., Narayanan, R. (2009) "Finite Element Simulation of Residual Stresses and Their Measurement by Contour Method," *Materials & Design*, 30(6):2067–2071.
[17] Richter-Trummer, V., Moreira, P., Ribeiro, J., and de Castro, P. (2011) "The Contour Method for Residual Stress Determination Applied to an AA6082-T6 Friction Stir Butt Weld," *Materials Science Forum* 681:177–181.
[18] Traore, Y., Paddea, S., Bouchard, P., and Gharghouri, M. (2013) "Measurement of the Residual Stress Tensor in a Compact Tension Weld Specimen," *Experimental Mechanics* 53(4):605–618.
[19] Hosseinzadeh, F., Toparli, M. B., Bouchard, P. J. (2012) "Slitting and Contour Method Residual Stress Measurements in an Edge Welded Beam," *Journal of Pressure Vessel Technology* 134(1):011402–011406.
[20] Pagliaro, P., Prime, M. B., Clausen, B., Lovato, M. L., Zuccarello, B. (2009) "Known Residual Stress Specimens Using Opposed Indentation," *Journal of Engineering Materials and Technology* 131:031002.

[21] Savaria, V., Hoseini, M., Bridier, F., Bocher, P., Arkinson, P. (2011) "On the Measurement of Residual Stress in Induction Hardened Parts," *Materials Science Forum* 681:431–436.

[22] Cheng, W., Finnie, I., Gremaud, M. and Prime, M. B. (1994) "Measurement of Near-surface Residual-stresses Using Electric-discharge Wire Machining," *Journal of Engineering Materials and Technology* 116(1):1–7.

[23] Johnson, G. (2008) Residual stress measurements using the contour method, Ph.D. Dissertation, University of Manchester. (Currently available at http://www.lanl.gov/contour).

[24] DeWald, A. T., Hill, M. R. (2006) "Multi-axial Contour Method for Mapping Residual Stresses in Continuously Processed Bodies," *Experimental Mechanics* 46(4):473–490.

[25] Hatamleh, O., Lyons, J., Forman, R. (2007) "Laser Peening and Shot Peening Effects on Fatigue Life and Surface Roughness of Friction Stir Welded 7075-T7351 Aluminum," *Fatigue and Fracture of Engineering Material and Structures* 30(2):115–130.

[26] Prime, M. B., Sebring, R. J., Edwards, J. M., Hughes, D. J., and Webster, P. J. (2004) "Laser Surface-contouring and Spline Data-smoothing for Residual Stress Measurement," *Experimental Mechanics* 44(2):176–184.

[27] DeWald, A. T., Ranki, J. E., Hill, M. R., Lee, M. J., Chen, H. L. (2004) "Assessment of Tensile Residual Stress Mitigation in Alloy 22 Welds Due to Laser Peening," *Journal of Engineering Materials and Technology* 126(4):465–473.

[28] Kartal, M. E., Liljedahl, C. D. M., Gungor, S., Edwards, L., Fitzpatrick, M. E. (2008) "Determination of the Profile of the Complete Residual Stress Tensor in a VPPA Weld Using the Multi-axial Contour Method," *Acta Materialia* 56(16):4417–4428.

[29] Zhang, Y., Ganguly, S., Edwards, L., Fitzpatrick, M. E. (2004) "Cross-sectional Mapping of Residual Stresses in a VPPA Weld Using the Contour Method," *Acta Materialia* 52(17):5225–5232.

[30] Frankel, P., Preuss, M., Steuwer, A., Withers, P. J., Bray, S. (2009) "Comparison of Residual Stresses in Ti6Al4V and Ti6Al2Sn4Zr2Mo Linear Friction Welds," *Materials Science and Technology* 25:640–650.

[31] Prime, M. B., and Hill, M. R. (2006) "Uncertainty, Model Error, and Order Selection for Series-Expanded, Residual-Stress Inverse Solutions," *Journal of Engineering Materials and Technology*, 128(2):175–185.

[32] Pagliaro, P., Prime, M. B., Swenson, H., and Zuccarello, B. (2010) "Measuring Multiple Residual-Stress Components Using the Contour Method and Multiple Cuts," *Experimental Mechanics*, 50(2):187–194.

[33] Wilson, G. S., Grandt Jr,, A. F., Bucci, R. J., and Schultz, R. W. (2009) "Exploiting Bulk Residual Stresses to Improve Fatigue Crack Growth Performance of Structures," *International Journal of Fatigue*, 31(8–9):1286–1299.

[34] DeWald, A. T., and Hill, M. R. (2009) "Eigenstrain Based Model for Prediction of Laser Peening Residual Stresses in Arbitrary 3D Bodies." Part 2: model verification, *Journal of Strain Analysis for Engineering Design*, 44(1):13–27.

[35] Evans, A., Johnson, G., King, A., and Withers, P. J. (2007) "Characterization of Laser Peening Residual Stresses in Al 7075 by Synchrotron Diffraction and the Contour Method," *Journal of Neutron Research*, 15(2):147–154.

[36] DeWald, A. T. and Hill, M. R. (2009) "Eigenstrain Based Model for Prediction of Laser Peening Residual Stresses in Arbitrary 3D Bodies." Part 1: model description, *Journal of Strain Analysis for Engineering Design*, 44(1):1–11.

[37] Pagliaro, P., Prime, M. B., Robinson, J. S., Clausen, B., Swenson, H., Steinzig, M., and Zuccarello, B. (2011) "Measuring Inaccessible Residual Stresses Using Multiple Methods and Superposition," *Experimental Mechanics*, 51(7):1123–1134.

[38] Moat, R. J., Pinkerton, A. J., Li, L., Withers, P. J., and Preuss, M. (2011) "Residual Stresses in Laser Direct Metal Deposited Waspaloy," *Materials Science and Engineering: A*, 528(6):2288–2298.

[39] Rangaswamy, P., Griffith, M. L., Prime, M. B., Holden, T. M., Rogge, R. B., Edwards, J. M., and Sebring, R. J. (2005) "Residual Stresses in LENS (R) Components Using Neutron Diffraction and Contour Method," *Materials Science and Engineering A*, 399(1–2): 72–83.

[40] Prime, M. B., Gnaupel-Herold, T., Baumann, J. A., Lederich, R. J., Bowden, D. M., and Sebring, R. J. (2006) "Residual Stress Measurements in a Thick, Dissimilar Aluminum Alloy Friction Stir Weld," *Acta Materialia*, 54(15):4013–4021.

[41] Woo, W., Choo, H., Prime, M. B., Feng, Z., and Clausen, B. (2008) "Microstructure, Texture and Residual Stress in a Friction-stir-processed AZ31B Magnesium Alloy," *Acta materialia*, 56(8):1701–1711.

[42] Zhang, Y., Ganguly, S., Stelmukh, V., Fitzpatrick, M. E., and Edwards, L. (2003) "Validation of the Contour Method of Residual Stress Measurement in a MIG 2024 Weld by Neutron and Synchrotron X-ray Diffraction," *Journal of Neutron Research*, 11(4):181–185.

[43] Richter-Trummer, V., Tavares, S. M. O., Moreira, P., de Figueiredo, M. A. V., and de Castro, P. (2008) "Residual Stress Measurement Using the Contour and the Sectioning Methods in a MIG Weld: Effects on the Stress Intensity Factor," *Ciência & Tecnologia dos Materiais*, 20(1–2):114–119.

[44] Turski, M., and Edwards, L. (2009) "Residual Stress Measurement of a 316L Stainless Steel Bead-on-plate Specimen Utilising the Contour Method," *International Journal of Pressure Vessels and Piping*, 86(1):126–131.

[45] Bouchard, P. J. (2008) "Code Characterisation of Weld Residual Stress Levels and the Problem of Innate Scatter," *International Journal of Pressure Vessels and Piping*, 85(3):152–165.

[46] Zhang, Y., Pratihar, S., Fitzpatrick, M. E., and Edwards, L. (2005) "Residual Stress Mapping in Welds Using the Contour Method," *Materials Science Forum*, 490/491:294–299.

[47] Kartal, M., Turski, M., Johnson, G., Fitzpatrick, M. E., Gungor, S., Withers, P. J., and Edwards, L. (2006) "Residual Stress Measurements in Single and Multi-pass Groove Weld Specimens Using Neutron Diffraction and the Contour Method," *Materials Science Forum*, 524–525:671–676.

[48] Withers, P. J., Turski, M., Edwards, L., Bouchard, P. J., and Buttle, D. J. (2008) "Recent Advances in Residual Stress Measurement," *The International Journal of Pressure Vessels and Piping*, 85(3):118–127.

[49] Thibault, D., Bocher, P., Thomas, M., Gharghouri, M., and Côté, M. (2010) "Residual Stress Characterization in Low Transformation Temperature 13%Cr-4%Ni Stainless Steel Weld by Neutron Diffraction and the Contour Method," *Materials Science and Engineering: A*, 527(23):6205–6210.

[50] Brown, D. W., Holden, T. M., Clausen, B., Prime, M. B., Sisneros, T. A., Swenson, H., and Vaja, J. (2011) "Critical Comparison of Two Independent Measurements of Residual Stress in an Electron-Beam Welded Uranium Cylinder: Neutron Diffraction and the Contour Method," *Acta Materialia*, 59(3):864–873.

[51] DeWald, A. T., Legzdina, D., Clausen, B., Brown, D. W., Sisneros, T. A., and Hill, M. R. (2013) "A Comparison of Residual Stress Measurements on a Linear Friction Weld Using the Contour Method and Neutron Diffraction," *Experimental and Applied Mechanics*, Volume 4, C. E. Ventura, W. C. Crone, and C. Furlong, eds., Springer: New York, pp. 183–189.

[52] DeWald, A. T., Hill, M. R., and Willis, E. (2011) Measurement of Welding Residual Stress in Dissimilar Metal Welds Using the Contour Method, Proc. ASME 2011 Pressure Vessels and Piping Conference (PVP2011) ASME, pp. 1599–1605.

[53] Kelleher, J., Prime, M. B., Buttle, D., Mummery, P. M., Webster, P. J., Shackleton, J., and Withers, P. J. (2003) "The Measurement of Residual Stress in Railway Rails by Diffraction and Other Methods," *Journal of Neutron Research*, 11(4):187–193.

[54] Schleinzer, G., and Fischer, F. D. (2001) "Residual Stress Formation During the Roller Straightening of Railway Rails," *International Journal of Mechanical Sciences*, 43(10):2281–2295.

[55] Ismonov, S., Daniewicz, S. R., Newman, J. J. C., Hill, M. R., and Urban, M. R. (2009) "Three Dimensional Finite Element Analysis of a Split-Sleeve Cold Expansion Process," *Journal of Engineering Materials and Technology*, 131(3): 031007.

[56] Stuart, D. H., Hill, M. R., Newman, J. C., and Jr. (2011) "Correlation of One-dimensional Fatigue Crack Growth at Cold-expanded Holes Using Linear Fracture Mechanics and Superposition," *Engineering Fracture Mechanics*, 78(7):1389–1406.

[57] Rubio-González, C., Felix-Martinez, C., Gomez-Rosas, G., Ocaña, J. L., Morales, M., and Porro, J. A. (2011) "Effect of Laser Shock Processing on Fatigue Crack Growth of Duplex Stainless Steel," *Materials Science and Engineering: A*, 528(3):914–919.

[58] Lillard, R. S., Kolman, D. G., Hill, M. A., Prime, M. B., Veirs, D. K., Worl, L. A., and Zapp, P. (2009) "Assessment of Corrosion Based Failure in Stainless Steel Containers Used for the Long-term Storage of Plutonium Base Salts," *Corrosion*, 65(3):175–186.

[59] Cuellar, S. D., Hill, M. R., DeWald, A. T., and Rankin, J. E. (2012) "Residual Stress and Fatigue Life in Laser Shock Peened Open Hole Samples," *International Journal of Fatigue*, 44:8–13.

[60] Richter-Trummer, V., Suzano, E., Beltrao, M., Roos, A., dos Santos, J. F., and de Castro, P. M. S. T. (2012) "Influence of the FSW Clamping Force on the Final Distortion and Residual Stress Field," *Materials Science and Engineering A*, 538:81–88.

[61] Carlone, P., and Palazzo, G. S. (2012) "Experimental Analysis of the Influence of Process Parameters on Residual Stress in AA2024-T3 Friction Stir Welds," *Key Engineering Materials*, 504–506:753–758.

[62] Boyce, B. L., Reu, P. L., and Robino, C. V. (2006) "The Constitutive Behavior of Laser Welds in 304L Stainless Steel Determined by Digital Image Correlation," *Metallurgical and Materials Transactions A*, 37A(8):2481–2492.

[63] Shin, S. H. (2005) "FEM Analysis of Plasticity-induced Error on Measurement of Welding Residual Stress by the Contour Method," *Journal of Mechanical Science and Technology*, 19(10):1885–1890.

[64] Dennis, R. J., Bray, D.P., Leggatt, N.A., Turski, M. (2008) Assessment of the influence of plasticity and constraint on measured residual stresses using the contour method, Proc. 2008 ASME Pressure Vessels and Piping Division Conference, Chicago, IL, USA, ASME, pp. PVP2008-61490.

[65] Edwards, L., Smith, M., Turski, M., Fitzpatrick, M., and Bouchard, P. (2008) "Advances in Residual Stress Modeling and Measurement for the Structural Integrity Assessment of Welded Thermal Power Plant," *Advanced Materials Research*, 41–42:391–400.

[66] Prime, M. B. (2010) "Plasticity Effects in Incremental Slitting Measurement of Residual Stresses," *Engineering Fracture Mechanics*, 77(10):1552–1566.

[67] Pagliaro, P., Prime, M. B., and Zuccarello, B. (2007) Inverting multiple residual stress components from multiple cuts with the contour method, Proc. Proceedings of the SEM Annual Conference and Exposition on Experimental and Applied Mechanics 2007, pp. 1993–2005.

[68] Wong, W., and Hill, M. R. (2012) "Superposition and Destructive Residual Stress Measurements," *Experimental Mechanics*, 53(3):339–344.

[69] Prime, M. B., Newborn, M. A., and Balog, J. A. (2003) "Quenching and Cold-work Residual Stresses in Aluminum Hand Forgings: Contour Method Measurement and FEM Prediction," *Materials Science Forum*, 426–432:435–440.

[70] Olson, M. D., Wong, W., and Hill, M. R. (2012) Simulation of Triaxial Residual Stress Mapping for a Hollow Cylinder, Proc. ASME 2012 Pressure Vessels & Piping Division Conference, Toronto, Ontario, CANADA, pp. PVP-2012-78885.

[71] Prime, M. B. (2011) "Contour Method Advanced Applications: Hoop Stresses in Cylinders and Discontinuities," *Engineering Applications of Residual Stress*, Volume 8, T. Proulx, ed., Springer: New York, pp. 13–28.

[72] de Swardt, R. R. (2003) "Finite Element Simulation of Crack Compliance Experiments to Measure Residual Stresses in Thick-walled Cylinders," *Journal of Pressure Vessel Technology*, 125(3):305–308.

6

Applied and Residual Stress Determination Using X-ray Diffraction

Conal E. Murray[1] and I. Cevdet Noyan[2]
[1]*IBM T.J. Watson Research Center, New York, USA*
[2]*Columbia University, New York, USA*

6.1 Introduction

All diffraction techniques for strain/stress measurement utilize the same basic framework. Diffraction, constructive interference of a wave scattered by a periodic array of atoms (Figure 6.1(a)), is used to measure the atomic spacing. To observe constructive interference peaks in the scattered intensity, the wavelength of the incident waves must be similar to the spacing of the atoms. For most materials this limits the wavelengths that can be used to the X-ray region of the electromagnetic spectrum; typically photons with energies between 5–120 keV are used. Elastic tractions acting on the ensemble of atoms change the potential energy of the system and the atoms move to new equilibrium positions, resulting in a shift of the diffraction peaks (Figure 6.1(b)). The changes in the positions of the diffraction peaks can then be used to calculate the strain and/or stress tensor components of interest in the diffracting regions by using the appropriate formulations of solid mechanics.

While the basic framework, as posed above, is fairly straightforward, proper application of the technique can be non-trivial. Various formulations can be chosen to analyze the scattering data. These formulations may yield divergent definitions of the information volume from which the displacement data is obtained. Different *a priori* assumptions of

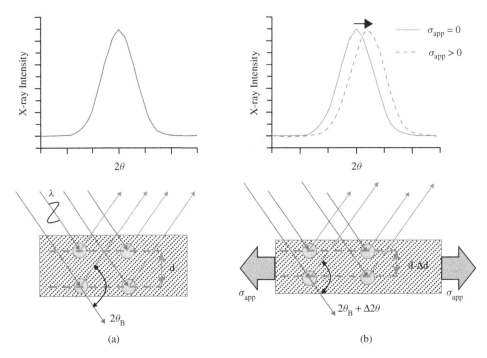

Figure 6.1 Schematic of diffraction emanating from an atomic array in an (a) unstrained state and (b) under tensile strain due to an applied load

the dimensionality of the strain/stress field within the information volume can yield very different results. In this chapter we discuss these issues and their implications on the practice of X-ray stress–strain analysis.

There are two techniques that can be used to determine homogeneous strain/stress profiles using X-ray diffraction:

1. A diffractometer is used to determine the elastic strain by measuring the lattice spacings in the material along various directions. These strains are then used, through the transformation law for 2nd rank tensors to compute the strain tensor in the sample coordinates. The stresses are then evaluated using the appropriate expression of Hooke's law. This technique is applicable to both polycrystalline and single crystal samples.
2. The local and global curvatures of a single-crystal sample can be determined by tracking the orientation of a crystal direction as a function of position within the sample using a goniometer equipped with a translation stage. If the curvature is caused by elastic constraint within the sample, it is possible to calculate the stresses due to the elastic constraint using various equations such as the Stoney formula. The diffraction system for this purpose can use a double-crystal diffractometer or a Lang camera, commonly used for X-ray topography.

In addition to these two techniques, the breadth of X-ray diffraction peaks can be used to get information about the RMS strain, $<\varepsilon^2>^{1/2}$. However, this value describes the distribution of elastic strains and cannot be used to calculate individual stresses. In the following sections we will discuss these techniques and their applications.

6.2 Measurement of Lattice Strain

This technique uses the spacing of atomic planes of the crystalline lattice as an internal strain gage and has been in use for almost 90 years; the first application was by Lester and Aborn in 1925 [1]. Figure 6.2 shows the coordinate systems used in the following discussion. $\vec{S}_1$ and $\vec{S}_2$ define the surface of the specimen. The measured plane spacing, $(d_{hkl})_{\phi\psi}$, is along the $\vec{L}_3$ axis of the laboratory coordinate system, $\vec{L}_i$. The coordinate systems $\vec{S}_i$, and $\vec{L}_i$ are related through the (rotation) angles ϕ and ψ. In what follows, primed tensor quantities refer to the laboratory system and unprimed tensor quantities refer to the sample coordinate system, following the usage by Dölle [2].

The plane spacing $(d_{hkl})_{\phi\psi}$ is obtained from the position of the diffraction peak through Bragg's law, $\lambda = 2d_{hkl} \sin\theta_B$, where λ is the X-ray wavelength, θ_B is the Bragg angle and hkl are the indices of the diffracting planes. Then the strain $(\varepsilon'_{33})_{\phi\psi}$ along $\vec{L}_3$ can be obtained from:

$$(\varepsilon'_{33})_{\phi\psi} = \frac{(d_{hkl})_{\phi\psi} - d_0}{d_0} \qquad (6.1)$$

Here d_0 is the unstressed lattice spacing. This strain can be expressed in terms of the strains, ε_{ij}, in the sample coordinate system by the tensor transformation [3]:

$$\varepsilon'_{33} = a_{3k}a_{3l}\varepsilon_{kl} \qquad (6.2)$$

where a_{3k}, a_{3l} are the direction cosines between $\vec{L}_3$ and the $\vec{S}_i$ axes, and summation over repeated indices is indicated. The direction cosine matrix corresponding to Figure 6.2 is:

$$a_{ik} = \begin{bmatrix} \cos\phi\cos\psi & \sin\phi\cos\psi & -\sin\psi \\ -\sin\phi & \cos\phi & 0 \\ \cos\phi\sin\psi & \sin\phi\sin\psi & \cos\psi \end{bmatrix} \qquad (6.3)$$

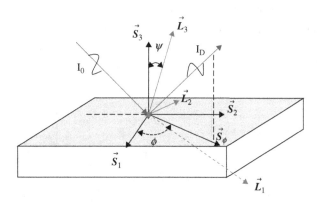

Figure 6.2 Definition of the sample, $\vec{S}_i$, and laboratory, $\vec{L}_i$, coordinate systems. The sample surface contains the $\vec{S}_1$ and $\vec{S}_2$ vectors. The incident and diffracted beams, I_0, I_D, and the normal to the diffracting planes, $\vec{L}_3$, are in the same plane as the surface direction $\vec{S}_\phi$. The Bragg angle, θ_B (not shown), is one half of the angle between the transmitted beam (the extension of I_0 through the sample) and the diffracted beam

The angles ϕ, ψ are settable on the diffractometer and are assumed to be known exactly. From Equations (6.1), (6.2) and (6.3) one obtains:

$$(\varepsilon'_{33})_{\phi\psi} = \frac{(d_{hkl})_{\phi\psi} - d_0}{d_0} = \varepsilon_{11}\cos^2\phi\sin^2\psi + \varepsilon_{12}\sin 2\phi\sin^2\psi$$
$$+ \varepsilon_{22}\sin^2\phi\sin^2\psi + \varepsilon_{33}\cos^2\psi$$
$$+ \varepsilon_{13}\cos\phi\sin 2\psi + \varepsilon_{23}\sin\phi\sin 2\psi \quad (6.4)$$

which is the fundamental equation for X-ray strain determination.

It is important to note that, Equations (6.2) and (6.3) have no explicit or implicit assumptions about the type of material. However, it is assumed that the strain data are obtained from a geometric point located at the origin of Figure 6.2. The use of Equation (6.1) limits the measurement to crystalline materials since diffraction is used to measure the atomic plane spacing. If the sample is a random (untextured) polycrystal, then there will be diffracted intensity at all $\vec{L}_3$ vectors since some crystallites will be oriented properly for diffraction for all Euler angles. For a textured polycrystal the $\vec{L}_3$ vectors for which $d_{\phi\psi}$ can be measured depends on the orientation distribution function of the sample. For a single crystal sample $d_{\phi\psi}$ can only be measured for the various (permitted) reflections; these depend on the symmetry of the unit cell. Consequently the rotation angles, ϕ, ψ, are not arbitrary, but are set by crystal symmetry.

For polycrystalline samples where it is possible to obtain a diffracted beam, and thus a plane spacing value, $d_{\phi\psi}$ for arbitrary ϕ, ψ rotations, three basic types of $d_{\phi\psi}$ vs. $\sin^2\psi$ behavior are observed, where the ψ tilts are carried out at fixed ϕ. These are shown in Figure 6.3a,b,c respectively. Figures 6.3a,b depict regular $d_{\phi\psi}$ vs. $\sin^2\psi$ behavior which can be predicted by Equation (6.4). When the strain components ε_{13} and/or ε_{23} are zero, Equation (6.4) predicts a linear variation of $d_{\phi\psi}$ vs. $\sin^2\psi$. When either or both of these shear strains (ε_{13}, ε_{23}) are finite, $d_\psi|_\phi$ measured at positive and negative ψ will be different due to the $\sin 2\psi$ term. This causes a "split" in $d_{\phi\psi}$ vs. $\sin^2\psi$ plots. This effect is termed ψ-splitting. Data exhibiting regular behavior can, thus, be analyzed by methods based on Equation (6.4). On the other hand, oscillatory $d_{\phi\psi}$ vs. $\sin^2\psi$ plots (irregular $d_{\phi\psi}$ vs. $\sin^2\psi$ behavior) cannot be predicted by Equation (6.4) without further modification. It should be noted that Equation (6.4) predicts regular $d_{\phi\psi}$ vs. $\sin^2\psi$ plots for

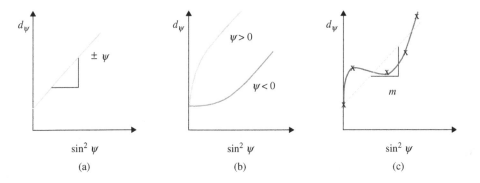

Figure 6.3 Plots of regular (a,b) and oscillatory d vs. $\sin^2\psi$ data. The regular data can be predicted by, and analyzed with, formulations based on Equation (6.4)

textured materials or single crystals. For these systems the ψ-angles at which diffraction can be obtained at a given ϕ are determined by the orientation-distribution function or by unit cell symmetry. However, those points that are measurable must still fall on the regular curves.

6.3 Analysis of Regular $d_{\phi\psi}$ vs. $\sin^2\psi$ Data

In the most general case Equation (6.4) is a linear equation with six unknown strain terms and may be solved exactly if $d_{\phi\psi}$ is measured along six independent directions $(\vec{L}_3)_{\phi\psi}$. It is better, however, to have more points to reduce statistical errors. There are two techniques that can be used to analyze such data.

6.3.1 Dölle-Hauk Method

In this approach [4] two terms based on Equation (6.4) are defined:

$$a_1 = \frac{1}{2}[(\varepsilon'_{33})_{\phi\psi+} + (\varepsilon'_{33})_{\phi\psi-}] = \{\varepsilon_{11}\cos^2\phi + \varepsilon_{12}\sin 2\phi + \varepsilon_{22}\sin^2\phi - \varepsilon_{33}\}\sin^2\psi + \varepsilon_{33}$$
(6.5-a)

$$a_2 = \frac{1}{2}[(\varepsilon'_{33})_{\phi\psi+} - (\varepsilon'_{33})_{\phi\psi-}] = \{\varepsilon_{13}\cos\phi + \varepsilon_{23}\sin\phi\}\sin|2\psi|$$
(6.5-b)

Equation (6.5-a) predicts a linear variation of a_1 vs. $\sin^2\psi$, where the slope and intercept are given by:

$$m_{a1}|_\phi = \{\varepsilon_{11}\cos^2\phi + \varepsilon_{12}\sin 2\phi + \varepsilon_{22}\sin^2\phi - \varepsilon_{33}\}$$
$$I_{a1}|_\phi = \varepsilon_{33}$$
(6.6)

Similarly, a_2 varies linearly with $\sin|2\psi|$. The slope in this case is:

$$m_{a2}|_\phi = \{\varepsilon_{13}\cos\phi + \varepsilon_{23}\sin\phi\}$$
(6.7)

If $d_{\phi\psi}$ vs. $\sin^2\psi$ data are obtained over a range of $\mp\psi$ at three ϕ rotations (0°, 45° and 90°), the unknown strain terms ε_{11}, ε_{12}, ε_{22} can be obtained from the slopes of the a_1 vs. $\sin^2\psi$ plots while the strain normal to the surface, ε_{33}, can be obtained from their intercepts. This value should be the same for all rotations; this serves as a check of the alignment of the system. The strain terms ε_{13}, ε_{23} can be obtained from the slopes of the a_2 vs. $\sin|2\psi|$ plots for $\phi = 0°$ and 90° respectively.

6.3.2 Winholtz-Cohen Least-squares Analysis

In this procedure Equation (6.4) is written in matrix form to facilitate a least-squares analysis [5]:

$$\varepsilon_1 = \varepsilon_{11}, \quad f_1(\phi, \psi) = \cos^2\phi \sin^2\psi$$
$$\varepsilon_2 = \varepsilon_{22}, \quad f_2(\phi, \psi) = \sin^2\phi \sin^2\psi$$

$$\varepsilon_3 = \varepsilon_{33}, \quad f_3(\phi, \psi) = \cos^2\psi$$

$$\varepsilon_4 = \varepsilon_{23}, \quad f_4(\phi, \psi) = \sin\phi \sin^2\psi$$

$$\varepsilon_5 = \varepsilon_{13}, \quad f_5(\phi, \psi) = \cos\phi \sin^2\psi$$

$$\varepsilon_6 = \varepsilon_{12}, \quad f_6(\phi, \psi) = \sin 2\phi \sin^2\psi \qquad (6.8)$$

The residual between calculated and measured strains (e'_i) along the $(\vec{L}_3)_i$ axis for the ith measurement (tilt-rotation combination) is:

$$r_i = \sum_{j=1}^{6} \varepsilon_j f_j(\phi_i, \psi_i) - e'_i \qquad (6.9)$$

The total weighted sum of the squared error, R, for n measurements of e' is given by:

$$R = \sum_{i=1}^{n} \frac{1}{\mathrm{var}(e'_i)} \left[\left(\sum_{j=1}^{6} \varepsilon_j f_j(\phi_i, \psi_i) \right) - e'_i \right]^2 \qquad (6.10)$$

where

$$\mathrm{var}(e') = \left(\frac{1}{d_0}\right)^2 \left(\frac{\pi}{180}\right)^2 \left(\frac{\lambda \cos\theta_B}{2\sin^2\theta_B}\right)^2 \frac{\mathrm{var}(2\theta_B)}{2} \qquad (6.11)$$

Here the variance of the peak position can be obtained from the $2\theta/\Omega$ scan by means of a particular peak-fitting algorithm. To find the solution with the minimum error one takes the partial derivatives of Equation (6.10) with respect to each strain ε_j and sets these equal to zero. This yields:

$$\sum_{i=1}^{n} \left[\left(\sum_{k=1}^{6} \varepsilon_k f_k(\phi_i, \psi_i) \right) - e'_i \right] \frac{f_j(\phi_i, \psi_i)}{\mathrm{var}(e'_i)} = 0 \qquad (6.12)$$

To formulate a matrix equation the B matrix and the $\vec{E}$ vector are defined as:

$$B_{jk} = \sum_{i=1}^{n} f_j(\phi_i, \psi_i) f_k(\phi_i, \psi_i)/\mathrm{var}(e'_i)$$

$$\vec{E}_j = \sum_{i=1}^{n} e'_i f_j(\phi_i, \psi_i)/\mathrm{var}(e'_i) \qquad (6.13)$$

For a non-singular B matrix, the strain matrix in the sample coordinate system that is associated with the minimum least squared error is given from the solution of

$$\varepsilon = B^{-1}\vec{E} \qquad (6.14)$$

The calculated strain matrix, ε should then be transformed into the strain tensor ε_{ij} (in the sample coordinates) through the left hand of Equation (6.8).

When the solutions given by Equations (6.5) and (6.10) are compared, it is seen that the least-squares analysis results in a more efficient use of available data since in

Applied and Residual Stress Determination Using X-ray Diffraction

Table 6.1 Comparison of the counting statistical errors in the strain tensor, computed by (a) the Dolle-Hauk technique, (b) the generalized least-squares method of Winholtz and Cohen. The (parentheses) show the variances associated with each strain value. Adapted from reference [5]

$$(\varepsilon_{ij})_{D-H} = \begin{bmatrix} 1.649\,(0.088) & -0.139\,(0.087) & -0.226\,(0.026) \\ -0.139\,(0.087) & 1.721\,(0.080) & 0.013\,(0.021) \\ -0.226\,(0.026) & 0.013\,(0.021) & -1.001\,(0.064) \end{bmatrix} \times 10^{-3} \quad \text{(a)}$$

$$(\varepsilon_{ij})_{W-C} = \begin{bmatrix} 1.515\,(0.036) & -0.045\,(0.043) & -0.234\,(0.010) \\ -0.045\,(0.043) & 1.888\,(0.031) & 0.029\,(0.009) \\ -0.234\,(0.010) & 0.029\,(0.009) & -0.936\,(0.010) \end{bmatrix} \times 10^{-3} \quad \text{(b)}$$

Equation (6.10), each error r_i is weighted by the inverse variance of the corresponding strain, e'_i, ensuring that the most reliable measurements are more heavily weighted in the analysis. Table 6.1 shows the strain tensors ε_{ij} obtained from the same specimen, a normalized plain carbon steel ground along the $\vec{S}_1$ direction, by the two techniques, showing that the counting statistical errors associated by the general least-squares method of Winholtz and Cohen are about half of the Dölle-Hauk analysis.

It must be noted that the generalized least-squares analysis described here can only be applied to data that form regular $d_{\phi\psi}$ vs. $\sin^2\psi$ plots. Consequently, even though it is possible to optimize the rotation and tilt angles, ϕ_i, ψ_i to minimize statistical counting errors and do the measurement more efficiently, the regularity of the data must be checked by plotting at least one plot of d_ψ vs. $\sin^2\psi$ at constant ϕ_i.

6.4 Calculation of Stresses

Once the full strain tensor, ε_{ij}, in the sample coordinate system is determined from the diffraction data, the stresses can be calculated using Hooke's law:

$$\sigma_{ij} = C_{ijkl}\varepsilon_{kl} \qquad (6.15\text{-a})$$

For an elastically isotropic specimen, this equation can be written as:

$$\sigma_{ij} = \frac{1}{\frac{1}{2}S_2}\left[\varepsilon_{ij} - \delta_{ij}\frac{S_1}{\frac{1}{2}S_2 + 3S_1}\varepsilon_{kk}\right] \qquad (6.15\text{-b})$$

where δ_{ij} is the Kronecker delta and summation over repeated indices is indicated. The terms S_1 and $\frac{1}{2}S_2$ are also termed X-ray elastic constants and have the following representation:

$$\frac{1}{2}S_2 = \left(\frac{1+\nu}{E}\right)_{hkl}, \quad S_1 = \left(-\frac{\nu}{E}\right)_{hkl} \qquad (6.15\text{-c})$$

where E is the Young's modulus, ν is the Poisson's ratio, and hkl refers to the Miller indices of the reflection under investigation. For an ideal isotropic specimen, these terms are independent of hkl but for quasi-isotropic polycrystalline materials, these terms depend on the reflection used and are best determined experimentally. Calculation of their values will be detailed in the next section.

In most of past literature, instead of first determining the strain tensor and then calculating the stresses through the appropriate formulation of Hooke's law, Equation (6.4) is re-written in terms of the stresses in the sample coordinate system. For an isotropic material this yields [6]:

$$(\varepsilon'_{33})_{\phi\psi} = \frac{(d_{hkl})_{\phi\psi} - d_0}{d_0} = \frac{1+\nu}{E}(\sigma_{11}\cos^2\phi + \sigma_{12}\sin 2\phi + \sigma_{22}\sin^2\phi - \sigma_{33})\sin^2\psi$$
$$+ \frac{1+\nu}{E}\sigma_{33} - \frac{\nu}{E}(\sigma_{11} + \sigma_{22} + \sigma_{33})$$
$$+ \frac{1+\nu}{E}(\sigma_{13}\cos\phi + \sigma_{23}\sin\phi)\sin 2\psi \quad (6.16)$$

Both the Dölle-Hauk and Winholtz-Cohen techniques can be used for analysis of Equation (6.16) by appropriately modifying Equations (6.5) to (6.7) and (6.8) to (6.14), respectively.

The most-used variant of Equation (6.16) is the case when a bi-axial stress state exists in the plane of the sample surface. In this case all stresses normal to the sample surface are zero, $\sigma_{3j} = 0$, $j = 1, 3$. For such a stress tensor Equation (6.16) becomes:

$$d_\psi = \frac{1+\nu}{E}d_0\sigma_\phi\sin^2\psi - \frac{\nu}{E}(\sigma_{11} + \sigma_{22})d_0 + d_0, \quad (6.17)$$

Here σ_ϕ is the surface stress along the $\vec{S}_\phi$ direction (Figure 6.2):

$$\sigma_\phi = \sigma_{11}\cos^2\phi + \sigma_{12}\sin 2\phi + \sigma_{22}\sin^2\phi \quad (6.18)$$

Equation (6.17) shows that the variation of the plane spacing with $\sin^2\psi$ is linear, with a slope proportional to:

$$m = \frac{1+\nu}{E}d_0\sigma_\phi \quad (6.19)$$

Equation (6.17) is known as the "$\sin^2\psi$" technique. If the unstressed plane spacing, d_0, and the elastic constant, $\frac{1+\nu}{E}$, are known, the stress along $\vec{S}_\phi$ can be determined directly from the slope of the d_ψ vs. $\sin^2\psi$ data. A positive slope indicates a tensile surface stress, σ_ϕ. In this procedure, since d_0 is a multiplier, and the maximum elastic strain would be less than 2%, the value of d_0 can be taken as the plane spacing at zero tilt, $d_{\psi=0}$. This would result in an error less than 2%. This is not the case, however, if there is a triaxial stress state in the sampled volume. Then the strain values, $(\varepsilon'_{33})_{\phi\psi} = \frac{(d_{hkl})_{\phi\psi} - d_0}{d_0}$, must be used in the analysis and, since a very small difference is being calculated, any error in d_0 would cause large errors in the strain and stress results obtained by using the general formulations previously discussed.

6.5 Effect of Sample Microstructure

The microstructure of the region irradiated by X-rays has important implications in diffraction strain/stress analysis. All of the equations discussed above are based on the second-rank tensor transformation, Equation (6.2), which is strictly applicable when both

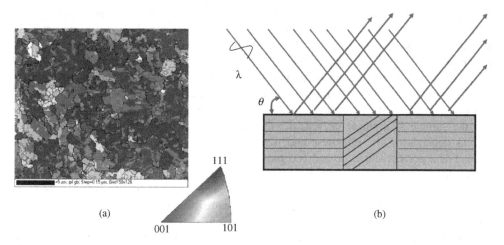

Figure 6.4 For random or incompletely-textured polycrystalline specimens, the crystallographic orientation of contiguous grains in (a) are not identical. Thus, the rays collected by the detector from various grains diffracting at a given ψ-tilt in (b) do not necessarily originate from contiguous grains

sets of axes and the strain tensors associated with them refer to the same geometric point. X-ray diffraction, on the other hand, is a volume probe. The rays that form the X-ray peak originate from the diffracting regions of volume V_D, contained in the irradiated volume V_{IR}. The plane spacing measured is thus an absorption-weighted diffraction average of the plane spacing distribution in V_D. For a polycrystalline sample only the grains that are in the Bragg condition will contribute to the diffraction peak at a given ψ tilt. These grains are not necessarily contiguous (Figure 6.4). For this case the diffracting volume for a particular ψ tilt, V_D^{ψ}, is:

$$V_D^{\psi} = \sum_{i=1}^{n} (V_g^{\psi})_i \qquad (6.20\text{-a})$$

Here $(V_g^{\psi})_i$ is the volume of the i-th grain diffracting at the particular ψ. Due to the selectivity of the Bragg condition, different grains scatter at different ψ tilts. Thus, for a set of plane spacing measurements for a given reflection carried over various ψ tilts, the total volume sampled will be:

$$(V_D^M)_{hkl} = \sum_{j=1}^{m} (V_D^{\psi})_j \qquad (6.20\text{-b})$$

The strain/stress values computed by Equations (6.4) and (6.16) will be averages referred to the total diffracting measurement volume $(V_D^M)_{hkl}$. A regular (linear or ψ-split) d_ψ vs. $\sin^2\psi$ indicates that the average strain tensor, $\langle \varepsilon_{ij} \rangle_\psi$, in the sample coordinates, for each V_D^{ψ} is the same. Thus the tensor transformation works as expected. Since the grains within a general, plastically deformed polycrystalline sample subjected to an applied load are expected to change from grain to grain due to elastic anisotropy and heterogeneous plastic flow, strain states that yield regular d_ψ vs. $\sin^2\psi$ plots are termed quasi-homogeneous, that is the (average) strain tensor, averaged over a representative number of grains, is homogeneous. However this average cannot be assumed to represent the strain state at a given point. If the grain size of the sample is large enough compared to the incident X-ray beam size such that an insufficient number of grains contribute to the X-ray peaks at various ψ tilts, a quasi-homogeneous average may not be achieved. In this case the sampling statistics will be inadequate to yield a proper volume average.

If the sample is a single crystal, the formulation used for describing the diffraction process depends on the thickness and perfection of the region irradiated by the X-ray beam. For a perfect single crystal (no mosaic regions or other dislocation distributions) thicker than the extinction depth[1] of the crystal the dynamic diffraction theory [7,8] should be used to describe the measured X-ray peaks. In such a case, because of the perfection of the lattice, interference of multiply-scattered beams becomes important. Consequently the width of the diffraction peak (FWHM) is no longer proportional to the thickness of the sample along the scattering vector and the intensity of the peak is not proportional to the diffracting volume. The Bragg angle will still yield the lattice spacing, but its measurement requires special instrumentation with very high angular resolution. Analysis of dynamically scattering samples for strain analysis is non-trivial and beyond the scope of this review. References [9] and [10] discuss this topic in more detail.

If the perfect crystal sample is thinner along the scattering vector than the relevant extinction depth, or has severe mosaic and dislocation distributions, kinematical diffraction theory [11] can adequately represent the diffraction process. This approach is also adequate for most polycrystalline samples and is the basis of almost all commercial diffraction analysis codes.

This approach has two fundamental assumptions: 1) an X-ray photon that is scattered once does not scatter again; multiple scattering is unimportant. 2) Scattering removes a negligible amount of energy from the transmitted X-ray beam. Under these assumptions the dependence of scattered intensity on the scattering angle depends only on the structure factor of the reflection in use (which depends on the symmetry of the particular unit cell) and on the size and shape distributions of the diffracting volumes [11]. The penetration depth of the X-rays into the sample in such cases depends on the diffraction geometry and the linear absorption coefficient, μ, of the X-rays of the particular energy within the sample under investigation. The diffraction geometry determines the path length of X-rays within the sample. For polycrystalline specimens, two common configurations are employed for tilting the sample with respect to the diffractometer plane (the plane containing the incident and the diffracted beams, along with the normal to the diffracting planes $\vec{L}_3$). In the Ω-goniometer geometry (Figure 6.5(a)), the rotation axis is normal to the diffractometer plane; in the ψ-goniometer geometry, the tilt axis is in the diffractometer plane (Figure 6.5(b)). The ψ-goniometer is advantageous since the sample rotation does not block the incident or diffracted beams during tilts.

---

[1] Extinction is caused by multiple scattering events and limits the penetration of the X-ray beam into the material to a very thin surface layer. This layer is usually much thinner than the penetration in the kinematical scattering regime.

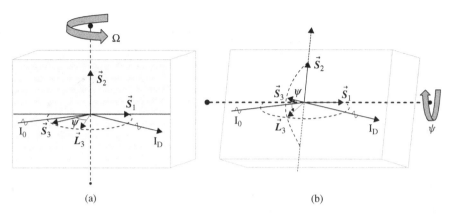

Figure 6.5 (a) The Ω-goniometer geometry. In this case the diffractometer plane (bounded by the dashed curve) contains the incident and diffracted beams, I_0, I_D as well as the specimen surface normal, $\vec{S}_3$ and the normal to the diffracting planes, $\vec{L}_3$. The axis of rotation is normal to the diffractometer plane. In this picture the rotation angle ϕ is zero, thus the measurement direction $\vec{S}_\phi$ is parallel to $\vec{S}_1$. For the ψ-goniometer (Figure 6.5(b)) the axis used for rotating the normal to the diffracting planes $\vec{L}_3$ with respect to $\vec{S}_3$ is in the diffractometer plane. In this case the measurement direction $\vec{S}_\phi$ is parallel to $\vec{S}_2$. The locus of the surface normal $\vec{S}_3$ during various tilts is a plane perpendicular to the diffractometer plane

The (kinematic) penetration depth for Ω and Ψ goniometers are given by:

$$\tau_\Omega = \frac{\sin^2\theta_B - \sin^2\psi}{2\mu \sin\theta_B \cos\psi} \quad (6.21\text{-a})$$

$$\tau_\psi = \frac{\sin\theta_B \cos\psi}{2\mu} \quad (6.21\text{-b})$$

Figure 6.6(a) shows the variation of penetration depth with tilt angle for α-Fe, Cu, and Al samples irradiated with Cu-Kα radiation at 145° 2θ on an Ω-goniometer. Figure 6.6(b) shows the variation of penetration depth with tilt angle for α-Fe on both Ω and Ψ goniometers. In all cases the penetration depth decreases with increasing tilt. The limited penetration depth (of the order of several μm for high atomic number materials) can be an advantage for the analysis of thin film samples such as surface coatings. However, this limited penetration depth also makes the strains highly sensitive to surface treatment and specimen handling. Special care must be taken to avoid handling artifacts.

6.6 X-ray Elastic Constants (XEC)

As indicated in the previous section, the link between the volume-averaged stress and strains measured using diffraction greatly depends on the microstructure of the specimen. In particular, the crystallographic nature of the sample dictates the expected mechanical

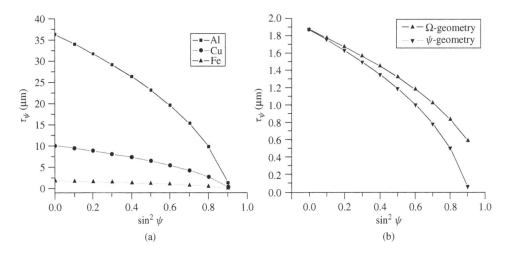

Figure 6.6 (a) Variation of the X-ray penetration depth with $\sin^2 \Psi$ for Fe, Cu, and Al samples irradiated with Cu–Ka radiation at 145° 2Ψ on an Ω-goniometer. Figure 6.6(b) shows the variation of penetration depth for the Fe-sample on both Ω and Ψ goniometers

response. If the system is linear elastic, then Equation (6.16) describes this representation for an elastically isotropic sample, or an ensemble of elastically isotropic crystallites. For samples composed of elastically anisotropic crystals, a more general form must be established. In polycrystalline ensembles, the elastic response of the subset of diffracting grains (i.e. those that satisfy the Bragg condition at a particular combination of angles, ϕ and Ψ) must be considered. The following section details how these aspects are incorporated into stress analysis by X-ray diffraction.

6.6.1 Constitutive Equation

The linear, elastic constitutive equation relates the strain tensor, ε_{ij}, as measured using X-ray diffraction to the stress tensor, σ_{ij}:

$$\langle \varepsilon_{ij} \rangle = \langle S_{ijkl} \sigma_{kl} \rangle \tag{6.22}$$

where S_{ijkl} is the fourth-rank compliance tensor of each crystal in laboratory coordinates and bracketed terms refer to an average over the particular crystallites that satisfy the diffraction condition. This diffraction average can be determined by integrating the quantity of interest, x, over all possible grains whose (hkl) normals are aligned with the diffraction vector, L_3, within the penetration depth of the X-rays. Using the angle, ξ, to represent a rotation of a crystallite about L_3 (see Figure 6.2), the form of the averaging can be written as [12]:

$$\langle x \rangle \equiv \frac{\int_0^{2\pi} x\, f(\phi, \psi, \xi, h, k, l)\, d\xi}{\int_0^{2\pi} f(\phi, \psi, \xi, h, k, l)\, d\xi} \tag{6.23}$$

where f represents the orientation distribution function (ODF) of grains within the penetration depth of the material under investigation. In the case of a random distribution, the averaging procedure is greatly simplified:

$$\langle x \rangle = \frac{1}{2\pi} \int_0^{2\pi} x \, d\xi \tag{6.24}$$

Note that the diffraction average differs from the bulk average, which is calculated by integrating over every crystallite in the sample:

$$\bar{x} \equiv \frac{\int_0^{2\pi} \int_0^{2\pi} \int_0^{\pi} x \, g(\alpha, \beta, \gamma) \sin(\beta) d\beta \, d\alpha \, d\gamma}{\int_0^{2\pi} \int_0^{2\pi} \int_0^{\pi} g(\alpha, \beta, \gamma) \sin(\beta) d\beta \, d\alpha \, d\gamma} \tag{6.25}$$

where g is the crystal ODF expressed in terms of the Euler angles, α, β and γ [12]. For a randomly oriented polycrystalline ensemble, the corresponding bulk average is:

$$\bar{x} = \frac{1}{8\pi^2} \int_0^{2\pi} \int_0^{2\pi} \int_0^{\pi} x \, \sin(\beta) d\beta \, d\alpha \, d\gamma \tag{6.26}$$

6.6.2 Grain Interaction

From Equation (6.22), it is clear that the diffraction-averaged compliance tensor and stress tensor are coupled in the most general form of the constitutive equation, requiring *a priori* knowledge of the sample ODF to solve for the stress state. Only for single-crystal samples does this relation possess a unique solution. However, for a polycrystalline aggregate, assumptions are often employed as to the interaction among the individual grains. For example, if all of the crystallites possess identical stress tensors, the diffraction-averaged stress tensor is equal to bulk averaged stress tensor, $\bar{\sigma}_{ij}$:

$$\langle \sigma_{ij} \rangle = \bar{\sigma}_{ij} \tag{6.27}$$

Though an unlikely occurrence, termed as the Reuss limit [13], this assumption affords significant simplification in the averaging procedure. In this case, the stress tensors of the diffracting grains can be decoupled from the averaging procedure, which can be written as:

$$\langle \varepsilon_{33}^L \rangle = A_{33ij}^L a_{ik}^{LS} a_{jl}^{LS} \bar{\sigma}_{kl}^S \tag{6.28}$$

where A_{ijkl}^L refers to a general compliance tensor expressed in laboratory coordinates based on the assumed mechanical model. Its specific representations are displayed below:

$$A_{33ij}^L = \begin{cases} \langle S_{33ij}^L \rangle & \text{Reuss} \\ \bar{C}_{33ij}^L{}^{-1} & \text{Voigt} \\ \langle C_{33ij}^L \rangle^{-1} & \text{Modified Voigt} \\ (\langle S_{33ij}^L \rangle + \bar{C}_{33ij}^L{}^{-1})/2 & \text{Neerfeld – Hill} \\ \langle S_{33ij}^L + t_{33ij}^L \rangle & \text{Kröner} \end{cases} \tag{6.29}$$

For the case of the Voigt limit [14], all grains possess strain tensors that are equal to the macroscopic strain tensor: $\langle \varepsilon_{ij} \rangle = \bar{\varepsilon}_{ij}$ so that the constitutive equation can be written as:

$$\langle \sigma_{ij} \rangle = \langle C_{ijkl} \rangle \bar{\varepsilon}_{kl} \qquad (6.30)$$

where C_{ijkl} is the stiffness tensor of each crystallite. The traditional formulation of the Voigt limit, in which Equation (6.30) is averaged over all grains in the material:

$$\bar{\sigma}_{ij} = \overline{C}_{ijkl} \bar{\varepsilon}_{kl} \qquad (6.31)$$

or

$$\bar{\varepsilon}_{ij} = \overline{C}_{ijkl}^{-1} \bar{\sigma}_{kl} \qquad (6.32)$$

also separates the macroscopic stress tensor from the averaging of the crystal compliance tensors. The arithmetic average of the XEC derived under the Voigt and Reuss limits is termed the Neerfeld-Hill model [15]. In the application of the Kröner model [16]:

$$\langle \varepsilon_{ij} \rangle = \langle S_{ijkl} + t_{ijkl} \rangle \bar{\sigma}_{kl} \qquad (6.33)$$

where t_{ijkl}, termed the elastic susceptibility tensor, represents the elastic interaction of the crystal in an elastic matrix.

The calculation of the compliance and stiffness tensors in the laboratory coordinate system, L_i, for each crystal involves a fourth-order transformation from the single crystal coordinate system, C_i:

$$S_{ijkl}^L = a_{im}^{LC} a_{jn}^{LC} a_{ko}^{LC} a_{lp}^{LC} S_{mnop}^C \qquad (6.34)$$

because the relevant non-zero compliance coefficients are described in crystal coordinates. For cubic crystals, the matrix representation of the compliance tensor, S^C possesses the form:

$$S^C = \begin{bmatrix} S_{1111}^C & S_{1122}^C & S_{1122}^C & 0 & 0 & 0 \\ S_{1122}^C & S_{1111}^C & S_{1122}^C & 0 & 0 & 0 \\ S_{1122}^C & S_{1122}^C & S_{1111}^C & 0 & 0 & 0 \\ 0 & 0 & 0 & 4S_{1212}^C & 0 & 0 \\ 0 & 0 & 0 & 0 & 4S_{1212}^C & 0 \\ 0 & 0 & 0 & 0 & 0 & 4S_{1212}^C \end{bmatrix} \qquad (6.35)$$

The diffraction-averaged compliance tensor possesses the following form:

$$\langle S^L \rangle = \begin{bmatrix} \langle S_{1111}^L \rangle & \langle S_{1122}^L \rangle & \langle S_{1133}^L \rangle & 0 & 0 & 0 \\ \langle S_{1122}^L \rangle & \langle S_{1111}^L \rangle & \langle S_{1133}^L \rangle & 0 & 0 & 0 \\ \langle S_{1133}^L \rangle & \langle S_{1133}^L \rangle & \langle S_{3333}^L \rangle & 0 & 0 & 0 \\ 0 & 0 & 0 & 4\langle S_{1313}^L \rangle & 0 & 0 \\ 0 & 0 & 0 & 0 & 4\langle S_{1313}^L \rangle & 0 \\ 0 & 0 & 0 & 0 & 0 & 4\langle S_{1212}^L \rangle \end{bmatrix} \qquad (6.36)$$

where $2\langle S_{1212}^L \rangle = \langle S_{1111}^L \rangle - \langle S_{1122}^L \rangle$. Note that this tensor exhibits transverse isotropy due to the averaging procedure about all orientations rotated around L_3, even though the constituent crystallites possess cubic symmetry. The bulk-averaged compliance tensor possesses full isotropy:

$$\overline{S}^L = \begin{bmatrix} \overline{S}^L_{1111} & \overline{S}^L_{1122} & \overline{S}^L_{1122} & 0 & 0 & 0 \\ \overline{S}^L_{1122} & \overline{S}^L_{1111} & \overline{S}^L_{1122} & 0 & 0 & 0 \\ \overline{S}^L_{1122} & \overline{S}^L_{1122} & \overline{S}^L_{1111} & 0 & 0 & 0 \\ 0 & 0 & 0 & 4\overline{S}^L_{1212} & 0 & 0 \\ 0 & 0 & 0 & 0 & 4\overline{S}^L_{1212} & 0 \\ 0 & 0 & 0 & 0 & 0 & 4\overline{S}^L_{1212} \end{bmatrix} \quad (6.37)$$

where again $2\overline{S}^L_{1212} = \overline{S}^L_{1111} - \overline{S}^L_{1122}$. Using Equation (6.36), Equation (6.28) can now be simplified [17]:

$$\begin{aligned} \langle \varepsilon^L_{33} \rangle &= \langle S^L_{3311} \rangle (a^{LS}_{1m} a^{LS}_{1n} + a^{LS}_{2m} a^{LS}_{2n}) \overline{\sigma}^S_{mn} + \langle S^L_{3333} \rangle a^{LS}_{3m} a^{LS}_{3n} \overline{\sigma}^S_{mn} \\ &= \langle S^L_{3311} \rangle \delta_{mn} \overline{\sigma}^S_{mn} + [\langle S^L_{3333} \rangle - \langle S^L_{3311} \rangle] a^{LS}_{3m} a^{LS}_{3n} \overline{\sigma}^S_{mn} \\ &= S_1 \overline{\sigma}^S_{ii} + \frac{1}{2} S_2 a^{LS}_{3m} a^{LS}_{3n} \overline{\sigma}^S_{mn} \end{aligned} \quad (6.38)$$

where $\frac{1}{2} S_2 = \langle S^L_{3333} \rangle - \langle S^L_{3311} \rangle$ and $S_1 = \langle S^L_{3311} \rangle$ are the common terminology used to express the quasi-isotropic XEC. Let us define m as the slope of the measured d_{hkl} with respect to $\sin^2(\psi)$. If the film is under isotropic, biaxial stress, then Equation (6.16) can be reduced to:

$$m \equiv \frac{\partial d_{hkl}}{\partial \sin^2(\psi)} = d_0 \frac{1}{2} S_2 \overline{\sigma}^S_{11} \quad (6.39)$$

6.7 Examples

6.7.1 Isotropic, Biaxial Stress

For the first example, a 1 μm thick Cu film deposited on a Si single-crystal substrate was analyzed using 8.6 keV radiation. Single crystal values of the Cu compliance tensor used to calculate XEC are $S^C_{1111} = 15.0$ TPa^{-1}, $S^C_{1122} = -6.3$ TPa^{-1} and $S^C_{1212} = 3.3$ TPa^{-1} [6]. Measurements of the Cu lattice spacing, d, for the (220), (311) and (222) reflections were conducted over a large range of both positive and negative ψ-tilts. The fitted slopes and XEC calculated under the Neerfeld-Hill and Kröner limits, and the corresponding in-plane stresses are contained in Table 6.2.

Figure 6.7 depicts the X-ray diffraction data, normalized by the square root of the squares of their respective Miller indices ($a_{hkl} = d_{hkl} \sqrt{h^2 + k^2 + l^2}$), for the three different reflections as a function of $\sin^2(\psi)$. The slopes of the three sets of data, normalized by their respective unstrained lattice parameters d_0 and listed in Table 6.2, vary by approximately 60%, which confirms that the elastic strain measured by X-ray diffraction greatly depends on the mechanical response of the diffracting grains. The Kröner XEC and Neerfeld-Hill (N-H) XEC also reflect a similar variation among the reflections under investigation

Table 6.2 Experimentally determined d vs. $\sin^2\Psi$ slopes, XEC and in-plane stress values as calculated under the Neerfeld-Hill and Kröner limits for 1 μm thick Cu film

	m/d_0 [%]	$1/2S_2^{N-H}$ [TPa^{-1}]	$1/2S_2^{K}$ [TPa^{-1}]	$\sigma_\parallel^{N-H}$ [MPa]	$\sigma_\parallel^{K}$ [MPa]
(311)	0.216(11)	11.786	11.094	183(9)	195(10)
(220)	0.171(7)	9.743	9.592	176(7)	178(7)
(222)	0.135(3)	7.912	8.246	171(4)	164(4)
average				177	179

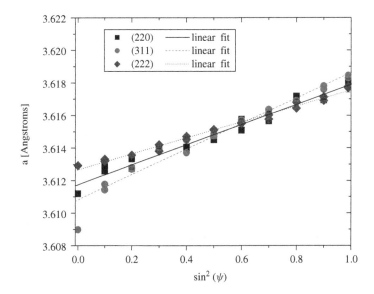

Figure 6.7 Normalized Cu (220), (311) and (222) lattice spacing values and corresponding linear fits measured from a 1 μm thick Cu film. Reproduced with permission from [17], Copyright 2011 American Institute of Physics

so that the corresponding in-plane stress values lie within 9% of the average. This fact illustrates the point that XEC, rather than one set of bulk elastic constants for all X-ray reflections, must be used to calculate the correct value of stress from strain determined using diffraction.

6.7.2 Triaxial Stress

For patterned features, the assumption of an isotropic biaxial stress state may no longer be valid. In fact, capped features can often possess finite out-of-plane stresses. Samples under investigation consisted of arrays of lines lithographically patterned within a low-k organosilicate dielectric film, which was deposited on 300 mm diameter Si (001) substrates possessing a 500 nm thick SiO_2 layer. A cross-sectional schematic is shown in Figure 6.8. The Cu (220) interplanar spacing was measured at 21 values of ψ and four different values of the in-plane rotation angle, ϕ, (0^0, 45^0, 90^0 and 135^0) for a total of 84 measurements

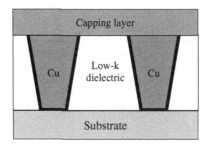

Figure 6.8 Cross-sectional schematic of 250 nm wide Cu line arrays

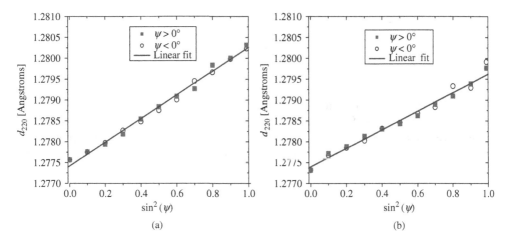

Figure 6.9 Cu (220) lattice spacings measured using X-ray diffraction of the 250 nm wide lines in the (a) longitudinal direction and (b) in the transverse direction

per sample. Figure 6.9 depicts the d vs. $\sin^2(\psi)$ data along the 250 nm wide lines. The data follow a linear trend with respect to $\sin^2(\psi)$, confirming that the Cu lines can be treated as a quasi-isotropic elastic aggregate. The open and closed circles show the values obtained for positive and negative values of ψ. Two procedures were used to calculate the six independent components of the strain tensor. The first involved a linear least-squares minimization of an overconstrained system of equations [5] for all 84 measurements. The second approach followed that of Dölle and Hauk, where components that are linear with $\sin^2(\psi)$ and those linear with $\sin(2\psi)$ [3] were generated. The value of the out-of-plane lattice strain, ε_{33}, was averaged among the four measurements performed at $\psi = 0$ with different ϕ.

The unstressed lattice spacing, d_0, was determined through measurements on 150 × 150 µm² Cu pad regions manufactured on the same wafer and adjacent to the Cu line arrays, where an isotropic, in-plane biaxial stress ($\sigma_{11} = \sigma_{22}$) was assumed so that the out-of-plane stress components (σ_{i3}) were taken to be zero. The three principal strain values are calculated by subtracting d_0 from the principal lattice spacings, so that all six independent components of the strain tensor are found. The corresponding stress tensor values are solved using tensor multiplication of the stiffness tensor and the strain tensor.

Table 6.3 Normal stress values in Cu line array using linear least-squares refinement vs. Dölle-Hauk approach

	σ_{11} [MPa]	σ_{22} [MPa]	σ_{33} [MPa]
Least-squares fit	366 ± 11	306 ± 12	128 ± 10
Dölle-Hauk	358 ± 23	324 ± 29	133 ± 25

Table 6.4 Shear stress values in Cu line array using linear least-squares refinement vs. Dölle-Hauk approach

	σ_{23} [MPa]	σ_{13} [MPa]	σ_{12} [MPa]
Least-squares fit	5.4 ± 2.4	0.2 ± 2.1	−1.9 ± 2.5
Dölle-Hauk	−2.6 ± 3.7	−2.6 ± 3.2	−4.1 ± 6.2

The XEC were calculated using the Neerfeld-Hill limit, where for Cu (220), $\frac{1}{2}S_2 = 9.743 \times 10^{-3}$ GPa^{-1} and $S_1 = -2.448 \times 10^{-3}$ GPa^{-1}. To extract individual components of the stress tensor, we convert the fitted slopes of the d_{220} vs. $\sin^2(\psi)$ plots measured along and transverse to the lines to deviatoric stress values with the assistance of Equation (6.38):

$$m_0 = \frac{\partial d_{220}}{\partial[\sin^2(\psi)]} = d_0 \frac{1}{2} S_2 (\overline{\sigma}_{11}^S - \overline{\sigma}_{33}^S) \quad (6.40)$$

$$m_{90} = \frac{\partial d_{220}}{\partial[\sin^2(\psi)]} = d_0 \frac{1}{2} S_2 (\overline{\sigma}_{22}^S - \overline{\sigma}_{33}^S) \quad (6.41)$$

Tables 6.3 and 6.4 contain the normal and shear stress components, respectively, in the Cu array for the two different methods. The normal stress component values lie within 18 MPa of those calculated using the full linear, least-squares analysis. The three principal stress tensor components are all tensile, indicating that a triaxial stress state is present. The out-of-plane stress, σ_{33}, is generated by the constraint imposed by the barrier layer materials along the sidewalls of the Cu lines. Since the Ta-based barrier layers possess a lower CTE than that of Cu, the cooling of the structure from processing temperature introduces a residual, tensile stress in the Cu lines [18]. The transverse in-plane stress, σ_{22}, is lower than σ_{11} due to the load sharing that develops between the Cu and adjacent dielectric material. As shown in Table 6.4, the magnitudes of the shear stress components are 5 MPa or less, suggesting that the principal directions are aligned with the sample axes.

6.7.3 Single-crystal Strain

Strain distributions generated by stressor elements within silicon-on-insulator (SOI) layers are deliberately introduced to enhance carrier mobility within neighboring devices because of piezoresistivity within silicon [19]. Through microbeam diffraction, we can map strain

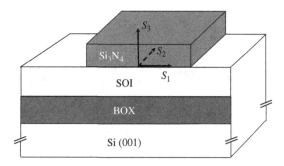

Figure 6.10 Cross-sectional schematic of Si_3N_4 stressor feature overlying an SOI substrate

within the SOI as a function of proximity to stressor features. Because of the crystallographic offset between the SOI layer and the underlying Si substrate, which are separated by a layer of SiO_2 in bonded wafers, diffraction information can be isolated from either the SOI layer or the substrate by proper sample and detector slit orientation. An example of a compressively stressed Si_3N_4 feature, approximately 105 nm thick and 1 μm wide, deposited on a 200 mm diameter substrate possessing a 140 nm thick SOI layer is depicted in Figure 6.10. X-ray micro-diffraction measurements, conducted at Argonne National Laboratory's APS 2-ID-D beamline, were used to map the Si (008) reflection within the SOI layer, whose surface normal was aligned with the SOI [001] orientation, at 0.2 μm increments across the Si_3N_4 features. A plane strain assumption could be made due to the geometry of the Si_3N_4 features along the S_2 direction. For the SOI region, the depth-averaged lattice spacings, c_{Si}, can directly be transformed into out-of-plane strain values:

$$\overline{\varepsilon}_{33} = (\overline{c}_{Si} - a_{Si})/a_{Si} \tag{6.42}$$

where a_{Si} refers to the unstrained lattice spacing of Si, which was determined from a measurement of the substrate at approximately 1 mm from the Si_3N_4 feature. The corresponding constitutive equation has the form:

$$\overline{\varepsilon}_{33} = S^S_{33ij}\overline{\sigma}_{ij} \tag{6.43}$$

Note that the sample axis S_3 is aligned with the crystal axis C_3 of the single-crystal SOI layer. Because the edges of the Si_3N_4 feature, which define S_1 and S_2, are parallel to the {110} SOI orientations, C_1 and C_2 are rotated 45° about C_3 from S_1 and S_2, respectively. Equation (6.34) can be used to transform the relevant compliance tensor coefficients to those in crystal coordinates:

$$S^S_{3333} = S^C_{1111}, \; S^S_{3311} = S^S_{3322} = S^C_{1122} \tag{6.44}$$

and all other $S^S_{33ij} = 0$, so that Equation (6.43) may be simplified to form:

$$\overline{\varepsilon}_{33} = S^C_{1122}(\overline{\sigma}_{11} + \overline{\sigma}_{22}) + S^C_{1111}\overline{\sigma}_{33} \tag{6.45}$$

A triaxial stress state must be considered in the SOI due to the elastic relaxation of the overlying Si_3N_4 feature edges.

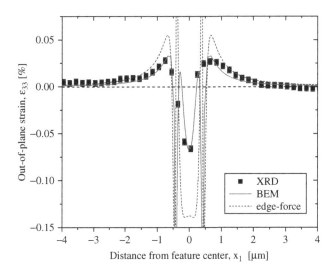

Figure 6.11 Depth-averaged out-of-plane strain measured in the SOI layer underneath the 1.0 μm wide Si_3N_4 feature and comparison to mechanical modeling results. Reproduced with permission from [20], Copyright 2008 American Institute of Physics

Figure 6.11 depicts the depth-averaged SOI out-of-plane strain across the interaction region imposed by the overlying Si_3N_4 stressor feature. In-plane tensile stress in the SOI along S_1, $\bar{\sigma}_{11}$, underneath the compressively stressed feature transitions to in-plane compressive stress outside of the Si_3N_4 feature. Poisson contraction and expansion generates the corresponding out-of-plane compressive strain underneath the feature and out-of-plane tensile strain in the bare SOI, respectively. The magnitude of the out-of-plane strain values exhibit a maximum underneath the center of the Si_3N_4 feature due to the overlapping strain distributions induced by both feature edges.

To compare these results to those based on continuum mechanics, the stress and strain tensors were simulated using an anisotropic, edge-force model and an elastically isotropic Boundary Element Method (BEM) model [20], where perfect bonding was assumed at the interfaces. The Young's modulus, E, and Poisson's ratio, v, were assumed to be 162.5 GPa and 0.28 for Si and 160 GPa and 0.3 for Si_3N_4 in the BEM modeling, respectively. A compressive, in-plane film stress of −2.5 GPa, based on wafer curvature measurements performed on blanket films, was assumed prior to elastic relaxation of the features. As shown in Figure 6.11, the BEM simulations reproduce the depth-averaged, out-of-plane SOI strain whereas the edge-force simulations over-predict the magnitude of SOI strain, particularly under the Si_3N_4 feature. Note that the large strain gradients under the Si_3N_4 feature edges, as predicted by linear elasticity, occur over an area too fine to be captured by the microbeam (250 nm FWHM) so that an average value is measured. Finite out-of-plane stress, $\bar{\sigma}_{33}$, and shear stress, $\bar{\sigma}_{13}$, values along the Si_3N_4/SOI interface, which are included in the BEM simulations but not in the edge-force model, are key to predicting the mechanical behavior of these composite systems.

6.8 Experimental Considerations

In the discussion so far we have not treated the errors associated with the measurement. In general there are three categories of errors: (1) instrumental errors; (2) errors due to counting statistics; (3) errors due to sampling statistics. Extended treatment of these issues is beyond the scope of this review [6]. Furthermore, most practitioners use pre-aligned instruments for the measurement and canned programs for data analysis and have little control over the actual procedure. However, the following points might be helpful in planning an experiment.

6.8.1 Instrumental Errors

This category includes specimen displacement from goniometer center, horizontal and vertical divergence issues, beam instabilities, diffractometer misalignment and similar issues. The best technique to check for such errors is to coat the surface of the specimen with a very thin, fine-grained, non-toxic, crystalline powder (such as Si powder) where the powder peaks are close to, but do not interfere with, the sample reflections. Then the stress measurement is conducted on one or more powder reflections. Since a powder sample, by definition, cannot support a long-range stress, the strain/stress values measured from the powder are the "error" associated with the measurement. In general stress values smaller than ± 10 MPa range are desirable. If the powder measurement yields unacceptable stress values, the source of the errors must be identified and remedied.

6.8.2 Errors Due to Counting Statistics and Peak-fitting

The intensity measured at any 2θ position on the line contains a finite statistical error since the arrival of X-ray photons at the detector is random in time. Consequently the number of pulses, $N(2\theta)$, counted for a fixed time, t, at a specific location, 2θ, will have an error, $\pm \delta N(2\theta)$, proportional to $\sqrt{N}$. The peak fitting algorithm used to determine peak positions should propagate these errors and yield the statistical error associated with the computed peak positions, $\pm \delta N(2\theta)$ for all ψ-tilts. The statistical error in strain/stress is then computed by propagating $\delta(2\theta)$ through the analysis algorithm. This procedure, if carried out completely without user supervision, can be quite error-prone. It is important to make sure that the peak profile function assumed by the analysis software actually fits the data and carefully examines the difference between measured and fitted profiles. Relying on simple metrics like χ^2 can be problematic.

6.8.3 Errors Due to Sampling Statistics

Since the maximum intensity scattered by relatively large, perfect, grains will be much stronger than smaller, imperfect grains, analysis of large-grained samples, or samples with heterogeneous grain-size distributions will be problematic. In the first case, just a few grains may contribute to the diffraction peak for each ψ-tilt and rocking the sample over a few degrees in $\pm \psi$ can be helpful. In the second case the scattering from the small grains will not be sampled whether one rocks the sample or not and only the data from the larger grains will be analyzed. Such issues can be immediately identified if pinhole X-ray patterns are obtained from the samples by using two-dimensional detectors such as film or CCD devices.

6.9 Summary

X-ray diffraction provides us with a very effective way to interrogate crystalline samples non-destructively to ascertain information about their residual stress states. The procedures to obtain lattice spacing information as a function of orientation are straightforward as are the methods to link this data to the sample stress. As highlighted in this chapter, an understanding of how the assumptions inherent to these procedures apply to the sample is a critical aspect of the analysis. Is the sample homogeneous or heterogeneous, what is the relationship between the beam size and grain size, what components of the stress tensor should be non-zero are but a few of the questions that must be addressed to arrive at both accurate and meaningful results. Although stress measurement techniques by X-ray diffraction have existed for decades, advances in equipment and computational analysis have produced "turn-key" systems that can provide answers with less user input. In such an environment, it becomes increasingly important for the user to understand the fundamental attributes and limitations of these methods to ensure that a proper analysis has been performed.

Acknowledgments

Use of the Advanced Photon Source, an Office of Science User Facility operated for the U.S. Department of Energy (DOE) Office of Science by Argonne National Laboratory, was supported by the U.S. DOE under Contract No. DE-AC02-06CH11357.

References

[1] Lester, H. H., Aborn, R. H. (1925) "The Behavior Under Stress of the Iron Crystals in Steel," *Army Ordnance* 6:120–127, 200–207, 283–287, 364–369.
[2] Dölle, H., Hauk, V. (1977) "System for Possible Lattice Strain Distributions for Mechanically Stressed Metallic Materials," *Zeitschrift für Metallkunde* 68(11):725–728.
[3] Dölle, H. (1979) "Influence of Multiaxial Stress States, Stress Gradients and Elastic Anisotropy on the Evaluation of (Residual) Stresses by X-rays," *Journal of Applied Crystallography* 12:489–501.
[4] Dölle, H., Cohen, J. B. (1979) "Residual Stresses in Ground Steels," *Journal of Metals* 31(12):79–89.
[5] Winholtz, R. A., Cohen, J. B. (1988) "Generalized Least-Squares Determination of Triaxial Stress States by X-Ray-Diffraction and the Associated Errors," *Australian Journal of Physics* 41(2):189–199.
[6] Noyan, I. C., Cohen, J. B. (1987) *Residual Stress: Measurement by Diffraction and Interpretation*, Springer: New York.
[7] Pinsker, Z. G. (1978) *Dynamical Scattering of X-Rays in Crystals*, Springer-Verlag: Heidelberg.
[8] Authier, A. (1998) "Dynamical Diffraction at Grazing Incidence," *Crystal Research and Technology* 33(4):517–533.
[9] Yan, H., Kalenci, O., Noyan, I. C. (2007) "Diffraction Profiles of Elastically Bent Single Crystals with Constant Strain Gradients," *Journal of Applied Crystallography* 40:322–331.
[10] Kalenci, O., Murray, C. E., Noyan, I. C. (2008) "Local Strain Distributions in Silicon-on-insulator/stressor-film Composites," *Journal of Applied Physics* 104(6):063503.

[11] Ying, A. J., Murray, C. E., Noyan, I. C. (2009) "A Rigorous Comparison of X-ray Diffraction Thickness Measurement Techniques Using Silicon-on-insulator Thin Films," *Journal of Applied Crystallography* 42:401–410.
[12] Leoni, M., Welzel, U., Lamparter, P., Mittemeijer, E. J., Kamminga, J-D. (2001) "Diffraction Analysis of Internal Strain–stress Fields in Textured, Transversely Isotropic Thin Films: Theoretical Basis and Simulation," *Philosophical Magazine A* 81:597–623.
[13] Reuss, A. (1929) "Calculation of Flow Limits of Mixed Crystals on the Basis of Plasticity of Single Crystals," *Zeitschrift für Agnewandte Mathematik und Mechanik* 9:49–58.
[14] Voigt, W. (1928) *Lehrbuch der Kristallphysik*, Teubner, Leipzig/Berlin.
[15] Neerfeld, H. (1942) Zur Spannungsberechnung aus röntgenographischen Dehnungsmessungen, Mitteilungen aus dem Kaiser-Wilhelm-Institut für Eisenforschung zu Düsseldorf 24:61–70.
[16] Kröner, E. (1958) Berechnung der elastischen Konstanten des Vielkristalls aus den Konstanten des Einkristalls Zeitschrift für Physik 151:504–518.
[17] Murray, C. E. (2011) "Invariant X-ray Elastic Constants and Their Use in Determining Hydrostatic Stress," *Journal of Applied Physics* 110:123501.
[18] Murray, C. E., Besser, P. R., Ryan, E. T., Jordan-Sweet, J. L. (2010) "Triaxial Stress Distributions in Cu/low-k Interconnect Features," *Applied Physics Letters* 98(6):061908.
[19] Smith, C. S. (1954) "Piezoresistance Effect in Germanium and Silicon," *Physical Review* 94:42–49.
[20] Murray, C. E., Saenger, K. L., Kalenci, O., Polvino, S. M., Noyan, I. C., Lai, B., Cai, Z. (2008) "Submicron Mapping of Silicon-on-insulator Strain Distributions Induced by Stressed Liner Structures," *Journal of Applied Physics* 104:013530.

7

Synchrotron X-ray Diffraction

Philip J. Withers
University of Manchester, Manchester, UK

7.1 Basic Concepts and Considerations

7.1.1 Introduction

The basic concept underlying the non-destructive measurement of residual strain by synchrotron X-ray diffraction is fundamentally the same as for other diffraction techniques. The method rests on the fundamental relation formulated by W. L. Bragg in 1913 connecting the spacing, d_{hkl}, between certain lattice planes of index hkl to the diffraction angle, $2\theta_{hkl}$, at which the radiation is scattered coherently and elastically for a given wavelength of the radiation, λ

$$2d_{hkl} \sin \theta_{hkl} = \lambda \qquad (7.1)$$

Figure 1.13 in Chapter 1 illustrates the various geometrical quantities. For diffraction to occur, the lattice plane of the crystallites (grains) must be oriented such that the normal direction bisects the incident, $\mathbf{k}_i$, and the diffracted, $\mathbf{k}_f$, wavevectors ($k = 2\pi/\lambda$), as shown in Figure 7.1(a). In such a configuration the spacing is measured parallel to the scattering wavevector, $\mathbf{Q}$, as shown in Figure 7.1(b). If a tensile strain exists, as shown in Figure 7.1(c), the lattice spacing, d, increases and causes a decrease in the diffraction angle, θ, according to Equation (7.1). Conversely, a compressive strain would cause an increase in diffraction angle.

Conceptually, the crystal lattice, which is the characteristic structural element of crystalline solids, is exploited as a natural and omni-present atomic-scale strain gage embedded in each crystallite or grain. Although the technique does not probe the deformation with atomic spatial resolution, it does probe the average deformation of the lattice planes over a certain sampled (gage) volume, as discussed in Section 7.2.1.

Practical Residual Stress Measurement Methods, First Edition. Edited by Gary S. Schajer.
© 2013 John Wiley & Sons, Ltd. Published 2013 by John Wiley & Sons, Ltd.

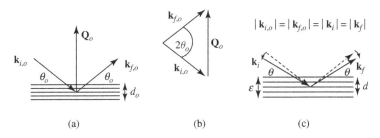

Figure 7.1 (a) Schematic showing the incident, $\mathbf{k}_i$, and diffracted, $\mathbf{k}_f$, wave vectors for Bragg diffraction from an unstrained crystal lattice plane, (b) the scattering vector $\mathbf{Q} = \mathbf{k}_f - \mathbf{k}_i$ is normal to the lattice planes and (c) change in Bragg angle when the lattice planes are under tension, the $\mathbf{Q}$ vector is parallel to the lattice strain measurement direction, ε

The three diffraction-based residual stress measurement techniques using, X-rays, synchrotron X-rays and neutrons differ in their application ranges. The X-ray technique can be used in-house or in the field with relatively modest laboratory equipment, but is limited to measurements within a few microns of the material surface. Synchrotron X-rays have much higher energy and can penetrate many mm or cm into materials. Their high intensity also allows rapid measurement times. However, access to a major synchrotron facility is required. Neutron measurements similarly require access to a major scientific facility, in this case with a neutron source. Neutrons have the greatest penetration depth (Figure 7.6(a)), but their lower intensity results in much longer measurement times.

In general practice, strain is measured by analyzing continuous diffraction rings (see Figure 7.2(b), as would be obtained from a powder sample, arising as the net result of diffraction from very many crystallites within the gage volume rather than from the individual diffraction spots arising from a single crystal grain. This means that there must be a statistically significant number of grains scattering from within the gage volume to obtain a 'powder' (uniformly distributed) diffraction pattern (Figure 7.2). This commonly places a limit on the smallest gage volume that can be employed.

The relationship between X-ray energy, E, and wavelength, λ, is $E = hc/\lambda$ where h is Planck's constant and c the speed of light. The wavelength is thus given by

$$\lambda \text{(in Å)} \approx 12.4/E \text{ (in keV)} \tag{7.2}$$

At the high X-ray energy end of the spectrum typical of engineering measurements in thick samples (say >100 keV), wavelengths are typically below 0.12 Å. In such cases Bragg's Law gives low scattering angles ($2\theta \sim 4°$) for low index (hkl) crystal planes. If high penetrating capability is not required, X-ray energies around 30–40 keV can be used. This leads to scattering angles similar to those of conventional X-ray diffraction.

There are essentially two different ways to exploit Bragg's Law to measure strain, ε (Equation 7.3), either to use a single wavelength (monochromatic) beam and measure shifts in scattering angle, 2θ, for one or more hkl diffraction peaks, or to use a polychro-

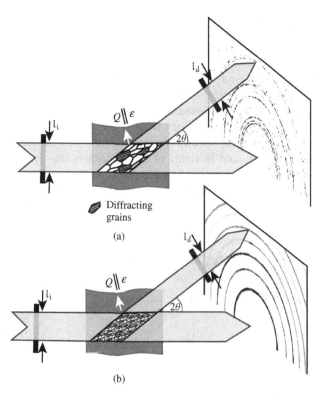

Figure 7.2 (a) Schematic showing diffraction from those crystal grains that satisfy Bragg's Law and are correctly oriented for diffraction within the gage volume. In (a) there are only a few grains giving rise to a spotty diffraction pattern while in (b) there are sufficient grains for continuous rings (powder diffraction). The strain measurement direction bisects the incident and diffracted beams

matic (white) beam keeping the angle fixed and to measure shifts in the wavelength $\Delta\lambda$ at which the maximum in the Bragg diffraction peak is located.

At constant λ: $\quad\varepsilon = \Delta d/d_0 = (d-d_0)/d_0 = -\cot\theta \, (\theta - \theta_0)$ (7.3a)

At constant θ: $\quad\varepsilon = \Delta d/d_0 = (d-d_0)/d_0 = (\lambda - \lambda_o)/\lambda_o$ (7.3b)

These are often termed angular and energy dispersive methods. Figures 7.21 and 7.22 show diffraction patterns collected as a function of scattering angle and wavelength. These methods are discussed in Sections 7.3 and 7.4 respectively through a series of examples. Irrespective of the method chosen, there are several basic principles to be considered and important issues that need to be addressed. These are treated in Section 7.2.

7.1.2 Production of X-rays; Undulators, Wigglers, and Bending Magnets

X-rays are radiated whenever a charged particle is accelerated or decelerated. Consequently, the centripetal accelerations that occur while using magnets to maintain a 'bunch' of electrons moving around a circle (called the storage ring) at a synchrotron creates X-ray beams tangential to the curved electron flight path, see Figure 7.3.

Greater beam intensity can be achieved by more drastically bending the electron beam. Figure 7.4 shows the three main classes of insertion devices used to generate X-rays and Figure 7.5 shows indicative X-ray spectra obtainable from such devices. Insertion device performance is sometimes quoted in terms of flux, which is the number of photons per second passing through a defined aperture, and is the appropriate measure for experiments that use the entire, unfocused X-ray beam, but more commonly in terms of brilliance (expressed as the flux per unit area of the radiation source per unit solid angle of the

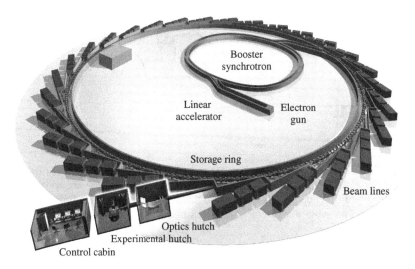

Figure 7.3 Insertion devices (magnets) are used to deflect the electron beam in the storage ring of a synchrotron creating an intense beam of X-rays essentially tangential to the ring. The beamlines are located along these tangents and include optical elements to control the, wavelength, beam dimensions, and focus. Images courtesy of Diamond Light Source

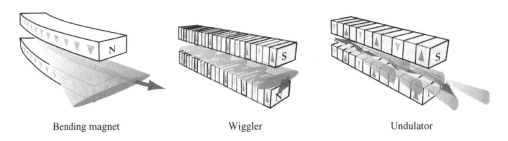

Figure 7.4 Schematic showing the operation of a bending magnet, wiggler, and undulator

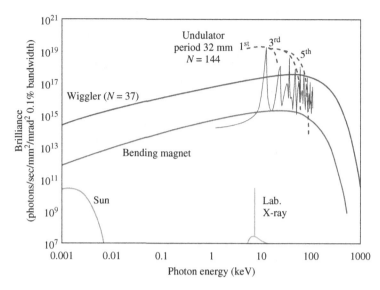

Figure 7.5 The spectral brilliance of a Spring-8 bending magnet, wiggler and undulator; the solid curve for the undulator shows the output at a fixed gap between top and bottom poles, the dashed lines the variation in the harmonic peaks as the gap is varied from 25 to 8 mm. Reproduced with permission from [1], Copyright 1998 John Wiley & Sons

radiation cone per unit spectral bandwidth – see Figures 7.5 and 7.6), which is a measure of the intensity and divergence of an X-ray beam.

Bending magnet: Bending magnets are placed strategically around the storage ring to maintain the circular path of the electrons. Each set of bending magnets creates a splay of X-rays having a broad range of wavelengths emanating tangentially from the point at which the beam was bent.

Wiggler: By placing special insertion devices in the straight segments between bending magnets it is possible to "wiggle" the beam creating much more intense X-ray sources. At the extremities of the "wiggle," where the acceleration is greatest, a bright beam of light is emitted. These beams of light add to produce a broad spectrum of incoherent light centred on the direction of travel of the electron bunch.

Undulator: Similar to a wiggler, but in this case the magnet poles deflect the beam less significantly. Here the light produced is semi-coherent and constructive interference occurs at particular frequencies controlled by the gap between the poles giving rise to significant peaks in intensity (see Figure 7.5).

7.1.3 The Historical Development of Synchrotron Sources

Laboratory X-ray stress measurement methods (see Chapter 6) have developed since the 1920s and neutron stress measurement techniques (see Chapter 8) since the 1980s. Nevertheless, both have their limitations; laboratory X-rays typically sample only a very shallow (typically a few microns) surface layer, while neutron strain measurements are

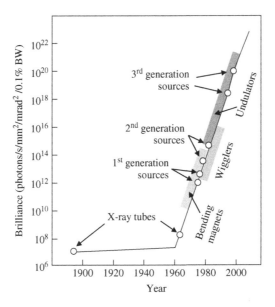

Figure 7.6 The historical increase in brilliance of X-ray sources. After [2]

characterized by relatively low intensities giving rise to slow rates of data acquisition and a spatial resolution of around 1 mm. By contrast, hard (high energy) synchrotron X-ray beams can be very penetrating (see Section 7.1.3) and incredibly intense, providing a probe with high spatial and time resolution. Consequently, whereas neutron diffraction is usually used to measure strains along specific lines at millimeter resolution, synchrotron diffraction can measure the strain over large areas at sub-mm spatial resolution, or can follow changes in strain over short timescales (much less than a second) making it a useful complement to laboratory X-ray and neutron strain measurement.

First-generation synchrotron light sources were basically beamlines exploiting bending magnets on synchrotrons designed for particle physics studies where the bending magnets were needed to maintain the motion of the near relativistic electrons in a circular trajectory. Daresbury, in the UK, was the first second-generation synchrotron light source (1981); second generation sources were especially designed to produce synchrotron radiation and employed bending magnets (see Figure 7.4). Subsequently, many were upgraded to host insertion devices too. Current (third-generation) synchrotron light sources optimize the intensity of the light through long straight sections that house 'insertion devices' such as undulators and wigglers (Section 7.1.2).

The low level of angular divergence and narrow energy bandwidth leads to diffraction peak widths that are symmetric and inherently very narrow ($\sim 0.01°$ diffraction peak full width half-maximum compared with a degree or so for neutron diffraction), and wavelengths can be selected down to below 0.1 Å. These attributes alongside the high flux and excellent penetration typical of hard (short wavelength) X-rays in most engineering materials make synchrotron X-ray diffraction an increasingly important way of mapping residual stresses.

7.1.4 Penetrating Capability of Synchrotron X-rays

Critical to measuring elastic strain deep inside engineering components is the distance through which synchrotron X-rays are transmitted because this places a limit on the depth from which strain data can be obtained. This property is usually quantified by the *attenuation length*, l_o, which is the distance through which 37% (e^{-1}) of the photons will pass. As electromagnetic radiation, X-rays are scattered by the electron cloud of an atom and hence the higher the atomic number, the greater the attenuation (see Figure 7.7(a)). Sudden increases in the attenuation (drops in the attenuation length) are observed at specific energies (see Figure 7.7(b)) corresponding to the point when there is sufficient energy for specific electronic transitions characteristic of the specific atom. Generally, penetration rises approximately as the third power of the X-ray energy (keV). The attenuation length, however, is only an indicator of the depth at which measurements can be made. Because synchrotron beamlines are so much more intense than conventional X-ray and neutron sources, the signal can still be sufficient for measurement after travelling as far as four to eight times the attenuation length, a distance through which only 1.80–0.03% (= e^{-4} to e^{-8}) of X-rays are transmitted. This equates to path lengths of around 17 cm, 6.5 cm and 2.8 cm for Al, Ti and Fe respectively at 100 keV. By comparison, path lengths of only two to four times the attenuation length are practically feasible for neutrons [3].

7.2 Practical Measurement Procedures and Considerations

Several practical matters must be considered when embarking on an experiment to measure residual stress by synchrotron diffraction. The first relates to the most appropriate method and the beamline to select. In this respect a number of key considerations come into play:

Penetration depth: Normally the greater the penetration depth the higher the X-ray energy required (Figure 7.7). However this results in lower scattering angles, 2θ, which can lead to difficulties in getting the beam in and out of the sample. Sometimes this difficulty

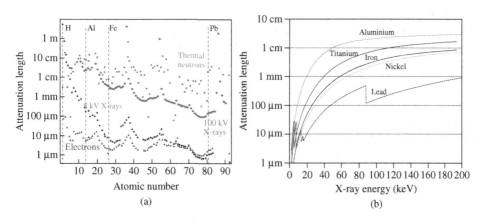

Figure 7.7 (a) The variation in penetrating capability for thermal neutrons, 8 keV (lab. Cu target) X-rays, 100 keV (hard synchrotron) X-rays, and 100 keV electrons and (b) the X-ray attenuation length (37% transmission) through various metals as a function of energy

can be obviated by using a lower energy so that a back-scattered reflection geometry can be employed.

Gage volume required: The smallest gage volume is usually limited more by the need to illuminate sufficient grains to achieve powder diffraction than by flux. One way of circumventing this limitation is to oscillate the sample angularly over a degree or so, so that more grains within the gage diffract. The gage length along the beam is sometimes difficult to confine, especially if the diffraction rings are collected on an area detector, or the scattering angle is very low (see Section 7.2.1). Consequently, if the gage length needs to be small ($<500\,\mu$m), it is usually best to use angle-dispersive scanning with a relatively low energy so that the scattering angle is as large as possible.

Need to scan through surfaces: Diffraction peak shifts independent of stress in the sample can arise if the gage volume straddles a surface; generally angular scanning with an analyzer crystal is the best way to limit this effect (see Section 7.2.6). In such cases it is essential to know accurately where the surface is – this is usually identified by surface scans of the integrated diffraction peak intensity (see Figure 7.14(a)).

Required time resolution: Energy dispersive methods, or angular dispersive methods using an area detector (see Section 7.3.1), are often a good choice if fast measurement timescales are required in order to track events dynamically because they involve no moving parts.

Single diffraction peak or multiple peaks? There are many situations where it is necessary to collect several diffraction peaks. These include cases where there is significant texture, where intergranular stresses are important so that the study of one peak may be misleading, in multiphase materials, or where it is likely that certain phases will appear or transform into other phases during the experiment. These favour energy-dispersive diffraction, or angle-dispersive diffraction using an area or line detector (Section 7.3.1).

How many directions of strain are required: Some techniques are well suited to measuring the strain in more than one direction simultaneously (see Figures 7.17(a) and 7.22(a)). For some methods it is difficult to measure the strain in three or more directions. These may be necessary in order to calculate the stresses from the strain components.

Variation of strain-free lattice spacing across the sample: Some methods are particularly suited to cases where the strain-free lattice spacing varies from point to point (see Section 7.2.3).

In view of the fact that synchrotron X-ray beamlines are usually accessible through peer reviewed access systems, it is important that the choice of beamline and the experimental procedure is well planned because it may be many months between application and experiment. Poorly planned experiments are rarely awarded beamtime and are even more rarely successful! Furthermore, experiments that can be done in the laboratory should not be done at a synchrotron, so the case for a synchrotron experiment needs to be well constructed.

7.2.1 Defining the Strain Measurement Volume and Measurement Spacing

While in mathematical terms strain is a pointwise quantity, such infinite resolution cannot be measured practically by any method. Although diffraction can map the strain within

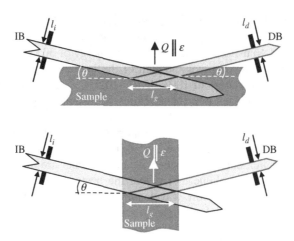

Figure 7.8 Definition of the instrumental gage volume (IGV) for reflection (a) and transmission (b) measurements using simple slits

individual grains, it is the Type I macrostresses (see Section 1.1.1 in Chapter 1) that vary over many grains that are usually of primary engineering interest. Consequently the strain must normally be averaged over many grains (Figure 7.2) to be sufficiently representative.

The gage volume over which the strain is measured is normally defined by ensuring that only a small volume is both illuminated by the incident beam and capable of diffracting photons to the detector. This is usually achieved by simple slits (apertures) on the incoming and diffracted beams, as shown in Figure 7.8(a). The incoming beam could also be defined by focusing optics. The diffracted beam cannot be defined by a simple aperture if an area detector is used; instead a conical slit can be used (see Figure 7.17) [4]. By tracing the trajectory of the diffracted beams it is possible to infer the location from which it originated.

Because of the low scattering angles characteristic of diffraction using hard (short wavelength) X-rays, gage definition is typically much better lateral to the incoming beam, where it is usually well approximated by the incoming beam dimensions, than along the beam direction (see Figure 7.8). For reflection geometries it is evident from Figure 7.8(b) that even for small gage depths the path length can be long, limiting the maximum depth that can be probed. Indeed, because of the very different path lengths over the gage volume as a function of depth, the diffracted signal will be weighted towards the surface. Consequently, the sampled gage volume (SGV) is different from the instrumental gage volume (IGV) defined simply in terms of the geometry of the slits. This must be corrected for (see Section 7.2.6) when accurate surface measurements are required. Conversely for transmission measurements (see Figure 7.8(b)), the path length is not much more than the thickness of the sample. Further because the diffraction angle, θ, is often less than 5° the sample is sometimes oriented normal to the incident beam (as shown in Figure 7.2) and the strain vector taken to be representative of the in-plane strain without serious error.

The length (l_g) of the gage normal to the scattering vector is given by:

$$l_g = (l_i/2\sin\theta + l_d/2\sin\theta) \approx (l_i + l_d)/2\theta \quad \text{(for small scattering angles)} \qquad (7.4)$$

where θ is in radians, l_i is the incident aperture width and l_d the width of the diffracted beam aperture (Figure 7.8(a)). Taking the Al (111) diffraction peak as an example (plane spacing $d(111) = 2.34$ Å) using a moderate incident energy of 30 keV gives a Bragg angle $(\theta) \approx 5°$ so that l_g is about 5.6 times the sum of the incident and diffracted slit widths. By contrast for an incident energy of 100 keV ($\theta \approx 1.5°$), l_g is about 18 times the sum of the aperture widths. This means that for apertures of 50 μm the gage is around 560 μm long at $2\theta = 10°$, but 1800 μm at 3°. Of course, the distinction between moderate and high energy is somewhat arbitrary, but the important point is that lower energy X-rays do allow for better spatial resolution along the beam path.

When making measurements where steep strain gradients are anticipated it is natural to want to use a small gage volume, but is that the best approach? While a small gage will clearly be sensitive to the short wavelength fluctuations in the strain field, if the gage is small then it will take a long time to acquire a sufficient number of counts for good precision (see Section 7.2.5). On the other hand, a large gage will tend to smear out the strain field, but will achieve good counting statistics in a short time. A deconvolution method for recovering the underlying strain profile from discrete diffraction measurements has been used to explore the optimal choice of gage size and measurement spacing in order to obtain the strain profile in as little time as possible [5]. This method provides an efficient way to recover the underlying strain profile by using a gage of length approximately equal to the wavelength of the expected strain variation, but with substantial overlap (~80%) between successive points (Figure 7.9). This procedure differs from current practice, where experimenters tend to choose a gage volume significantly smaller than the 'wavelength' of the strain feature they wish to capture along with a measurement spacing equal to the gage length (i.e., zero overlap). The efficacy of this large gage approach is demonstrated in Figure 7.9 where a sharp change in strain/stress state (~200 μm) near the surface has been recovered using reflection measurements to measure the out-of-plane strain where

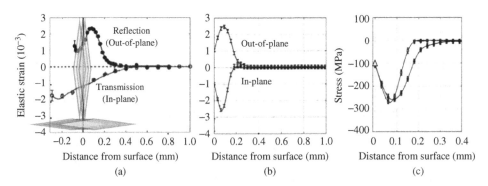

Figure 7.9 Strain measurement data recorded in reflection (filled circles) and transmission (open circles) as the nominal gage volume (illustrated) is translated through the surface of a shot-peened Al sample using increments of 0.02 mm (80% overlap), 0.1 mm (87%) respectively. The full lines are profiles reconstructed using a Bayesian fit to the data, (b) The deconvoluted residual strain profile versus depth below the peened surface. (c) The inferred residual stress profiles from the in-plane and out-of-plane strain measurements assuming a biaxial shot peen stress, a surface X-ray stress measurement (triangle) is also included for comparison. Reproduced with permission from [5], Copyright 2006 John Wiley & Sons

the gage extends over 100 μm depth, and in transmission where the gage extends 760 μm normal to the surface. In this case the sampled gage volume (SGV – see Section 7.2.6) gets smaller when the gage centroid is outside the surface, but nevertheless the point is that deconvoluting the profile from data collected using a large gage with extensive overlap between successive points is the best option from a time efficiency viewpoint.

7.2.2 From Diffraction Peak to Lattice Spacing

Whether the diffraction peaks are recorded as a function of scattering angle using a monochromatic beam (see Figure 7.17), or as a function of wavelength (more commonly plotted as a function of energy) (see Figure 7.22), the task is to determine the lattice spacing from Bragg's Law (Equation 7.1). Individual *hkl* diffraction peaks can be analyzed by fitting each to a Gaussian or some other function (e.g., Voigt or pseudo-Voigt) to determine a specific lattice spacing $d(hkl)$ as in Figure 7.10(a). Alternatively, a number of peaks can be analyzed simultaneously to refine the unit cell dimensions (a, b, c) as in Figure 7.10(b). The data in Figure 7.10(a) also illustrate the incredible single peak instrumental resolution capable with synchrotron diffraction (standard deviation in θ of 0.0034°).

One advantage of refining many peaks simultaneously is that the lattice parameter represents an average over many *hkl* reflections and therefore is more likely to be representative of the behavior of the bulk.

7.2.3 From Lattice Spacing to Elastic Strain

Bragg's Law provides a means of determining the lattice spacing, d. To do this accurately in an absolute sense would require extensive calibration of the incident wavelength and diffraction angles. However to determine the lattice strain using Equation (7.3) it is necessary only to evaluate the change in lattice spacing, Δd, relative to the strain-free lattice spacing, d_0.

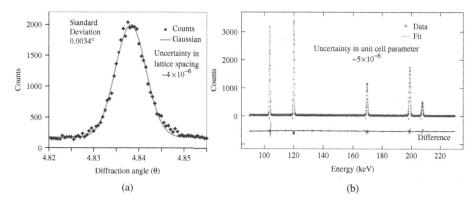

Figure 7.10 (a) A single Al(311) diffraction peak recorded with an analyzer crystal on ID31 at the ESRF at 60 kV alongside a Gaussian peak fit. (b) A typical energy dispersive spectrum recorded in transmission in a 25 mm thick stainless steel specimen on ID15 at the ESRF ($2\theta = 3.5°$). The difference between the data and the multiple peak fit has been displaced vertically for clarity

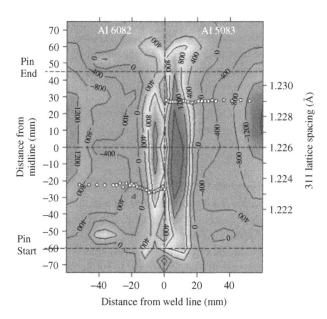

Figure 7.11 The variation in the longitudinal residual stress as a function of position from the weld line for a friction stir weld joining 2.8 mm thick dissimilar aluminium alloy plates. The variation in strain-free lattice spacing, d_{311}, was determined by the $\sin^2\psi$ method and is shown across the mid-line (circles/right hand axis) [9]. If not accounted for, this effect would swamp the elastic strains and lead to misinterpretation

Further, if the lattice strain is to be interpreted in terms of stress, any other mechanisms by which the lattice spacing may vary must be accounted for. An increase in temperature, or the movement of an alloying element into solution, may cause the lattice spacing to change in the absence of stress. In this respect it is helpful to note that, to first order, plastic deformation does not affect the atomic lattice spacing.

In planning an experiment, the task of measuring a representative value of the strain-free lattice spacing, d_0, is of critical importance. Indeed, in cases where a global strain-free lattice spacing is expected, it has been suggested that more time should be spent measuring the strain-free lattice spacing than on the individual strains because the initial error propagates through all the subsequent measurements as a systematic error. Special care is warranted in cases where d_0 might be expected to vary either across the component or over time. An important example of the former is provided in the welding of heat treatable Al alloys where alloy elements pass into or out of solution changing the strain free lattice spacing in the weld and heat affected zone (Figure 7.11), an example of the latter would be the development of internal stresses in a two-phase composite during cooling from manufacture.

The ubiquitous $\sin^2\psi$ laboratory X-ray stress measurement method (see Chapter 6) does not require a separate measurement of the strain free lattice spacing, d_0. Rather it exploits the fact that for laboratory X-rays, the diffracted signal comes from a region very near to the surface for which it is often possible to assume that the out-of-plane stresses are zero. The large penetration of synchrotron X-rays means that in many cases

Table 7.1 The merits and limitations of various strategies for measuring strain-free lattice spacing

Method	Local/Global	Determined or implied	Validity
Reference Standard	Global	Determined	Absolute lattice spacings are not normally accurate enough for strain measurement
Filings/powders	Global	Determined	Well suited to ceramics; only valid for metals if filings representative of alloy condition in the component
Cubes	Local	Determined	May retain intergranular stresses (Section 7.2.4), vulnerable to geometrical effects if poorly positioned
Far-field	Global	Determined	Only valid if no spatial variation in d_0^{hkl}: critical to ensure chosen locations are actually stress-free
$\sin^2\psi$	Local	Implied	Only valid if truly plane-stress, e.g., near surface, in thin plates/slices. Sensitive to intergranular and interphase stresses (Section 7.2.4) which may not be zero over the sampling volume even when macrostress is zero
Stress balance	Global	Implied	Only valid if no spatial variation in d_0^{hkl} and complete area normal to stress is mapped. Best held in reserve as a confirmation that a_0 has been correctly estimated.

this circumstance cannot be assumed and so a measurement or inference of d_0 is needed. The various strategies for determining d_0 have been reviewed in detail elsewhere [6] many of which have been borrowed from neutron or X-ray diffraction approaches. Some are applicable only to materials for which a global strain free lattice parameter is valid, while others can map local variations in d_0 (see Table 7.1). Some of the most suitable methods for synchrotron diffraction are briefly summarized below.

Powders, Filings, Cubes and Combs: Since every point within a fine powder is near a free surface, the material cannot sustain a long-range macrostress and thus can be used as macrostress-free reference. The powder should be measured at the same temperature and should completely fill the instrumental gage volume (Section 7.2.6). An elegant alternative approach, suitable in appropriate circumstances, is to manufacture so-called reference 'combs' from a notionally identical reference sample. The comb should be cut in an orientation such that the teeth are essentially free from the constraint of the surroundings, but left in registration through a small mechanical connection to the base material. Such structures circumvent the logistical problems associated with handling small reference cubes, while at the same time retaining the positional relationship between each 'tooth' and hence strain-free lattice spacing measurement.

Measurement of d_0 in a region known to be free of macrostress: A region must be identified where it is considered likely that the sample will be stress-free. Note that this method of globally fixing d_0 is not appropriate in cases where there may be localized heating, such as near a weld, due to possible local compositional/precipitation changes.

Exploitation of plane stress or plane strain conditions: There are many methods that rely on some assumption about the state of stress, or strain, to infer the strain-free lattice spacing or in-plane stress. Of these the $\sin^2\psi$-method is commonly applied, notably with X-ray measurements. It is based upon the general stress–strain relationship [7]:

$$\varepsilon_{\phi\psi} = \frac{d_{\phi\psi} - d_0}{d_0} = \frac{1+\nu}{E}\{\sigma_{11}\cos^2\phi + \sigma_{12}\sin 2\phi + \sigma_{22}\sin^2\phi - \sigma_{33}\}\sin^2\psi$$

$$+ \frac{1}{E}\{\sigma_{33} - \nu\sigma_{11} - \nu\sigma_{22}\} + \frac{1+\nu}{E}\{\sigma_{13}\cos\phi + \sigma_{23}\sin\phi\}\sin 2\psi \quad (7.5)$$

where ψ is the polar angle from the surface normal and ϕ the azimuthal angle to the in-plane 1-axis. In the presence of shear stresses σ_{13} or σ_{23} ψ-splitting occurs (different spacings recorded for $\pm\Psi$). For a plane stress field ($\sigma_{33} = \sigma_{23} = \sigma_{13} = 0$), it allows residual stress analysis without precise knowledge of the strain-free lattice spacing by plotting $\varepsilon_{\phi\psi}$ vs $\sin^2\psi$, for example by tilting about the two axis ($\phi = 0$):

$$\varepsilon_\psi = \frac{d_\psi - d_0}{d_0} = \frac{1+\nu}{E}\sigma_{11}\sin^2\psi - \frac{\nu}{E}\{\sigma_{11} + \sigma_{22}\} \quad (7.6)$$

The high penetrations available with high-energy synchrotron X-ray diffraction opens up the possibility of using the $\sin^2\psi$ method over a wider range of ψ angles than is available for laboratory X-rays. In fact the low scattering angles characteristic of high energy X-rays mean that low ψ angles cannot be accessed in reflection because of the large path lengths associated with the glancing angles (Figure 7.8). Of course in such a case the method still relies on the out-of-plane stress being zero through the thickness of the plate, or if the sampling gage spans the entire thickness it requires that the local through-thickness stresses average to zero (which is usually true for plates). A good example is provided by the mapping in Figure 7.11 of axial stress across a friction stir weld butt joint between 150×60 mm, 2.8 mm thick plates of non age-hardenable AA5083 and age-hardenable AA6082 alloy. In this case the $\sin^2\psi$ method was used (with ψ varied from $45°$ to $90°$ using five ψ-tilts) to account for variations in d_0 across the welded plate [8]. These changes were extensive across the weld line (equivalent to 5000×10^{-6}) because of the different alloy chemistry of the two plates being joined, but significant stress-free variation (600×10^{-6}) was also observed in the heat affected zone for the age-hardenable AA6082 side, presumably because of local repartitioning of the solute elements.(strain-free lattice spacing: mapping)

d_0 inferred from measurements on thin slices: A thin slice cut from a sample will tend to be in a condition of plane stress with the normal component averaging to zero through thickness ($\sigma_{33} = 0$). As a result the $\sin^2\psi$ technique can be applied to map d_0 across the slice. Following Equation (7.7) [10] a series of measurements are made at different ψ tilts, at each of two angles, ϕ and $\phi + 90°$.

$$d_0 = (d_1^\perp + d_2^\perp)/2 + [\nu/(1+\nu)](m_1 + m_2) \quad (7.7)$$

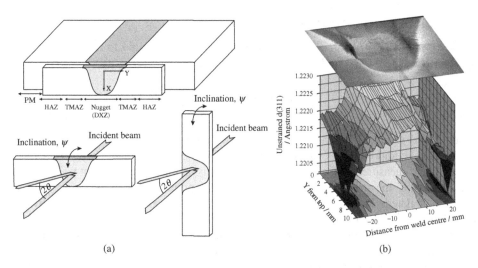

Figure 7.12 (a) schematic of the cross-sectional slice cut from the Al 7010 friction stir weld showing the set-up for the synchrotron X-ray diffraction d_0 measurements. (b) A graph showing of the variation of the strain-free lattice spacing of $d_0(311)$ calculated by the $\sin^2\psi$-method

where $d_i^\perp$ and m_i ($i = 1, 2$) are the intercept and gradient, respectively, of the $\sin^2\psi$ vs d plots in the two directions ($\varphi = 0°$ and $90°$) (see Figure 7.12). The $\sin^2\psi$ analysis can be carried out quickly and effectively on excised cross-sectional slices using high-energy synchrotron X-ray diffraction at low scattering angles in transmission. The $\varphi = 0°$ and $90°$ measurement process requires no prior knowledge of the in-plane principal stresses. A good example is provided by the measurement of the local variation in lattice parameter across a friction stir weld of 13 mm thick Al 7010 alloy plate [11]. A 1 mm thick cross-sectional slice was cut mid-way along the FSW in order to use the transmission $\sin^2\psi$ method, and the variation in d_0 mapped across the weld in Figure 7.12. It is evident that the changes in strain-free lattice parameter are significant. Age hardening alloys are particularly prone to changes in solute content with precipitation; if a single global value of d_0 were used as reference in this case, it would give an apparent strain of up to 2000×10^{-6}, corresponding to a stress error of up to 140 MPa.

Imposing Stress and Moment Balance: This method exploits the fundamental continuum mechanics-based requirements that force and moment must balance across selected cross-sections of the sample. The approach involves measuring the required field of d-spacings, or diffraction angles, in the sample, and, using a nominal d_{ref} to calculate the strain and stress in the sample. Subsequently the reference value is varied iteratively in order to infer the true strain free value; that is, the value that renders a stress field in which force and moment balance. In adopting this approach great care must be taken in selecting appropriate cross-sections over which to require the forces and moments to balance. It must be ensured that the experimental data set covers the entire cross-section, and not just a part of it, and also that a single global value of d_0 is appropriate. In practice, this method is probably best held in reserve to check the validity of the d_0 value obtained from one of the other methods listed above.

7.2.4 From Elastic Strain to Stress

Stress (σ) and strain (ε) are second rank tensor quantities related to one another by the elastic stiffness tensor C, and the elastic compliance tensor S;

$$\sigma_{ij} = \sum_{kl} C_{ijkl} \varepsilon_{kl} \quad \text{and} \quad \varepsilon_{ij} = \sum_{kl} S_{ijkl} \sigma_{kl} \qquad (7.8)$$

where σ and ε have 3×3 components, 6 of which are independent and C and S have $3 \times 3 \times 3 \times 3$ components, of which as many as 36 can be independent [12]. As a result the conversion of measured strain components to stress is an inherently difficult task, requiring measurement of the strain in many directions at each point in the sample before the stress can be calculated. Indeed, the low scattering angles typical of synchrotron diffraction mean it may not be possible to measure strain in sufficient directions. Systematic and statistical errors in the individual strain measurements combine to reduce the accuracy of the inferred stresses. For example the presence of intergranular stresses between grains of different orientations may mean that the strain is very different for grains of different orientations. This necessitates an understanding of the effects of elastic and plastically induced anisotropy (see below) of lattice strain in crystals so that the strain recorded for one or more *hkl* lattice planes in particular directions can be used to infer the continuum macrostress tensor.

Most engineering investigations are based on isotropic continuum mechanics. In this case, C can be written in terms of just two independent elastic components, for instance Young's modulus, E, and Poisson's ratio, v. Consequently, the relationship between stress and strain can be expressed using the generalized Hooke's law equations:

$$\sigma_{ij} = \frac{E}{(1+v)} \left[\varepsilon_{ij} + \frac{v}{(1-2v)} (\varepsilon_{11} + \varepsilon_{22} + \varepsilon_{33}) \right] \qquad (7.9)$$

where $i, j = 1, 2, 3$ indicate the components relative to chosen axes.

From this it is clear that in order to derive the stress tensor, the strain tensor must first be determined requiring at least six measurements. Considering the inherent measurement uncertainty associated with individual strain measurements, it is generally best to over-determine the problem by measuring more components than theoretically needed to solve the mathematical problem. To date, this significant measurement task has not yet been tackled using synchrotron X-rays, even for just a few key locations within a component. Instead, experimenters tend to invoke symmetry considerations to reduce the number of measurements necessary, or focus on obtaining the normal stress in three perpendicular directions. This exploits the form of Equation (7.9), and the invariance of the trace of a tensor under rotation, which means any three measured orthogonal normal strain components, ε_{11}, ε_{22}, and ε_{33} can form the basis for calculating the corresponding three orthogonal stress components, without any knowledge of the shear (off-diagonal) strain components or the principal stress directions.

In practice the applied stress *vs.* elastic strain response of each lattice plane family *hkl* is usually different. In the elastic regime this is because in general the stiffness of a single crystal is not isotropic (elastic anisotropy), and in the plastic regime this is because different grains deform plastically to different extents (plastic anisotropy), thereby

Table 7.2 Suitability of various diffraction planes for different crystal systems for strain measurement [14] for face-centered (fcc) and body centered (bcc) cubic and hexagonally close packed (hcp) systems of engineering interest

Material	Recommended planes – small intergranular strains	Problematic planes – large intergranular strains
fcc (Ni, Fe, Cu)	111, 311, 422	200
fcc (Al)	111, 311, 422, 220	200
bcc (Fe)	110, 211	200
hcp (zircaloy, Ti)	pyramidal ($10\bar{1}2$, $10\bar{1}3$)	basal (0002) prism ($10\bar{1}0$, $1\bar{2}10$)
hcp (Be)	2nd order pyramidal ($20\bar{2}1$, $11\bar{2}2$)	basal, prism and 1st order pyramidal ($10\bar{1}2$, $10\bar{1}3$)

generating stresses between them [13]. This means that different stresses will be obtained using different hkl planes if these effects are not accounted for.

At first glance it is tempting to replace the continuum elastic constants (E, ν) in Equation (7.9) with their single crystal hkl dependent ones to convert the hkl specific strain $\varepsilon(hkl)$ into the continuum macrostress σ^I. However the question then arises as to the most appropriate elastic constants. The single crystal values are not representative of the behaviour of grains within a polycrystal because of intergranular stresses (constraint) generated between the differently oriented grains. Here, a pragmatic approach is taken to derive representative elastic constants to relate the lattice strains to the macrostress during elastic loading. These are termed the diffraction peak specific elastic constants (DEC), E_{hkl} and ν_{hkl} for texture-free materials. If these are substituted in the generalized Hooke's law equations then the strain evaluated for each reflection, $\underline{\varepsilon}hkl$, can be converted to a single valued estimate of the macrostress, σ^I, where,

$$\sigma_{ij}^I = \frac{E_{hkl}}{(1+\nu_{hkl})}\left[\varepsilon_{ij}(hkl) + \frac{\nu_{hkl}}{(1-2\nu_{hkl})}(\varepsilon_{11}(hkl) + \varepsilon_{22}(hkl) + \varepsilon_{33}(hkl))\right] \quad (7.11)$$

The diffraction elastic constants can be measured from calibration experiments, in which a sample is subjected to known uniaxial loading. They can also be calculated using polycrystal models. The behavior of some diffraction peaks are less affected by plastic anisotropy and are thus more representative of the continuum stress. The best peaks for residual stress measurement are discussed in the draft standard developed for residual stress measurement by neutron diffraction [14] the conclusions of which are summarized in Table 7.2.

7.2.5 The Precision of Diffraction Peak Measurement

As discussed in Section 7.2.2, a key issue for diffraction methods of strain measurement is the precision, Δx, with which the diffraction peak position (in θ or E) can be determined. Normally the peak is approximately Gaussian in shape (Figure 7.10(a)), although other functions are also commonly fitted. The precision of peak location for peaks approximately Gaussian in shape can be expressed in terms of the standard deviation, u, the integrated number of counts under the peak, N, the height of the diffraction peak, H, and the background, B [15],

$$\langle \Delta x^2 \rangle = \frac{u^2}{N}\left[1 + \frac{2\sqrt{2B}}{H}\right] \qquad (7.12)$$

Normally the peak width for synchrotron diffraction peaks can be extremely narrow (a standard deviation in θ of $0.0034°$ in Figure 7.10(a)), the number of counts per second high (tens of thousands, say) and the background low such that the statistical strain measurement accuracy associated with the peak fit can be very high (equivalent to strain accuracies of 10^{-5} to 10^{-6}). In cases where the signal to noise ratio (H/B) becomes less than 1 measurement becomes increasingly impractical, involving long count times.

While the statistical precision of diffraction peak measurement is easily quantified and is often small, in practice it is often the case that other factors conspire to limit the accuracy of strain measurement achievable.

7.2.6 Reliability, Systematic Errors and Standardization

In addition to statistical uncertainties in determining the diffraction peak position and concerns over the most appropriate estimate of d_0, there are also systematic errors. These can be due to poor setting up of the beamline, but some errors can arise even for a well-conditioned setup. Most important among these are those relating to incomplete filling of the instrumental gage volume.

As the sample is moved into the gage volume at first the gage is only partially filled (Figure 7.13). In these circumstances both the position recorded for the measurement location and the angle recorded for the diffraction peak must be corrected. This effect is very well documented for neutron diffraction (see Chapter 8). The gage determined by the instrument (the instrumental gage volume (IGV)) is defined by the slits (a parallelogram in 2D – see Figure 7.8), whereas the sampled gage volume (SGV) is the multiplication of the IGV with the probability of scattering across it (this gives a triangle when the sample is less than half within the IGV) [13]. In this case the centroid of the SGV (marked by a white circle in Figure 7.13) is not coincident with the centroid of the IGV (marked by a cross). For strongly attenuating samples, where attenuation across the gage is significant, this effect must also be included when calculating the weighted average centroid of the SGV. This effect is significant for measurements in reflection where photons diffracted from the region of the gage nearest the surface travel shorter path lengths and thus are attenuated less giving a weight centroid nearer to the surface than the geometrical centroid of the IGV (Section 7.2.1). By contrast, for transmission measurments (see Figure 7.8) the path lengths are the same for all positions within the SGV and so in this case attenuation doesn't affect the position of the effective SGV.

Correcting for errors in position: As illustrated in Figure 7.13, when the sample begins to enter the IGV the sampled gage is a triangle and the centroid of the SGV lies just within the sample whereas the centroid of the IGV still lies outside the sample (see also Figure 7.14(a)). Given that the centroid of the IGV is usually taken to represent the location of the SGV (which is true for a fully buried gage provided attenuation

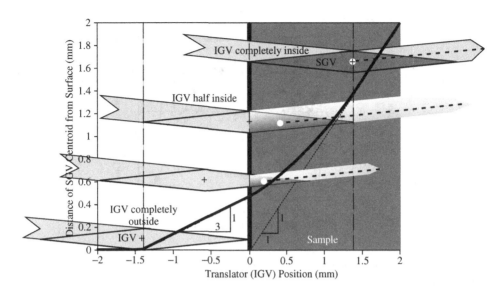

Figure 7.13 The relationship between the depth of the centroid (white circle) of the sampled gage volume (SGV) and the position (cross) representing the instrumental gage volume (IGV) for an IGV which is 2.8 mm long normal to the strain measurement direction (vertical) for transmission geometry. Note that the diffracted signal is weighted towards the lower region of the IGV until it is full, this would be recorded as a shift towards a lower diffraction angle unless an analyzer crystal is used

across the gage can be neglected) this must be corrected according to the curve given in Figure 7.13.

Correcting for errors in diffraction peak angle: As the sample enters the IGV in Figure 7.13 the fact that the effective centroid is not at the instrument reference point means that most of the diffracted signal comes from the region towards the bottom of the SGV. Unless an analyzer crystal is used (see Figure 7.14(a)) to ensure that the diffractometer is solely angularly sensitive, this shift in the weight of the signal towards the bottom of the IGV would be recorded as a shift to lower angles (equivalent to a tensile strain – see Figure 7.14(a)) if uncorrected. This spurious strain should be corrected for analytically or experimentally for surface entry and exit.

There are many other situations, besides partial gage filling, where a shift in the weighted average signal from the centroid of the IGV can lead to apparent changes in diffraction angle and hence apparent strains. The most commonly encountered case relates to poor grain size statistics. This is shown schematically in Figure 7.15. As the individual grains move into the gage volume they contribute a spot to the diffraction pattern. According to their location this can occur at a larger (apparent compressive strain) or smaller (apparent tensile strain) radius as shown schematically in Figure 7.15. For small scattering angles the variation is determined by the diffracting slit (or the sample thickness if no diffraction slit is used). This variation in the apparent scattering angle can

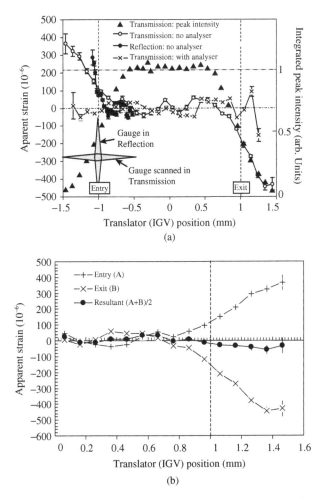

Figure 7.14 (a) Shifts in the d spacing recorded for a 2 mm thick Al powder phial as it is scanned in transmission and reflection on a $\theta/2\theta$ diffractometer without an analyzer and in transmission with an analyzer. The peak shifts are of geometric origin and occur up until the point at which the gage is totally filled. The edges of the samples can be accurately located from the variation in integrated diffraction peak intensity and (b) the surface effect can be accounted for by undertaking scans before and after rotating the sample by 180° and adding them together so that the surface is sampled on entry and exit in turn. From [26]

be no more than:

$$2\Delta\theta = l_d/l_c \quad \text{or} \quad 2\Delta\theta = t\sin\theta/l_c \quad \text{(if no diffracting slit)} \tag{7.13}$$

so for a slit of 100 μm and a camera length, l_c, of 1 m the error is 10^{-4} radians which equates to a strain uncertainty of $\pm 1.4 \times 10^{-3}$ for $2\theta = 4°$, which is much bigger than the statistical uncertainty in the peak position (typically $<10^{-5}$).

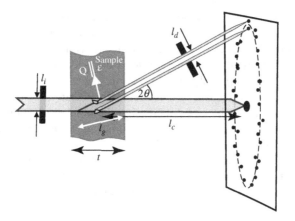

Figure 7.15 A schematic showing the difference in apparent scattering angle arising from diffraction from large grains either side of the gage width. If there are many grains present the average of all the scattering grains is coincident with the centroid of the IGV giving the diffraction ring shown by the dashed line

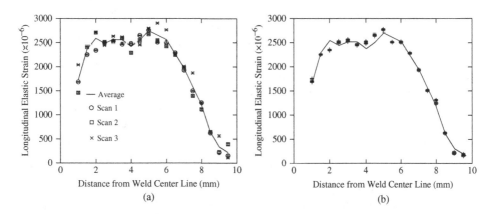

Figure 7.16 (a) three lines scans made laterally from a weld, each made 100 μm further down the weld line made using a gage defined by 100 μm slits, the line shows the average of the three sets, (b) three repeat linescans made at the position of Scan 1 compared to the average of Scans 1 to 3 [26]

For an area detector, as shown in Figure 7.15, insufficient grain sampling is evident as spotty diffraction rings. For a conventional $\theta/2\theta$ scan or when using an energy sensitive detector, it is evident as a sharp variation in diffraction peak intensity from measurement to measurement as individual grains move in and out of the gage volume. The effect on the strain profile is exemplified by the data in Figure 7.16a that shows three nominally identical linescans made side by side. The statistical uncertainties are smaller than the symbols ($\sim \pm 15 \times 10^{-6}$) and yet the point-to-point scatter between the measurements is as large as 500×10^{-6}. That this is due to grain sampling errors is demonstrated by the three repeat scans along the same scan line in Figure 7.16(b) which shows that three repeat measurements at the same location lie within $\sim 15 \times 10^{-6}$ (consistent with the expected statistical scatter), yet deviating significantly in places from the average of the three scans made side by side in Figure 7.16(a). As a general rule, if the peak intensity varies from point to point by more than 25% it is probably because grain size effects are important and special measures need to be taken to recover strains accurately. Clearly, this effect can be reduced by increasing the camera length, l_c; by increasing the number of grains sampled (e.g. by oscillating the sample or increasing the gage volume); or by using an analyzer crystal to ensure that the detector is only angularly discriminating and not position discriminating.

As mature measurement methods, both laboratory X-ray [16] and neutron diffraction [14] have best-practice guidelines for repeatable strain measurement, from which much can be learnt relevant to synchrotron X-ray measurement. To date, no such guidelines have developed for synchrotron methods.

7.3 Angle-dispersive Diffraction

7.3.1 Experimental Set-up, Detectors, and Data Analysis

There are several possible experimental set-ups used for mapping elastic strains using monochromatic radiation. Many use an area detector to collect whole Debye-Scherrer cones (Figure 7.17(a)). In many cases when a 2D detector is used no diffracted slit is employed to define the exit beam, which means that the strain is sampled over the whole sample thickness. Alternatively a conical slit system can be used in order to define the gage volume. In some cases rather than use incident slits to determine the incoming beam, beam focusing is used, either using refractive lenses or focusing monochromators. Area detectors have the clear advantage that poor grain sampling statistics become clearly evident in terms of spotty diffraction rings (see Figure 7.2(a)). An alternative approach is to use a two-axis diffractometer set-up whereby an analyzer crystal ensures that all detected photons originate from the sample at the same angle, 2θ, as shown in Figure 7.17(b). In this case, the diffraction pattern is collected by scanning the sample and the detector/analyser over θ and 2θ respectively to collect a diffraction profile.

If whole diffraction rings are collected, these are often "caked" so as to collapse the 360° data into a series of linear profiles corresponding to the center of the cake-slices (e.g., 0°, 10°, 20°, etc.) as shown in Figure 7.18. Because the diffraction angles are small, the diffraction vector, Q, is almost perpendicular to the incident beam, that is, the strain is sampled in the plane normal to the beam. Under stress, the Debye-Scherrer cones are distorted from producing circular rings to make ellipses on the detector. Strain can then be obtained in a number of ways from the data. For example, the strain in specific directions can be assessed by measuring the change in radius (or diameter), typically, in the horizontal (90° slice) and vertical (0° slice) directions. Sometimes, however, it is more useful to analyze the complete rings. By fitting the diffraction rings to ellipses it is possible to deduce the principal in-plane-strain directions (the major and minor axes) and their angles to the laboratory frame [27]. Another approach is to use a $\sin^2\phi$ plot to extract the axial and transverse responses as shown in Figure 7.19.

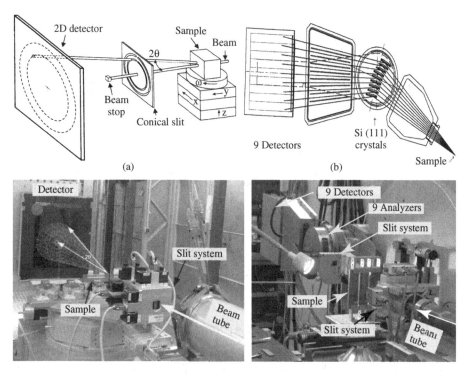

Figure 7.17 (a) Schematic (top) and photograph (HARWI II beamline-bottom) showing a 2D detector arrangement with a conical slit and (b) schematic (top) and photo (bottom) showing a $\theta/2\theta$ arrangement with an analyzer used on ID31 at ESRF. From [28]

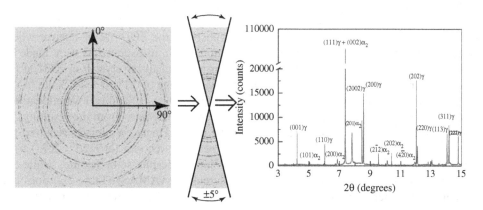

Figure 7.18 The diffraction pattern (left) can be 'caked' (center) into a series of 'slices' which can then be collapsed to give diffraction profiles (right) representative of the strain at 0°, 10°, and so on

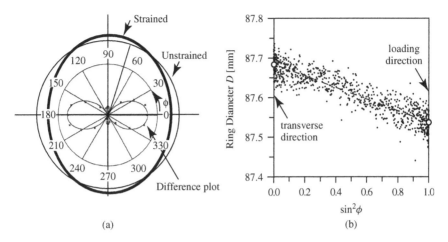

Figure 7.19 (a) The 2D distortion of the diffraction pattern due to biaxial in-plane strain (tensile vertical and compressive horizontally) [27] and (b) similar data analyzed using a $\sin^2\phi$ plot. Reproduced from [29], Copyright 1993 Elsevier

7.3.2 Exemplar: Mapping Stresses Around Foreign Object Damage

Aero-engine compressor blades are susceptible to foreign object damage (FOD) by small angular particles, particularly on the leading edge of the aerofoil. The combination of the geometric stress concentration introduced by foreign object damage, the local plastic work, the propensity for localized damage and the residual stress all affect the fatigue resistance of the blade. In order to study the residual stresses introduced by FOD, a FOD event was introduced by means of a 3 mm hardened steel cube, fired normal to the center of the edge of a fatigue test-piece at a velocity of 200 m s^{-1} using a light gas gun (see Figure 7.20) [17]. The steep local gradients in stress necessitate the use of synchrotron diffraction. Because the stresses around the periphery of the FOD are of interest where the gage may protrude from the surface (see Section 7.2.6), measurements were made by angle dispersive diffraction using 60 kV X-rays using an analyzer crystal on ID31 at the European Synchrotron Radiation Facility (ESRF) with a gage of $0.1 \times 0.1 \times 2.3$ mm in x, y and z. The measurement locations (crosses in Figures 7.20(a), (b) and (c)) were identified by measuring the shape using a coordinate measurement machine. From the resulting 3D model of the sample it was possible to extract the measurement coordinates required for each point. The $(10\bar{1}1)$ hcp titanium diffraction peak was chosen (at $\sim 2\theta = 5.3°$) as it showed good intensity for all the orientations of interest despite some texture. While the European draft standard for residual stress measurement (VAMAS) (Table 7.2) does not explicitly recommend this peak, it does fulfill the criteria set therein as it is neither a basal nor prism plane.

The strains parallel to the fatigue axis are well resolved because they are most tensile near the surface. While the measured elastic strain field is similar in form to finite element predictions [17], their magnitude is only 40% of those predicted by finite element modeling. This is either because the constitutive equation for high strain rate deformation is inadequate, or because damage was introduced local to the FOD site. Since the diffraction

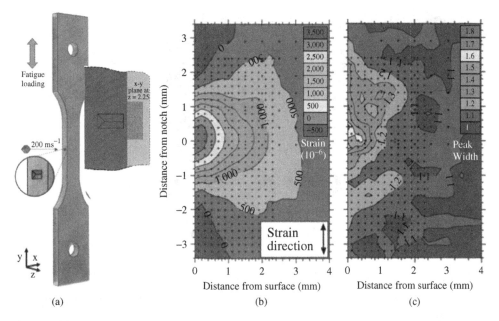

Figure 7.20 (a) Schematic showing sample geometry, location of impact and the measurement location, (b) elastic strain (10^{-6}) at the periphery of the FOD at mid-thickness of the remaining ligament measured on ID31 at the ESRF, (c) local variation in diffraction peak full width half maximum normalized by the average FWHM recorded far from the FOD (~0.03°). Since the instrumental peak width is actually very narrow it is very sensitive to plasticity in the sample. Reproduced from [17], Copyright 2012 Elsevier

peak width is sensitive to plastic strain it is clear that the ligaments lateral to the FOD have been plastically strained (Figure 7.20(c)). Only synchrotron X-ray diffraction is able to produce such high spatial and strain resolution maps comprising hundreds of points. Incidentally the same scan had previously been undertaken by energy dispersive scanning but the surface effects (see Section 7.2.6) were very large masking the peak shifts from elastic strain.

7.3.3 Exemplar: Fast Strain Measurements

Using an area detector enables phase changes to be monitored with a sub-second frame rate. This is important, for example, when looking at the phase changes and the associated weld residual strains that arise as steel weld filler metal cools, because the cooling naturally occurs over a short period of time (Figure 7.21). The large misfit between the austenite and the displacively formed martensite means that on cooling large misfit strains are generated which have a very significant effect on the final weld residual stresses. Indeed if the transformation occurs at low temperature (around 300–200 °C) it can change the weld stress from tension to compression [18]. Here the effect of the stress on the onset of the phase transformation during cooling has been quantified by diffraction. This is important because the weld filler metal cools in the presence of weld residual stresses.

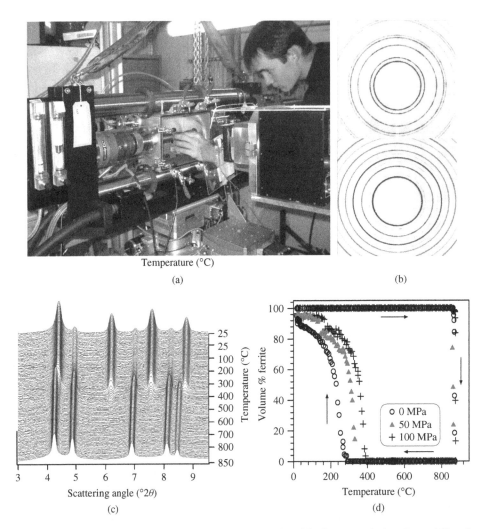

Figure 7.21 (a) Monitoring phase changes and associated residual stresses during the rapid heating and cooling of weld filler metal at the ESRF using a thermo-mechanical test facility. (b) Debye-Scherrer cones collected on an area detector at 900°C (top) and on cooling to 250° (bottom) using acquisition times of 30 ms to capture the fast cooling typical of weld cooling. (c) Caked segment (0°) of the diffraction pattern showing the transformation of the weld filler metal OK75.78 from austenite to ferrite as a function of temperature at a cooling rate of 10°C/s. (d) The phase fractions as a function of weld constraint (0 MPa, 50 MPa and 100 MPa) introduced at 550 °C and held to completion

7.4 Energy-dispersive Diffraction

Analogous to time of flight neutron diffraction (Chapter 8), it is possible to measure the diffraction profile at a fixed angle using a polychromatic incident beam. X-rays travel at the speed of light and so it is not possible to deduce their energy from their time of arrival, as one can for neutrons. Instead, an energy sensitive detector must be used.

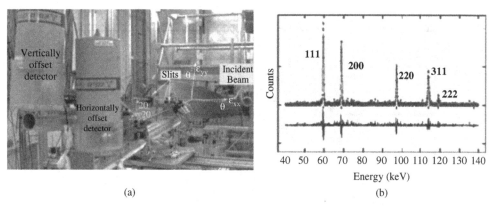

Figure 7.22 (a) Schematic showing how with two energy dispersive detectors one can measure the vertical and horizontal strains within a gage shown here for a compact tension specimen; note that the two strain measurement directions are essentially normal to the incident beam and (b) energy dispersive diffraction profile for Al alloy collected on ID15 at the ESRF with 2θ set at $5°$

One of the main advantages of this method is that one can measure a number of peaks simultaneously, just as one can when using a monochromatic beam with an area detector, but because the angles are fixed one can determine the gage volume using simple slits rather than the more geometrically involved conical slits (Figure 7.17(a)).

7.4.1 Experimental Set-up, Detectors, and Data Analysis

Central to the energy dispersive technique is the need to determine the energy of the detected X-ray. At present, gadolinium oxide detectors are used. Generally they have an energy discrimination of around $\Delta E \leq 200$ eV, which is sufficient for strain measurements. The experimental set-up, illustrated in Figure 7.22, has the advantage of having no moving parts.

As shown in Figure 7.22(b) the diffraction profile can be fitted with a multiple-peak fit (Rietveld refinement) to simultaneously derive all the lattice parameters representative of the polycrystal. Except when the grain size is very small, the main limit on the spatial resolution (gage dimensions) is not the diffracted X-ray intensity, but the limited number of diffracting grains sampled by the gage (see Figure 7.2).

7.4.2 Exemplar: Crack Tip Strain Mapping at High Spatial Resolution

One of the advantages of white beam experiments is that a number of peaks can be fitted simultaneously. This both improves statistical grain sampling and combats to some extent any texture changes across the sample. Further, with the set-up shown in Figure 7.22, two perpendicular components can be measured at the same time. Consequently, the method is very well suited to 2D strain mapping problems, for example the stresses across thin welded plates. Indeed the high brightness of synchrotron sources combined with an ability to define very small gage volumes means that significant areas can be mapped within feasible time periods.

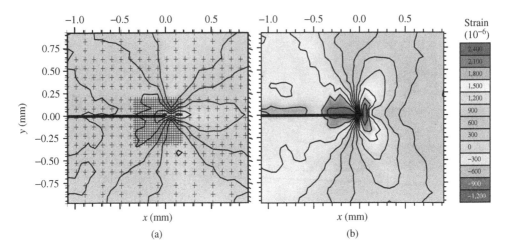

Figure 7.23 The elastic strain fields (in 10^{-6}) for (a) the crack growth (ε_{xx}) and (b) the crack opening (ε_{yy}) directions at a stress intensity $K_{IMax} = 6.6\,MPa\sqrt{m}$ in a Al-Li compact tension specimen. In (a) the crosses mark the measurement locations and the crack tip lies at (0,0). The lateral gage dimensions were $20 \times 20\,\mu m$. Reproduced from [19], Copyright 2010 Elsevier

A good example of the capability of synchrotron strain measurement for 2D strain mapping is provided by the measurement of stresses around a crack tip by Steuwer et al. [19]. They have exploited the sub-micron grain size of an Al-Li alloy to deliver extremely fine scale maps of the stresses around a crack tip (Figure 7.23). Further it is possible to infer the stress intensity at the crack tip by fitting the measured strain field to the linear elastic fracture mechanics solution for the stress local to the tip. In the current example the best fit corresponds to a crack-tip stress intensity of ($K_{IMax} = 6.2\,MPa\sqrt{m}$, $K_{IIMax} = 0.2\,MPa\sqrt{m}$) showing that in this ideal case the actual crack tip stress intensity is very close to that nominally applied ($K_{IMax} = 6.6\,MPa\sqrt{m}$).

7.4.3 Exemplar: Mapping Stresses in Thin Coatings and Surface Layers

It is possible to use the fact that X-rays are attenuated according to their energies to obtain surface depth profiles of residual stress. The penetration depth (e^{-1}) is given by:

$$\tau_{(hk.l)} = \sin\theta \cos\psi/2\mu \qquad (7.14)$$

In other words increasing the diffraction angle, θ, or increasing the inclination, ψ, of the scattering plane from the normal to the surface (inset in Figure 7.24), increases the depth due to the steeper entry angle into the surface, while for a given diffraction peak, (hkl), increasing the energy decreases the linear absorption coefficient, μ, (increasing the penetration depth approximately as $1/E^3$ as discussed in Section 7.1.3) but also decreasing the scattering angle, θ (decreasing the penetration depth).

This dependency lies at the heart of a number of different depth probing measurement strategies. The simplest is based around using the conventional $\sin^2\psi$ method to determine the stresses from the slope of the d vs $\sin^2\psi$ plot and then ascribing the stress determined

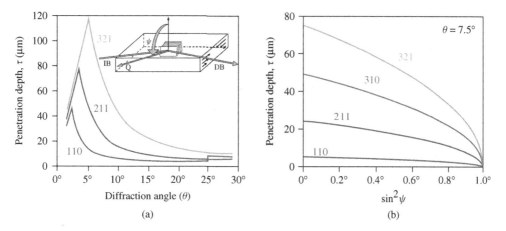

Figure 7.24 (a) The dependence of the penetration depth, τ, on the diffraction angle for a ferritic steel, (b) the variation in penetration depth as a function of inclination angle at a fixed Bragg angle for different reflections. Reproduced from [20], Copyright 2004 Elsevier

in this way to an average penetration depth $\tau(hk.l)$:

$$\tau_{(hk.l)} = \frac{\tau_{(hk.l)\,min} + \tau_{(hk.l)\,max}}{2} \quad (7.15)$$

where $\iota_{(hk.l)\,min}$ and $\tau_{(hk.l)\,max}$ are the minimum and maximum penetration depths corresponding to maximum and minimum ψ respectively. This procedure only provides good results if the non-linearity in the d vs $\sin^2\psi$ slope is small, in other words, if the stress does not vary significantly over the X-ray penetration depth [21], otherwise the method can underestimate the near surface gradient.

This method has been extended to deal with steeper stress gradients by assembling d vs. $\sin^2\psi$ plots at specific penetration depths [23] using plots such as Figure 7.25(b) to select the inclination angle, ψ, for each Bragg reflections that corresponds to a given depth. The resulting composite d vs. $\sin^2\psi$ plot should give a straight line slope. This method has been applied successfully to the analysis of the stress depth profile introduced into titanium alloys by laser peening for example (Figure 7.25(b)). The maximum information depth depends on the available energy range, but for a range of 10–80 kV the depth accessible in reflection mode experiments is about 100 µm for titanium [20]. Therefore in order to get a stress distribution over a deeper region, layer removal in steps of 100–150 µm must be applied by electropolishing.

More sophisticated methods are available, for example the $\sin^2\psi$ data obtained for the evaluation of perpendicular in plane stresses σ_{11} and σ_{22} can be combined to form a master plot of the individual stress profiles with depth [24]. Provided there is no variation in the stress free lattice parameter with depth can even be used to evaluate a triaxial residual stress state [20].

7.5 New Directions

Strain measurements are being used increasingly alongside other methods; in particular in combination with small angle X-ray, or neutron, scattering to explore the relationship

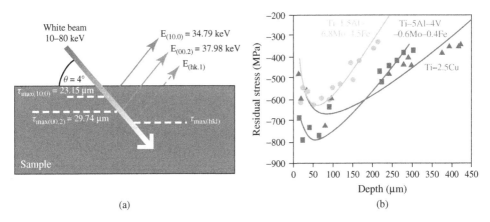

Figure 7.25 (a) Scheme showing a relation between energy [$E(hk.l)$] of a certain reflection ($hk.l$) of α phase and maximum penetration depth, $\tau_{\max(hk.l)}$ at $2\theta = 8°$ using white X-ray beam. $\tau_{\max(hk.l)}$ corresponds to the minimum tilting angle $= 0°$ and is calculated by $\tau_{(hk.l)} = (\sin\theta \cos\psi)/2\mu = \sin 4°/2\mu$, where μ is the linear absorption coefficient which depends upon $E(hk.l)$, b) In-plane residual stress-depth profiles in α phase of 3 Ti alloys after laser shock peening using the method shown in (a) [22]. At large depths layer removal in steps of $\sim 100\,\mu m$ was applied by electropolishing Reproduced from [22], Copyright 2012 Elsevier

between residual stress and damage populations, such as the size and density distributions of creep cavity defects.

Furthermore, recent developments at synchrotron sources make it possible to switch seamlessly between high spatial resolution X-ray tomography and diffraction modes analogous to the 2D imaging and diffraction modes of an electron microscope. This opens the way for the correlation of structure and stress. This is particularly useful when trying to understand the relationship between the state of stress and the initiation, growth and coalescence of defects. The bringing together of these techniques has been termed 3D "crack-tip microscopy" to probe the local condition of the crack-tip region [25]. The imaging mode can capture the level of damage, identify crack debris, closure, the crack-tip shielding mechanisms, quantify the crack tip opening displacement variations through the loading cycle, while the diffraction mode can quantify the crack-tip stress field, give a measure of local plasticity and any phase changes (e.g., phase transformations) (Figure 7.26). From this information it is possible to extract measures of the stress driving force actually experienced locally by the propagating crack-tip in complex microstructured materials.

7.6 Concluding Remarks

Synchrotron X-ray diffraction is fast becoming a practical tool for residual stress measurement. Its main advantages derive from the extremely high flux of hard X-rays available, enabling it to map elastic strain fields in 2D in the bulk of materials and components very quickly and at high spatial resolution, and to follow changes in elastic strain over time. However, the technique should be selected only in cases where other techniques fall short because of the need to access scarce beamtime at national and international facilities.

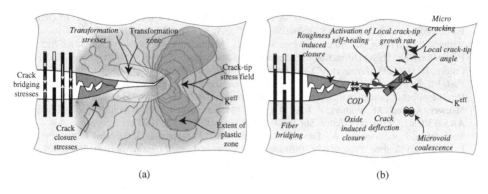

Figure 7.26 Schematic showing the qualitative and quantitative fracture mechanics information provided by (a) diffraction and (b) imaging

Consequently synchrotron experiments require a great deal of planning. Another practical aspect is that the measurement speed also means that large volumes of data are collected in a relatively short time meaning that in many cases data analysis procedures need to be automated.

It should also be noted that many current synchrotron beamlines put a premium on high brilliance (many photons with low divergence); this means that relatively few grains are likely to satisfy the diffraction condition within a sampling volume compared to neutron and laboratory X-ray methods. As a consequence, the very high spatial resolutions that might otherwise be achievable are often compromised to achieve good statistical sampling of sufficient grains. Strategies for offsetting this restriction, such as "wobbling" the sample around the gage volume, have only been partially successful and do add considerable complexity to experiments with bulky samples.

As a result of the high energies usually employed, the scattering angles are characteristically low ($<5°$), which means that the in-plane strain tensor can be mapped quickly and easily. It does however mean that the gage volume is typically extended along the beamline and it can be difficult to obtain measurements in other directions so as to obtain the stress tensor.

In summary, synchrotron diffraction is proving a valuable addition to the residual stress measurement toolbox, amongst other things, providing one of the few methods capable of tracking changes in strain at millisecond timescales (say to follow the effect on stress of phase transformations occurring during weld cooling) and of mapping strain fields in 2D at the 100 μm length scale.

References

[1] Kamitsubo, H. (1998) "SPring-8 Program." *Journal of Synchrotron Radiation* 5(3): 162–167.
[2] Winick, H. (1987) "Synchrotron Radiation." *Scientific American* 257: 88–99.
[3] Withers, P. J. (2004) "Depth Capabilities of Neutron and Synchrotron Diffraction Strain Measurement Instruments: Part II – Practical Implications." *J. Appl. Cryst.* 37: 607–612.
[4] Martins, R. V. and Honkimäki, V. (2003) "Depth Resolved Strain and Phase Mapping of Dissimilar Friction Stir Welds Using High Energy Synchrotron Radiation." *Textures and Microstructures* 35: 145–152.

[5] Xiong, Y. S. and Withers, P. J. (2006) "A Deconvolution Method for the Reconstruction of Underlying Profiles Measured Using Large Sampling Volumes." *J. Appl. Cryst.* 39 (3): 410–424.

[6] Withers, P. J., Preuss, M., Steuwer, A., et al. (2007) "Methods for the Obtaining the Strain-free Lattice Parameter When Using Diffraction to Determine Residual Stress." *J. Appl. Crystallography* 40: 891–904.

[7] Cullity, B. D. (1978) *Elements of X-ray Diffraction*. Addison-Wesley.

[8] Peel, M., Steuwer, A. and Withers, P. J. (2005) "Dissimilar Friction Stir Welds in AA5083-AA6082. Part III: The effect of process parameters on residual stress." *Metal. Mat. Trans* 37A: 2195–2206.

[9] Steuwer, A., Peel, M., Withers, P. J., et al. (2003) "Measurement and Prediction of Residual Stresses in AA5083/6082 Dissimilar Friction Stir Welds." *J. Neutron Res.* 11: 267–272.

[10] Santisteban, J. R., Steuwer, A., Edwards, L., et al. (2002) "Mapping of Unstressed Lattice Parameters Using Pulsed Neutron Transmission Diffraction." *Journal of Applied Crystallography* 35(4): 497–504.

[11] Steuwer, A., Dumont, M., Peel, M., et al. (2007) "The Variation of the Unstrained Lattice Parameter in an AA7010 Aluminium Friction Stir Weld." *Acta Mater.* 55: 4111–4120.

[12] Nye, J. F. (1985) *Physical Properties of Crystals – Their Representation by Tensors and Matrices*. Clarendon: Oxford.

[13] Hutchings, M. T., Withers, P. J., Holden, T. M., et al. (2005) *Introduction to the Characterisation of Residual Stresses by Neutron Diffraction*. CRC Press, Taylor & Francis, London.

[14] Webster(Editor) G. A. (2001) "Polycrystalline Materials – Determinations of Residual Stresses by Neutron Diffraction." Geneva 20, Switzerland, ISO/TTA3 Technology Trends Assessment.

[15] Withers, P. J., Daymond, M. R. and Johnson, M. W. (2001) "The Precision of Diffraction Peak Location." *J. App. Cryst* 34: 737–743.

[16] Fitzpatrick, M. E., Fry, A. T., Holdway, P., et al. (2005) "A National Measurement Good Practice Guide: Determination of Residual Stresses by X-ray Diffraction." *NPL*. London. 52.

[17] Frankel, P. G., Withers, P. J., Preuss, M., et al. (2012) "Residual Stress Fields After FOD Impact on Flat and Aerofoil-shaped Leading Edges." *Mech. Mater.* 55: 130–145.

[18] Dai, H., Francis, J. A., Stone, H. J., et al. (2008) "Characterizing Phase Transformations and Their Effects on Ferritic Weld Residual Stresses with X-rays and Neutrons." *Metal. Mater. Trans. A* 39A: 3070–3078.

[19] Steuwer, A., Rahman, M., Shterenlikht, A., et al. (2010) "The Evolution of Crack-tip Stresses During a Fatigue Overload Event." *Acta. Mater.* 58(11): 4039–4052.

[20] Genzel, C., Stock, C. and Reimers, W. (2004) "Application of Energy-dispersive Diffraction to the Analysis of Multiaxial Residual Stress Fields in the Intermediate Zone Between Surface and Volume." *Materials Science and Engineering a-Structural Materials Properties Microstructure and Processing* 372(1–2): 28–43.

[21] Noyan, I. C. and Cohen, J. B. (1987) *Residual Stress – Measurement by Diffraction and Interpretation*. Springer-Verlag: New York.

[22] Maawad, E., Sano, Y., Wagner, L., et al. (2012) "Investigation of Laser Shock Peening Effects on Residual Stress State and Fatigue Performance of Titanium Alloys." *Materials Science and Engineering A* 536: 82–91.

[23] Somers, M. A. J. and Mittemeijer, E. J. (1990) "Devlopment and Relaxation of Stress in Surface-layers-substrates and Residual-stress Profiles in g'-Fe_4N_{1-x} Layers on a-Fe Substrates." *Metallurgical Transactions a-Physical Metallurgy and Materials Science* 21(1): 189–204.

[24] Ruppersberg, H. and Detemple, I. (1993) "Evaluation of the Stress-field in a Ground Steel Plate from Energy-dispersive X-ray-diffraction Experiments." *Materials Science and Engineering a-Structural Materials Properties Microstructure and Processing* 161(1): 41–44.

[25] Withers, P. J. (2011) "3D Crack-tip Microscopy: Illuminating Micro-Scale Effects on Crack-Tip Behavior." *Adv. Eng. Materials* 13(12): 1096–1100.

[26] Withers, P. J. *Use of Synchrotron X-ray Radiation for Stress Measurement*, in *Analysis of Residual Stress by Diffraction using Neutron and Synchrotron Radiation*, M. E. Fitzpatrick and A. Lodini, Editors. 2003, Taylor & Francis: London. p. 170–189.

[27] Korsunsky, A. M., Wells, K. E., and Withers, P. J. (1998) "Mapping Two Dimensional State of Strain Using Synchrotron X-ray Diffraction." *Scripta Mater.* 39: 1705–12.

[28] Hodeau, J.L., Bordet P., Anne M., Prat A., Fitch A. N., DooryheeE., Vaughan G., and Freund A., *Nine crystal multi-analyser stage for high resolution powder diffraction between 6 and 40 keV*, in *Crystal and Multilayer Optics*, A. T. Macrander, et al., Editors. 1998. p. 353–361.

[29] Wanner, A. and Dunand D. C., (2000) "Synchrotoron x-ray Study of Bulk Lattice Strains in Externally-loaded Cu-Mo Composites." *Metal. & Mater. Trans.* 31A: 2949–2962.

8

Neutron Diffraction

Thomas M. Holden
National Research Council of Canada, Ontario, Canada (Retired)

8.1 Introduction

8.1.1 Measurement Concept

Neutron diffraction joins with X-ray and synchrotron diffraction to form a family of residual stress evaluation methods based on diffraction as a measurement technique for crystalline materials. All three methods involve directing a beam of the chosen radiation on the sample material and measuring the angular distribution of the radiation diffracted from the material. This angular variation is governed by Bragg's Law

$$\lambda = 2d \sin \theta \tag{8.1}$$

where λ is the wavelength of the radiation used, d is the crystal lattice plane spacing, and 2θ is the angle between the incident and diffracted beams corresponding to the maximum diffracted beam intensity. Figure 1.13 in Chapter 1 illustrates these dimensions. For residual stress measurements, the crystal lattice is used as an intrinsic "strain gage," where changes in the lattice spacing "d" indicate the state of strain in the sample. These changes are determined through measurements of the diffraction angle 2θ.

While all three diffraction techniques share the same conceptual basis, they differ dramatically in their capabilities, applications, and practical details. X-rays have wavelengths of around several (Å), 10^{-10} m and can diffract within metals through a depth of a few microns. X-ray synchrotron radiation has wavelengths of around 10^{-11} m, which can penetrate through a depth of a few mm. Typical neutron beams have wavelengths of several Å and they can diffract through a depth of several tens of mm, sometimes hundreds. Of the three measurement techniques, X-rays have the simplest equipment needs, just an X-ray tube, detectors, and associated hardware, but they are limited to

---

Practical Residual Stress Measurement Methods, First Edition. Edited by Gary S. Schajer.
© 2013 John Wiley & Sons, Ltd. Published 2013 by John Wiley & Sons, Ltd.

2D near-surface measurements. In contrast, neutrons can penetrate substantial material depths, and with appropriate specimen positioning hardware can identify 3D stresses. The various neutron and synchrotron X-ray sources are user facilities where a peer-review process determines access.

The present discussion focuses on practical measurements of residual stress using neutron diffraction. Several publications also deal with particular aspects. Krawitz [1] gives a good general introduction to diffraction, and Reimers [2] summarizes many recent practical advances in diffraction and imaging. Noyan and Cohen [3] describe X-ray methods specifically for measuring residual stresses. Hutchings et al. [4] give a comprehensive overview of the neutron method, with the theory of neutron scattering described in more detail by Squires [5]. For practical measurements a standard test method for determining residual stresses by neutron diffraction has been developed by ISO/TS 21432 [6].

8.1.2 Neutron Technique

A nuclear reactor produces high-energy neutrons by fission of U-235 nuclei after neutron capture. In a spallation source neutrons are produced following the break up of target nuclei by high-energy protons. The high-energy neutrons are brought into the useful thermal range by collisions near ambient temperature with hydrogenous material, termed the moderator. The wavelength and direction of travel of the neutrons emerging from the moderator are selected either by a single crystal or by time-of-flight through a known distance. The beams of defined wavelength impinge on the sample set up on the instrument and are scattered by it. Diffraction occurs, whereby the neutron wavelets scattered by the nuclei of the sample add up in phase to give sharp peaks at defined angles. The sharp peaks permit strain measurements through the use of the Bragg relationship, Equation (8.1). Both the incident neutron beam falling on the sample and the diffracted beam may be defined in height and width by slits in masks in the beams as indicated schematically in Figure 8.1. The overlap of the incident and diffracted beams is known as the gage volume and is fixed in space at the center of rotation of the instrument. A computer-controlled table capable of motion in three orthogonal directions plus rotation can bring into the gage volume any sample location and direction. The lattice spacing measured in the test is the average over the gage volume, and may be mapped out comprehensively.

The basis of the neutron diffraction technique of stress measurement is a medium-resolution determination of diffraction peak position and hence lattice spacing as a function of location in the measured component. The advantage of using neutrons comes from their feature that they are scattered weakly by atomic nuclei so that the depth penetration of the neutron beam is high for most engineering materials.

8.1.3 Neutron Diffraction

The Bragg relationship in Equation (8.1) can be generalized to apply to multiple different crystal planes

$$\lambda = 2d_{hkl} \sin \theta_{hkl} \tag{8.2}$$

where the subscripts $\{hkl\}$ are the Miller indices that specify the planes of the atomic lattice. For example, $\{002\}$ indicates the cube edge planes of the cubic lattice and $\{111\}$ the

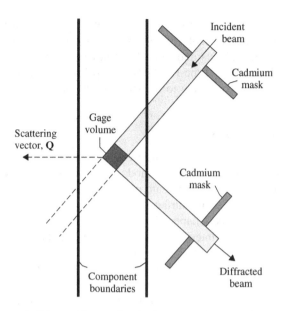

Figure 8.1 Incident and diffracted beams impinging on a plate sample. The gage volume is the intersection of the incident and diffracted beams defined by the apertures of the cadmium masks. The scattering vector **Q** (direction of measurement) bisects the incident and diffracted beams

cube diagonal planes. The corresponding lattice plane spacing is d_{hkl} and the angular position of the associated diffraction peak is $2\theta_{hkl}$. Section 8.2.1 describes the modifications to Equation (8.2) for time-of-flight diffraction.

Equation (8.2) requires that the normal $<hkl>$ to the diffracting planes lies along the bisector of the angle between the incident and diffracted beams. This is called the scattering vector for the diffracting grains. Thus the lattice "strain gage" is also directional. Grains whose plane normals $<hkl>$ do not lie along the scattering vector do not contribute to the diffracted signal. The strain from a single measurement is deduced from only a subset of the grains in the sample, namely, those with an $<hkl>$ normal along the scattering vector. This is unlike mechanical methods, which sample the behavior of all the grains in the sample irrespective of their crystallographic orientation. This selectivity reveals details of the anisotropy of strain and stress at the grain length scale.

Stresses and strains within a component exist on three length scales. The macroscopic or Type-I stress, of primary interest to the engineer, has the length scale of the part. For a butt weld, a tensile stress is expected within and nearby the weld, with a balancing compressive stress remote from the weld. The measurements determine the size and direction of the macroscopic stress field in a location selected by the experiment. The macroscopic stress in a particular sample direction is the average stress for all the grains, numbering perhaps thousands, in a small region of a few mm^3 in volume. But the stresses in the grains of differing crystallographic orientations $<hkl>$ are not equal. They change from grain to grain because the elastic and plastic properties vary with crystallographic direction. The deviations from the average for a particular $<hkl>$, say $<111>$ or $<002>$, are termed intergranular or Type-II stresses. Their scale is on the order of the grain size. The sum of the intergranular stresses over all the grains of various <hkl> in the small region is zero. Finally, there are stresses that vary from point to point within the grains due to dislocations, local alloying effects and grain boundaries. These are termed intragranular or Type-III stresses. Type-II and Type-III stresses contribute to diffraction line-widths and Type-I and Type-II stresses contribute to the shifts in the positions of diffraction lines.

8.1.4 3-Dimensional Stresses

Force equilibrium dictates that the normal stress at a surface is zero, thus the surface stresses are two-dimensional. For X-ray methods, which make use of conventional X-ray tubes with wavelengths around 2Å, the beam penetration is only a few μm. The condition that the normal stress is zero at the surface provides a reference point and becomes the basis of the $\sin^2\Psi$ method (see Chapter 6). However, there is no requirement that any stress component is zero in the interior. Thus, a three-dimensional state of stress must be considered. This is where penetrating beams of neutrons or high energy X-rays are useful because they allow measurements to be made in several different sample orientations. These provide measurements of the lattice spacings d_{hkl}. Thus, a separate sample free from macroscopic stress is required to provide a zero datum for evaluating the strain associated with the macroscopic stress. Small cubes cut from the material by an appropriate method can be used as a macroscopic stress-free reference. However, such samples may still retain the Type-II stresses and strains because of the grain scale and lattice parameter changes due to chemistry.

8.1.5 Neutron Path Length

Neutron diffraction measurements can be carried out on most industrial materials, the main requirement is that they are crystalline. The most fundamental limitation is the total path length of the neutron beam through the material of the sample. For materials like steels that display strong diffraction, the depth of penetration is limited by depletion of the beam by diffraction. The maximum path lengths in steels are in the order of 50–60 mm. Some metals, like vanadium, have practically no diffraction peaks so neutron stress measurements are not possible. Titanium alloys have low coherent scattering and also absorb neutrons and show appreciable "incoherent scattering" which contributes to the background rather than the peak thus worsening the peak-to-background ratio. These are therefore difficult to measure. Aluminum alloys also have low scattering but have no competing absorption or incoherent scattering. In this case, while the diffraction signal is never strong, usable path lengths of 250 mm are possible. In general, neutron diffractometers are large and robust with sample handling capabilities up to 1500 kg. It is usual to be able to test segments of line-pipe or large aircraft wing-stiffeners. At the other end of the sample size scale, a few facilities have the means to define beams as small as 0.2 mm in width so these represent the upper and lower limits of the capabilities. Table 8.3 lists the path lengths for 95% attenuation of neutron beams within various metals.

8.2 Formulation

8.2.1 Determination of the Elastic Strains from the Lattice Spacings

When a specimen is illuminated by a monochromatic beam of neutrons of known wavelength, the lattice spacing d_{hkl} may be determined from the observed Bragg angle using Equation (8.2). The lattice spacings corresponding to the macroscopic stress- (and strain-) free values for the material are denoted by $d_{0,hkl}$. The Bragg angle for the stress free material is denoted $\theta_{0,hkl}$. The elastic strains are given by:

$$\varepsilon_{hkl} = \frac{d_{hkl} - d_{0,hkl}}{d_{0,hkl}} = \frac{\sin\theta_{0,hkl}}{\sin\theta_{hkl}} - 1 \tag{8.3}$$

In a time-of flight instrument, short pulses of neutrons comprising a range of wavelengths illuminate the sample. From the measured time-of-flight from the source to the counter, t_{hkl}, the wavelength of the diffracted neutrons is calculated from the de Broglie relationship:

$$\lambda_{hkl} = \frac{h_p t_{hkl}}{m_n (L_0 + L_1)} \tag{8.4}$$

where m_n is the neutron mass, L_0 is the path length from the moderator to the sample, L_1 is the path length from the sample to the counter and h_p is Planck's constant. Substituting this into the Bragg equation gives

$$d_{hkl} = \frac{h_p t_{hkl}}{2m_n (L_0 + L_1)\sin\theta} \tag{8.5}$$

for a detector positioned at a scattering angle of 2θ. With a similar time-of-flight measurement of $d_{0,hkl}$ the determination of strain follows from Equation (8.3).

8.2.2 Relationship between the Measured Macroscopic Strain in a given Direction and the Elements of the Strain Tensor

It is customary [3] to display the relationship between measured strain and the elements of the strain tensor $\varepsilon_{i,j}$ by:

$$\frac{\Delta d}{d} = \sin^2\Psi\cos^2\Phi\,\varepsilon_{11} + \sin^2\Psi\,\sin^2\Phi\,\varepsilon_{22} + \cos^2\Psi\,\varepsilon_{33}$$
$$+ 2\varepsilon_{12}\sin^2\Psi\,\sin\Phi\cos\Phi + 2\varepsilon_{23}\sin\Psi\cos\Psi\sin\Phi + 2\varepsilon_{31}\sin\Psi\cos\Psi\cos\Phi \tag{8.6}$$

The sample coordinate system (X_1, X_2, X_3) is displayed in Figure 8.2 and is thought of as being attached to the sample, which in this example case represents a weld in a plate. ε_{11} is the element of the strain tensor along axis 1 and ε_{12} is the off-diagonal element in the (1,2) plane. Ψ is the angle between the scattering vector and the X_3 axis and Φ is the angle of the projection of the scattering vector onto the $(X_1 - X_2)$ plane and the X_1 axis.

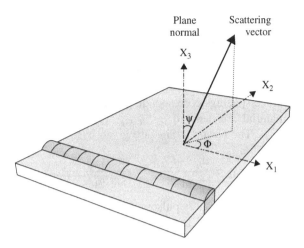

Figure 8.2 Plate sample showing Cartesian axes, (X_1, X_2, X_3) superposed, the scattering vector, and the conventional labeling of the angular deviation from the normal direction, X_3, Ψ, and the angular deviation from the X_1 axis, Φ

If all the elements of the strain tensor are required then at least six measurements of strain in independent directions are needed. If the chosen axes are known to be the principal axes, only the three principal strains are needed. In any case, three orthogonal stresses may always be obtained from three orthogonal strains. For laboratory X-rays, where the penetration is a few μm, measurements are made at a number of angles Ψ with respect to the normal direction and the values of ε_{11}, ε_{12} etc. are deduced by extrapolation to $\Psi = 90°$. Because of the high penetration of neutrons and high-energy X-rays, the strain in a given sample direction can be measured directly in transmission. Thus, the sample is oriented so that the desired directions are in turn aligned along the bisector of the incident and diffracted beams.

The relationship between strain and the elements of the strain tensor may also be written compactly as:

$$\frac{\Delta d}{d}(\theta_1, \theta_2, \theta_3) = \sum_{i,j} l_i l_j \varepsilon_{i,j} \quad (8.7)$$

where $\frac{\Delta d}{d}(\theta_1, \theta_2, \theta_3)$ is the measured strain in the direction making angles of $\theta_1, \theta_2, \theta_3$ with three orthogonal axes 1,2,3 chosen to coincide with the expected principal axes of the sample. l_1 represents the direction cosine $\cos\theta_1$ and so on. For a rolled plate, these axes likely align with the rolling direction, the transverse direction and the normal direction. For a cylinder they might be the radial, hoop, and axial directions.

8.2.3 Relationship between the Stress $\sigma_{i,j}$ and Strain $\varepsilon_{i,j}$ Tensors

In three dimensions, the stress/strain relationship is given by:

$$\sigma_{11} = \frac{E_{hkl}}{(1+\nu_{hkl})(1-2\nu_{hkl})}((1-\nu_{hkl})\varepsilon_{11} + \nu_{hkl}\varepsilon_{22} + \nu_{hkl}\varepsilon_{33}) \quad (8.8)$$

Table 8.1 Diffraction elastic constant, E_{hkl} (GPa) calculated from the Kröner model [7]

{hkl}	200	311	420	531	220	422	331	111
Al	67.6	70.2	70.3	71.2	71.9	71.9	72.3	73.4
Cu	101.1	122.0	122.5	131.5	139.1	139.1	144.3	159.0
Ni	160.0	185.0	185.6	195.6	203.9	203.9	209.5	224.6
304L	152	184	185	199	211	211	219	242

Table 8.2 Diffraction elastic constant, v_{hkl}, calculated from the Kröner model [7]

{hkl}	200	311	420	531	220	422	331	111
Al	0.35	0.35	0.35	0.35	0.34	0.34	0.34	0.34
Cu	0.38	0.35	0.35	0.34	0.33	0.33	0.32	0.31
Ni	0.36	0.33	0.33	0.33	0.33	0.33	0.31	0.30
304L	0.33	0.294	.293	.278	0.265	0.265	0.256	0.23

and

$$\sigma_{12} = \frac{E_{hkl}}{(1 + v_{hkl})} \varepsilon_{12} \quad (8.9)$$

for σ_{11} and σ_{12}. Similar expressions for the other components apply with appropriate permutation of the subscripts. These relations are a generalized form of Hooke's Law. However, instead of the macroscopic elastic constants E and v, which are average values of Young's modulus and Poisson's ratio over all grain orientations in the sample, the constants of proportionality E_{hkl} and v_{hkl} relate elastic strains in the grains that have an $<hkl>$ direction along the scattering vector to the macroscopic stress at that location. Known as diffraction elastic constants, they are empirical constants measured in calibration experiments where the {hkl} strains are measured with known applied stresses. They may also be calculated from the single crystal elastic constants by the Kröner method [7] if there is no crystallographic texture. If the sample does exhibit significant crystallographic texture, the elasto-plastic self-consistent model [8,9] can be used instead.

Tables 8.1 and 8.2 gives representative values [4] of the diffraction elastic constants for various materials calculated using the Kröner model. The presence of strong texture will modify the constants and then it is appropriate to measure them on the material of interest.

8.3 Neutron Diffraction

8.3.1 Properties of the Neutron

The nuclear attractive interaction between protons and neutrons and between neutrons, as opposed to the coulomb repulsion between protons is responsible for the stability of the elements up to the Actinides. The neutron is charge-neutral [5] with a mass

$m_n = 1.675 \times 10^{-27}$ kg. It exhibits wave-particle duality; it behaves like a wave in two-slit interference and has spin $1/2$ and therefore possesses a magnetic moment. Thermal neutrons have a wavelength in order of the spacing of atoms and is then well suited to interference effects from the crystal lattice leading to diffraction. They have an equivalent temperature around 300 K that is of the same magnitude as the energies of vibration of the crystal lattice. The momentum of a thermal neutron of velocity, v, is related to the wavelength λ by:

$$m_n v = \frac{h_P}{\lambda} \qquad (8.10)$$

If the velocity is expressed in m/sec and wavelength in Å, then $\lambda = 3956.03/v$. The kinetic energy, E, of a thermal neutron is given by:

$$E = \frac{m_n}{2} v^2 = \frac{h_P^2}{2\lambda^2 m_n} \qquad (8.11)$$

If E is expressed in meV, or as an effective temperature T in degrees K, and the wavelength is in Å, then $E = 81.8896/\lambda^2$ and $T = 949.3/\lambda^2$. A neutron with wavelength 1.8Å has an effective temperature of 293 K and a velocity of 2198 m/sec.

8.3.2 The Strength of the Diffracted Intensity

It is important to be able to calculate the expected intensity in a diffraction experiment. Neutrons are primarily scattered by the interaction with the nuclei of atoms, as opposed to X-rays, which are scattered by the electrons. Because the nucleus is small relative to the electron cloud, the momentum dependence of the scattering, the form factor, is flat. The number of neutrons, I, scattered per second into all directions (4π) from the sample may be written [5]:

$$I = \sigma \Phi_0. \qquad (8.12)$$

where the incident flux Φ_0 is the number of neutrons incident on the sample per cm^2 per second. The cross section for scattering, σ, has dimensions of area, $[L]^2$, and order of magnitude 10^{-24} cm^2. This unit, the barn (bn), is so called because it was easy to measure in the 1940s and was "as big as a barn door"!

The interference between neutron wavelets scattered from nuclei situated on a periodic lattice is the origin of the diffraction peaks. For a polycrystalline sample with a random orientation of grains, the macroscopic coherent cross section per unit volume, Σ_{coh} for scattering into the diffraction peaks is given by the sum over all the allowed reflections

$$\sum_{coh} = \frac{\lambda^3}{4V_0^2} \sum_h \frac{M_h F_h^2}{\sin \theta_h} \qquad (8.13)$$

where V_0 is the volume of the unit cell of the periodic lattice, λ is the wavelength of the neutron beam, θ_h is the Bragg angle and h represents the Miller indices $\{hkl\}$. M_h is the multiplicity of the reflection, tabulated in [4]. For example, there are eight peaks of the $\{111\}$ type, corresponding to the combinations of ± 1. The structure factor of the

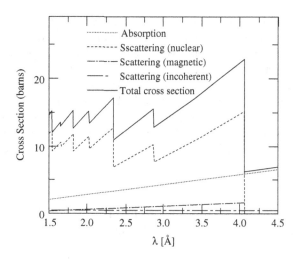

Figure 8.3 Total cross section (bn/atom) for body-centered cubic iron as a function of wavelength, showing the discontinuities in the coherent cross section at the Bragg edges, the absorption and the incoherent cross sections and the weak coherent magnetic diffraction. Reproduced with permission from Hsu, T. C., Marsiglio, F., Root, J. H., Holden, T. M. (1995) "Effects of multiple scattering and wavelength-dependent attenuation on strain measurements by neutron scattering." *Journal of Neutron Research* 3(1):27–39

reflection, F_h expresses the interference between sites within one unit cell and is given by the sum:

$$F_h = \sum_i b_i \exp(2\pi i (hx_i + ky_i + lz_i)) \tag{8.14}$$

where x_i are the fractional positions of the nuclei within one unit cell, and b_i is the coherent scattering length of the nucleus at site i. b_i is an experimentally known number for the element, alloy or compound. In a random polycrystal there are reflecting planes at all angles to the incident beam so the diffracted neutrons fall on a series of cones, the Debye-Scherrer cones, of semi angle $2\theta_h$. In Equation (8.13) the intensity in all the Debye-Scherrer cones is added for a given wavelength and for angles between 0 and 180°, as permitted by Bragg's law. However, $\lambda = 2d_{hkl} \sin \theta_{hkl}$ cannot be satisfied for $2\theta_{hkl} > 180°$. At a wavelength beyond $\lambda = 2d_{hkl}$ no further coherent scattering from the $\{hkl\}$ planes is possible and there is a rapid drop in intensity known as the Bragg edge. Figure 8.3 shows the coherent cross section for body-centered cubic iron and displays the Bragg edges and the λ^3 variation of the cross section.

Use can be made of the Bragg edges to obtain a through thickness average of the lattice spacing [10] and, with the advent of counters with very high (10 μm) spatial resolution, there is the possibility that Bragg edge tomography may be used to map strains in a part [11,12]. Reference [4] discusses these novel techniques.

8.3.3 Cross Sections for the Elements

There are three cross sections (barns/atom) competing to remove neutrons from the beam, namely coherent scattering, σ_{coh}, incoherent scattering, σ_{inc}, and absorption, σ_A. These, together with b_{coh}, have been measured for all the elements and their isotopes and are tabulated in [13]. The coherent cross section per atom is related to b_{coh} by $\sigma_{coh} = 4\pi b_{coh}^2$. The incoherent scattering arises from two sources. Firstly from the random distribution of isotopes on the crystal lattice, each isotope having a distinct value of b, and secondly from the interaction of the neutron spin with nuclear spins on the isotopes, if these are non-zero. The tabulated values of σ_{inc} include both these terms.

Nuclei can absorb neutrons, changing the isotope and creating an excited state that then decays often by γ-ray emission. The absorption cross section is wavelength dependent. It usually tabulated for 1.8Å and the cross section for any other wavelength is given by

$$\sigma_A(\lambda) = \frac{\sigma_A(1.8) \times \lambda}{1.8} \tag{8.15}$$

While very strong neutron absorbers such as Cd, B or Gd may prohibit neutron measurements, they do provide good shielding for instruments against unwanted background radiation.

The neutron magnetic moment also interacts in a dipole fashion with the unpaired magnetic moment of unpaired electrons in 3D elements such as Fe, Ni, and Co. The ferromagnetic cross section adds to the nuclear cross section for unpolarized neutrons and generally gives a small contribution in order of a few % to the low index $\{hkl\}$ planes. Figure 8.3 shows an example case for Fe.

8.3.4 Alloys

When elements are mixed to form an alloy, the result is a random distribution of nuclei of different elements on the lattice. The coherent scattering length for the alloy is the weighted sum of the constituent elements:

$$<b>_{alloy} = \sum_j c_j b_{coh,j} \tag{8.16}$$

where j labels the different elements in the alloy, $b_{coh,j}$ are the values of the coherent scattering lengths of the elements and c_j are the fractional atomic concentrations of the elements in the alloy. However, adding elements in random positions on the crystal lattice also adds an extra term to the incoherent scattering:

$$\sigma'_{inc} = 4\pi \left(<b^2>_{alloy} - <b>^2_{alloy} \right) \tag{8.17}$$

where $<b^2>_{alloy} = \sum_j c_j b_{coh,j}^2$. The total incoherent scattering for the alloy is the sum of two terms, the weighted sum of the incoherent scattering of the elements plus the additional term, Equation (8.17) as follows:

$$\sigma_{inc,alloy} = \sigma'_{inc} + \sum_j c_j \sigma_{inc,j} \tag{8.18}$$

The incoherent scattering impairs measurements because it adds an intrinsic background under the diffraction peaks as well as attenuating the incident and diffracted beams. For the common industrial alloy Ti6Al4V, the average coherent cross section is only 0.83 bn, the incoherent cross section is 3.21 bn and the absorption is 5.46 bn. In Ni-rich alloys, such as the special welding alloy In182/82, the coherent, incoherent, and absorption cross sections are 8.09, 4.73, and 5.98 bn respectively. These values make tests in In182/82 welds with long path lengths very difficult.

Finally the absorption cross section for an alloy is given by

$$\sigma_{A,alloy} = \sum_j c_j \sigma_{A,j} \qquad (8.19)$$

8.3.5 Differences with Respect to X-rays

The coherent cross section for X-rays is proportional to the square of the atomic number, Z, so light elements therefore contribute less to the alloy scattering than heavy elements. The X-ray analog of incoherent scattering is Thompson scattering from the individual electrons. Since the Thompson scattering is proportional to $Z^2\lambda^4$, laboratory X-rays are strongly depleted and this is the reason for the low penetration of laboratory X-rays into materials. Only by using very high X-ray energies, and hence short wavelengths, at synchrotron sources can the Thompson scattering be diminished to permit measurements at depth.

8.3.6 Calculation of Transmission

The transmission of neutrons through matter follows the exponential law:

$$I(t) = I(0)\, e^{-\Sigma t} \qquad (8.20)$$

where Σ, with units $[L]^{-1}$, is the macroscopic cross section, the cross section per unit volume, and t is the total path length through the material. Equation (8.20) allows a calculation of the feasibility of an experiment when there are long path lengths through the sample. All nuclear cross sections that deplete the beam, diffraction, incoherent scattering, and absorption contribute to Σ as:

$$\Sigma = \frac{N_{av}\rho}{AW}\left(\sigma_{coh} + \frac{\sigma_A \lambda}{1.8} + \sigma_{inc}\right) \qquad (8.21)$$

where $\frac{N_{av}\rho}{AW}$ is the number of atoms per unit volume, N_{av} is Avogadro's number, ρ is the density and AW is the atomic weight. Table 8.3 lists for several elements the macroscopic cross sections and the lengths, $l_{0.95}$, over which the incident beam is attenuated to 5% of its initial value.

The approximation to the coherent scattering per atom $\sigma_{coh} = 4\pi <b_{coh}>^2$ is often sufficient, but if the full wavelength dependence of the coherent scattering is required, including the effect of the Bragg edges, then the full expression in Equation (8.13) has to be used to give the coherent cross section per atom. In this case, the pre-factor before the summation is replaced by $\frac{\lambda^3}{4n_0 V_0}$, where n_0 is the number of atoms, or formulae units, in the unit cell.

Table 8.3 Macroscopic cross section and 95% attenuation length for a neutron beam penetrating within various metals

Element	Σ (cm^{-1})	$l_{0.95}$ (mm)
Al	0.104	287
Mg	0.163	185
Ti	0.597	50
Fe	1.204	25
Ni	2.099	14
Cu	1.003	30
Zr	0.285	105
W	1.447	21
Gd	1509	0.02
U	0.790	38

8.4 Neutron Diffractometers

8.4.1 Elements of an Engineering Diffractometer

For a continuous source at a reactor, a core flux of more than 10^{14} ncm^{-2} sec^{-1} is required and for a pulsed source the average flux of all wavelengths should be at least 5×10^{12} ncm^{-2} sec^{-1}. In monochromatic beam diffraction, a narrow band of wavelengths is selected from the Maxwell-Boltzmann thermal distribution of neutrons. In time-of-flight diffraction, a distribution of neutron wavelengths falls on the sample but the pulses have sharp edges to give good time and wavelength resolution. In both cases the incident beam falling on the sample is restricted in width and height by a slit in a mask or a radial collimator. The diffracted beam is similarly defined by a slit or radial collimator. The region of overlap between the incident and diffracted beams, the gage volume, is fixed in space and is arranged to be located over the reference point of the diffractometer, which is the center of rotation of the sample table. This is the region from which neutrons are diffracted out of the incident beam into the counter and over which the lattice spacings are averaged. The gage volume has its most compact section at a scattering angle of 90° and becomes an elongated diamond above and below 90°. The departure of the gage volume from its ideal square section at 90° depends on the angular divergences of the incident and diffracted beams and the distances of the beam defining slits or radial collimators from the gage volume. A computer controlled sample table with degrees of freedom in three orthogonal directions (X, Y, Z) and 360° of angular range, R, is required so as to be able to move any point in the sample into the gage volume. Ideally, the table should support 1500 kg with a setting precision smaller than 0.1 mm. Aids such as telescopes, theodolites or laser-trackers are needed to set up samples with this precision. Finally, well-shielded counters are required to measure the diffracted neutrons.

8.4.2 Monochromatic Beam Diffraction

Figure 8.4 shows the essential features of a monochromatic beam diffractometer. The neutron beam passes from the reactor to the diffractometer through a beam tube or a

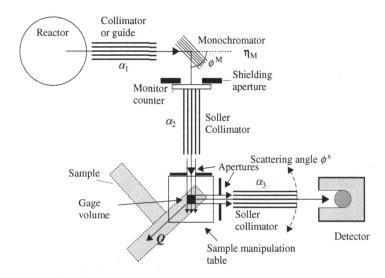

Figure 8.4 Monochromatic beam diffractometer showing diffraction from the monochromator to the right and diffraction from the sample to the left resulting in focusing Reproduced with permission from [4], Copyright 2005 Taylor and Francis

thermal guide. Guides [14], which use total internal reflection of the neutrons from the surface of the guide, avoid the inverse square loss in the beam intensity and allow beams to be brought many tens of metres from the reactor face to low background regions of a beam hall.

The single crystal monochromator is ideally made from materials with a tetrahedral atomic structure such Si or Ge. For this structure, reflections such as (226) are forbidden, so a neutron beam reflected from the (113) planes with wavelength λ is not contaminated with neutrons of wavelength $\lambda/2$. To optimize the intensity, the monochromator must exhibit a small angular or mosaic spread (of 0.2°) of the atomic planes. This can be achieved by plastically deforming the crystal or by bending a perfect crystal wafer elastically. The most widely available monochromators [15] are of the bent focusing type, which are bent in the horizontal plane to focus the beam at the sample position. The monochromator may be formed of several bent strips above and below the horizontal plane, which are tilted to direct the beam into the horizontal plane at the sample position. The resulting horizontal and vertical divergences have an impact on determining the actual shape of the gage volume and this requires careful design of the slits or radial focusing collimators [16] and their positions.

Figure 8.4 shows the diffraction at the monochromator to the right and the diffraction at the sample position to the left. This left-right asymmetry, known as the focusing condition, minimizes the instrumental diffraction line-width but also gives a variation of the line-width as a function of diffraction angle. The width variation is quite strong for bent focusing monochromators and may restrict the range of diffraction angles available for precise experiments before the instrumental line-width becomes too broad. Figure 8.5(a) shows the variation of line-width with diffraction angle for the KOWARI [17] instrument.

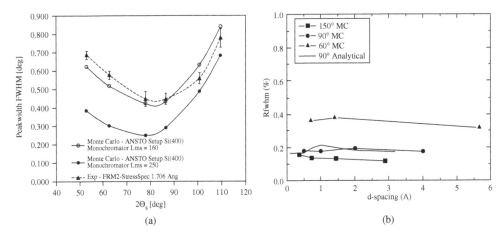

Figure 8.5 (a) The full-width at half maximum as a function of scattering angle for the KOWARI instrument showing the focusing minimum at 85°. Reproduced with permission from [17]. (b) The calculated full-width at half height as a function of lattice spacing for the VULCAN time-of-flight diffractometer at Oak Ridge National Laboratory showing relatively minor changes in width. Reproduced with permission from X.-L. Wang, T. M. Holden, G. Q. Rennich, A. D. Stoica, P. K. Liaw, H. Choo, C. R. Hubbard, "VULCAN – The engineering diffractometer at the SNS," *Physica B: Condensed Matter*, 385–386

Higher vertical and horizontal divergences, both before and after the sample, generally lead to higher count rates. If a slit is used in the diffracted beam it must be positioned close to the gage volume (say closer than 30 mm with a sample to counter distance of 1200 mm) so that the gage volume is well defined. A radial collimator in the diffracted beam may be positioned further from the reference position without spreading out the gage volume allowing more space for bulky samples. The counter is usually a multi-wire or area detector so that the whole diffraction peak can be collected at one angular setting of the diffractometer. Since the measurement is essentially an angle measurement, the relative positions of the wire elements have to be calibrated carefully. Since the counter may extend ±10 cm out of the horizontal plane, it is necessary to correct for the curvature of the Debye-Scherrer cones [18]. In North America there are monochromatic beam diffractometers at Oak Ridge National Laboratory, at the National Institute for Standards and Technology in Gaithersburg, at the University of Missouri and at Chalk River in Canada. There are also engineering diffractometers at Saclay and Grenoble in France, Berlin and Munich in Germany, at Tokai in Japan and Sydney in Australia.

To give some idea of the intensities involved, the flux of neutrons of all wavelengths at the end of a 6 m beam tube at a medium flux reactor is about 8×10^8 ncm^{-1}s^{-1}. A Ge or Si monochromator selects a narrow wavelength range that is about 1% of the Maxwell-Boltzmann distribution giving a flux of monochromatic neutrons of about 10^7 ncm^{-1}s^{-1} at the sample. The counter intercepts about 3% of the intensity in one Debye-Scherrer cone. The number of neutrons collected in the counter varies from about 1 per second for an easy experiment to about 1 per minute for a hard one. About 200 counts are sufficient to define a diffraction peak for stress measurements.

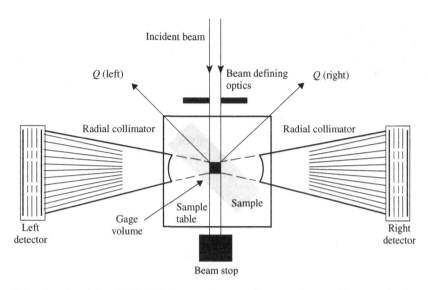

Figure 8.6 Sketch of the ENGIN-X time-of-flight instrument [20] showing the incident beam defined by a slit and diffraction to the right and to the left. The diffracted beams are defined by a pair of radial collimators. Reproduced with permission from [2], Copyright 2008 Wiley-VCH

8.4.3 Time-of-flight Diffractometers

The neutron source for time-of-flight diffraction is usually a short-pulsed spallation source, such as ISIS at the Rutherford Laboratory in the UK, LANSCE at Los Alamos National Laboratory, USA, the Spallation Neutron Source (SNS) at Oak Ridge National Laboratory, USA, or the J-PARC source in Tokai, Japan. The high-energy neutrons produced in the target are brought into the thermal range in a hydrogenous moderator, such as water or liquid methane. The pulse of thermal neutrons, about 30 μsec long, comprising a continuum of wavelengths then passes down a thermal guide to the diffractometer. Neutron choppers in the beam further shape the pulse and create an adjustable bandwidth, $\Delta\lambda$, typically about 1.3 Å wide. The incident beam is defined in height and width by adjustable slits and passes over the reference point of the instrument. A guide length of about 45 m gives the best compromise [19] between resolution and intensity. Figure 8.6 shows a typical time-of-flight diffractometer [20].

The divergence of the diffracted beam is usually restricted by a radial collimator chosen to define the gage volume that best matches the experimental requirements. The counter bank is much larger in height and width than for a monochromatic instrument and may extend ±15° about the center at ±90° and ±20° about the horizontal plane. The data in the individual counters in the bank are usually scaled and summed so as to be equivalent to single counters at ±90°.

Since there is no monochromator, there is little focusing in time-of-flight diffraction and two counter banks can be used at ±90°. The average variation of the instrumental width in time for the counter bank is also nearly wavelength independent over the bandwidth, $\Delta\lambda$ as Figure 8.5(b) shows for the VULCAN diffractometer at the SNS. The essential measurement in time-of-flight diffraction is an arrival time measurement in a fixed counter as opposed to an angle measurement. All grains whose crystallographic orientations <hkl> lie along the bisector of the incident and diffracted beams diffract from the continuum of wavelengths in the pulse and are registered as peaks in the time arrival spectrum.

8.5 Setting up an Experiment

The ISO/TS 21432 technical specification [6] gives detailed descriptions of the standard test method for determining residual stresses. Appendix 3 of [4] gives similar material.

8.5.1 Choosing the Beam-defining Slits or Radial Collimators

The slit size determines the size of the gage volume and hence the spatial resolution of the test. For both monochromatic and time-of-flight diffraction these also determine the angular resolution of the instrument. They are adjusted so that the incident and diffracted beams intersect at the reference point. Ideally, they are set up on kinematic mounts so that the position is reliably reproducible.

8.5.2 Calibration of the Wavelength and Effective Zero of the Angle Scale, $2\theta_0$

For the purposes of calibration, Bragg's Law can be written as follows

$$\lambda = 2d_{hkl} \sin(\theta_{hkl} - \theta_0) \tag{8.22}$$

where λ is the incident wavelength, $2\theta_0$ the zero of the angle scale, $2\theta_{hkl}$ are the scale settings for the reflections and d_{hkl} are the known lattice spacings of the standard. For the arrangement of slits or collimators selected, λ and $2\theta_0$ can be determined with high accuracy, ±0.0001Å for the wavelength and ±0.005° for the scale zero. This is done by measuring several diffraction peaks from standard powder samples such as Si, Ni, CaF_2 or CeO_2.

8.5.3 Calibration of a Time-of-flight Diffractometer

Combining Bragg's Law with the de Broglie relation gives:

$$(t_{hkl} - t_0) = \frac{2d_{hkl} m_n (L_0 + L_1) \sin\theta}{h_P} \tag{8.23}$$

where 2θ is the counter angle, L_0 and L_1 are the distances from the sample to the moderator and the counter, t_0 is the time origin of the neutron pulse in the moderator, t_{hkl} is the time of arrival of the {hkl} reflection and d_{hkl} are the plane spacings of the standard powder. By fitting to ten or more reflections from the standard, the combination $(L_0 + L_1) \sin\theta$ and t_0 may be determined. Using these calibration constants, the lattice spacing of the sample under study may be computed. For a counter bank made up of arrays of counters the relative efficiency of each counter must be calibrated with a vanadium sample that scatters isotropically. Then the scattered intensity falling on each counter should be identical and any differences are due to the relative efficiencies of the counters.

8.5.4 Positioning the Sample on the Table

The sample under study is usually attached to a sturdy base-plate fixture, which is in turn attached to the sample table. Typically, the X, Y, and Z axes of the sample are set up carefully to be parallel to the X, Y, and Z axes of the table. Deviations from parallel alignment may cause the gage volume to emerge from the sample and so generate a systematic error. These checks can be carried out with traveling telescopes, theodolites or laser trackers. The most elegant, and now widely adopted, solution for positioning the sample uses the SSCANS routine [21]. Fiducial spheres are attached to the sample and fixture, which are mapped off-line with a coordinate measuring machine to create a CAD-CAM image. With the sample attached to the table, the fiducial points are moved in turn to the reference point, the center of rotation of the sample table to establish their sample table coordinates. This establishes the sample table coordinates of every point in the sample. The CAD-CAM image is used to select the points where measurements are required in the sample and constructing the scan set. The SSCANS routine collates and records and keeps track of the experimental measurements made in the different sample orientations. In the absence of this sophisticated tool, fiducial marks are made on the sample, which are then measured carefully and a similar routine followed. The sample position may also be established using the neutron beam as an alternative to optics. This is done by measuring the characteristic variation of the intensity as a surface passes through the gage volume, providing what is known as an "entering curve" [4].

8.5.5 Measuring Reference Samples

The reference samples contain near zero macroscopic stress and may have the form of thin plates, combs or small cubes cut with the expectation that the stress normal to the surface in a thin section is zero. These are aligned on the sample table by the methods of the previous section. The reference samples must be sufficiently large so that the gage volume lies totally within them and they must be examined with exactly the same calibrated set-up, wavelength, collimators and slits as the intact sample and the standard powder calibration. In general it is necessary to ensure that the direction of measurement of the reference is the same as the direction of the intact sample because the lattice spacings can vary with sample direction due to intergranular effects.

8.6 Analysis of Data

8.6.1 Monochromatic Beam Diffraction

The basic data have the form of counts as a function of diffractometer setting angle. The peak is superposed on a flat or sloping background caused by intrinsic sample incoherent scattering, background neutrons near the reactor and neutrons generated by cosmic rays that penetrate the diffractometer shielding. The peak position, width and integrated intensity may be obtained by fitting the count data to a Gaussian line shape with a flat or sloping background. The precision of the fit depends on the number of points in the peak, the intensity, the instrumental and intrinsic widths and the setting accuracy of the angle scale. This fitting precision is often used as a preliminary measure of the accuracy and it may be as small as $\pm 0.005°$ in a peak around $90°$. A check on this preliminary value may be obtained by making several measurements at the same sample location. However, the precision may be considerably less than the true accuracy because of a number of systematic errors that are not revealed in a single peak measurement. With the aid of the calibration parameters, the corresponding lattice spacing, d_{hkl} and its uncertainty, $\pm \Delta d_{hkl}$, usually taken as the standard error, may be obtained from Equation (8.22). The reference lattice spacing $d_{0,hkl} \pm \Delta d_{0,hkl}$ may be obtained similarly.

The strain $\varepsilon_{L,hkl}$ and the uncertainty $\pm \Delta \varepsilon_{L,hkl}$ for example, the longitudinal L component are given by:

$$\varepsilon_{L,hkl} = \frac{d_{hkl} - d_{0,hkl}}{d_{0,hkl}} \quad \text{and} \quad \Delta \varepsilon_{L,hkl} = \sqrt{\frac{\Delta^2 d_{hkl} + \Delta^2 d_{0,hkl}}{(d_{0,hkl})^2}} \quad (8.24)$$

Finally the stress in the longitudinal direction, σ_L may be determined from the measured strains in three orthogonal, (and intermediate directions if available), following Equations (8.8) and (8.9). The uncertainty $\pm \Delta \sigma_L$ is given by:

$$\Delta \sigma_L = \frac{E_{hkl}}{(1+\nu_{hkl})(1-2\nu_{hkl})} \sqrt{(1-\nu_{hkl})^2 \Delta^2 \varepsilon_{L,hkl} + \nu_{hkl}^2 \Delta^2 \varepsilon_{T,hkl} + \nu_{hkl}^2 \Delta^2 \varepsilon_{N,hkl}} \quad (8.25)$$

with corresponding equations for $\Delta \sigma_T$ and $\Delta \sigma_N$.

8.6.2 Analysis of Time-of-flight Diffraction

The line shape for analysis of time-of-arrival spectra is more complex because of the nature of the pulse from the moderator. Typically it has leading and trailing exponential tails upon a Gaussian central component that are wavelength dependent [22]. Nevertheless, the powder calibration procedure determines the line shape parameters accurately. The whole spectrum, including the peaks and the background, may be fitted by the Rietveld method [23,24] to determine the lattice parameters. The usual procedure does not constrain the peak intensities to be the ideal powder averages, Equation (8.13), thus allowing for the non-random texture usually present in industrial samples. The Rietveld approach of fitting all the peaks averages over the intergranular strains and is deemed to be less sensitive to Type-II effects. The fits give $a \pm \Delta a$, where Δa is often an unreasonably small measure of the precision. With measurements of the reference samples the strains and their errors are given by

$$\varepsilon = \frac{a - a_0}{a_0} \quad \text{and} \quad \Delta \varepsilon = \frac{1}{a_0} \sqrt{\Delta^2 a + \Delta^2 a_0} \quad (8.26)$$

Since the Rietveld method averages over all the measured $\{hkl\}$, the stresses are usually calculated with Equations (8.7) and (8.8) using the macroscopic elastic constants rather than the diffraction elastic constants.

If the diffracted intensity is sufficiently strong, then individual peaks may be fitted. Typically the fitting precisions in single peak time-of-flight are about 50% more precise than fitting angular data. Single peak fits are made on the reference samples in the corresponding directions and the procedure outlined above is followed to obtain the sets of strains from the different $\{hkl\}$ reflections. Since the reference procedure takes into account both chemistry as well as intergranular effects, because of their length scales, the computed macroscopic stress, calculated with the appropriate diffraction elastic constants, E_{hkl} and ν_{hkl} for each reflection $\{hkl\}$, should be independent of the reflection used to within the calculated uncertainty. This process was followed for a weld problem in [25].

8.6.3 Precision of the Measurements

Lattice parameters can normally be measured to a precision between $\pm(0.5-1.0) \times 10^{-4}$, for both the intact sample and the reference sample. Assuming that the standard error of the strain is computed which is a combination of the lattice parameters of the intact sample and the reference, then each component of the strain tensor can be measured to between $\pm(0.7-1.4) \times 10^{-4}$. Assuming that $E = 200$ GPa and $\nu = 0.3$, appropriate for ferritic materials, this translates to a precision in stress of between $\pm(20-40)$ MPa. The material factors that decrease the precision include broad diffraction peaks, for example with martensitic materials, for which it is more difficult to determine the peak center, and weak signals, where a low peak-to-background ratio precludes a good determination of the center. Weak signals can be due to very weak texture associated with the $\{hkl\}$ reflection selected, small coherent scattering, as in Ti alloys, large absorption as in Co alloys or low transmission as in high Ni content alloys. In superalloys such as Waspaloy, the face-centered cubic reflections such as $\{111\}$, $\{002\}$ or $\{113\}$ have contributions from both the matrix and strengthening simple cubic phase so that the center of the composite peak is a measure of some combination of the strain in both phases.

Low measurement precision occurs when the signal size is low, for example with small gage volumes that are used when the stress is required close to a specific feature such as a crack tip. The use of long path lengths through the sample gives similar difficulties; the case of the hoop stress in a thick-walled cylinder is typical where the beam has to pass through the wall twice. Often a window is cut in the cylindrical sample to allow the beam to pass unimpeded through the first wall, but in spite of this, the path length may be near prohibitive. Sometimes, in an effort to minimize the size of the window, the component is measured precisely only along the hoop direction at the mid-thickness of the wall, and away from this position a combination of hoop and radial stress is measured.

8.7 Systematic Errors in Strain Measurements

8.7.1 Partly Filled Gage Volumes

Serious systematic errors will occur if the gage volume is not completely within the sample. Three effects [26–28] occur for monochromatic beam diffraction with an area detector and slit geometry. Firstly, the center-of-gravity of the diffraction from the sample is shifted away from the reference point of the instrument. This produces an error that is initially linear in partial filling. Secondly, there is always a wavelength distribution across the beam in monochromatic diffraction that is the source of the beam enhancement from focusing in the horizontal plane. For partial filling, part of this wavelength distribution does not diffract from the material so that the average wavelength of the beam is shifted.

Thirdly, the slit can cut off part of the angular distribution of diffracted neutrons and prevent it being counted. When the gage volume is filled, the "clipping" is symmetrical and does not shift the peak position. When the gage volume is partly filled, clipping occurs on one side of the beam and skews the fitting process to give a non-linear effect. For time-of-flight diffraction, the offset from the reference point is important since the Bragg angles, primarily, are slightly altered [29]. The uneven distribution of intensity across the counter can also lead to an error. The problem has been studied experimentally and also modeled [26–28]. Partial filling usually arises in attempts to measure near-surface stresses but can occur if the sample has been set up incorrectly or if slits have been bumped off alignment. Similar systematic errors occur with internal surfaces, such as different phases in a dissimilar metal weld, phase gradients, and rapid texture gradients such as are seen in the heat-affected zones in welds of hexagonal close packed metals [30]. Alloy concentration gradients and gradients in minor strengthening phases can affect the lattice parameter and also lead to errors. The use of radial collimators, as opposed to slits, may decrease the near surface errors to some degree, but does not remove them. The best use of models or simulations of the geometry is to design a set-up where the systematic errors from partly filled gage volumes are only of the order of the precision rather than to use them to correct basically faulty data.

8.7.2 Large Grain Effects

Large grains generate an error related to an incompletely filled gage volume. In this case, the few strongly scattering large grains will be randomly offset from the center of the gage volume rather than being distributed uniformly over the gage volume. Effects begin to be noticeable around an average grain size of $100\,\mu m$ and manifest as large changes in strain (5–10 times larger than the precision) from point to point in the sample. These effects are accompanied by large changes in intensity due to fluctuations in the small number of grains. The accuracy may be increased by improving the statistical sampling of the diffracting grains. This can be done by collecting data over a range of angles on either side of the desired direction or by translating the sample in small steps and averaging the results. Time-of-flight has an advantage over reactor measurements since the diffraction data are collected over an angle in the horizontal plane of, say, $\pm 15°$ about $\pm 90°$, corresponding to a spread of $\pm 7.5°$ in grain orientation as well as over the $\pm 20°$ vertical acceptance. It might appear initially that making the gage volume larger would solve the problem. Unfortunately, in this case, on average the large grains can be even further from the center of the gage volume so the systematic error and scatter get worse. A practical measure of the accuracy of the lattice spacing for large grains is the standard deviation about the average in a region where the stress is not varying strongly.

8.7.3 Incorrect Use of Slits

Slit geometry may lead to errors if the gage volume is far from the slit defining the diffracted beam. This is partly because the gage volume becomes poorly defined due to beam divergence, but also because the lines of sight from regions of the sample irradiated with neutrons to a 2D detector alter the line profiles and lead to inaccurate peak fitting as in the case of a partially filled gage volume. The problem arises if the slit is both an element of the resolution as well as a "window." If the beam is well collimated by a straight Söller collimator and the slit is merely a "window" and not a collimating element, then the error can be avoided. This approach was taken when examining a $25 \times 25 \times 25 \, \text{cm}^3$ aluminum forging [31]. A radial collimator will also avoid the problem.

8.7.4 Intergranular Effects

For face-centered cubic nickel alloys such as Inconel or stainless steels, the {002} reflection always gives rise to the largest tensile intergranular residual strains after uniaxial tensile loading. For example, in a relatively untextured material, grains with <002> along the loading direction unload in tension, whereas those with <220> grains along the loading direction unload in compression [32]. Orientations such as <111> and <113> tend to show very small residual strains, comparable with the experimental uncertainty, after unloading to zero macroscopic stress. For example, [33] in bent Incoloy-800 steam generator tubes, the {002} strains exceeded the {111} strains by almost a factor of two near the neutral axis and differed in sign at the top and bottom of the bend. The reference spacing had been taken with respect to straight tube that had not been plastically deformed by bending and displayed no intergranular effects. The stress at a single location at the top of the bend calculated from {111} and {002} reflections was opposite in sign indicating a serious contradiction. This arose from not recognizing initially that a sizeable fraction of the measured residual strain from {002} in the bent tube was a Type-II strain not Type-I.

The problem is often circumvented [4,5] by choosing reflections that appear to be less sensitive to intergranular effects such as {111} and {113} for nickel alloys and {110} and {112} for iron alloys. However, this choice is based on empirical observation for uniaxial stress, and may not necessarily apply to biaxial or triaxial stresses. The observation of which reflections show the biggest intergranular effects for a given material is also texture-dependent. For hexagonal close packed materials there are no {hkil} reflections that are always free from Type-II effects, mainly because of texture arising from plastic deformation generated during manufacture.

8.8 Test Cases

8.8.1 Stresses in Indented Discs; Neutrons, Contour Method and Finite Element Modeling

Figure 8.7 shows a schematic of a disc indentation process and the conceptual stress distribution expected [34]. The indented region corresponds to a compressive deformation of about -2% in the center of the disc. The lateral expansion of the indented region is constrained by the surroundings and generates compressive radial stresses and tensile circumferential (hoop) stresses. The in-plane stresses within the indented region are compressive.

The test sample was a 316 stainless steel disc, 60 mm in diameter and 10 mm thick, with a 0.085 mm indentation on the top and bottom surfaces, 15 mm in diameter. Finite element calculations of the stresses were made for both the indenting tool and disc. The principal axes are expected to be hoop, radial and axial from the symmetry, and the axial stress is expected to be near zero. Measurements of the hoop stress were additionally

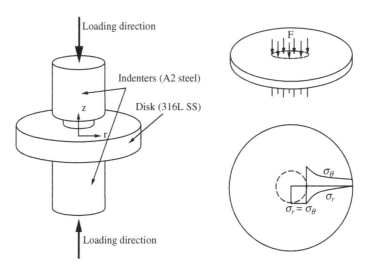

Figure 8.7 Schematic of the indenting tool generating the stress field in the 316 stainless steel plate and the resulting residual stress field (from [34]). Reproduced with permission from [34], Copyright 2009 ASME

made after the neutron experiments by the contour method [35], described in Chapter 5 of this volume. The assumptions involved in the interpretation of the contour method are completely different to diffraction, so that agreement of measurements by the two techniques enhances confidence in the results.

The diffraction experiments were carried out on the SMARTS time-of-flight diffractometer [36] at Los Alamos National Laboratory. The initial stress calculations indicated that a $2 \times 2 \times 2 \, mm^3$ gage volume would be adequate to describe the stress gradients expected while giving good counting statistics for the sample, which has an ideal thickness for neutron diffraction. SMARTS has a straight neutron guide from the moderator to the sample, L_0, of 31 m and a flight path, L_1, of 1.5 m from the sample to the detector. The repetition rate was 20 Hz, corresponding to a time interval between pulses of 50,000 μ sec. Figure 8.8 shows a time-of-flight spectrum, intensity vs. d-spacing for this sample, covering a range of d from 0.6 to 2.1 Å and therefore wavelengths (for 90° diffraction) from 0.85 to 2.97 Å. Table 8.4 shows the wavelengths, neutron velocities and time elapses from moderator to counter for this range of wavelengths. Note that there is no frame overlap because 0.85 Å neutrons from a subsequent pulse do not catch up with 2.97 Å neutrons from the preceding pulse.

The arrangement of the samples for the two cases of radial/axial and hoop/axial strains in the disc in Figure 8.9 shows the incident beam and the two diffracted beams. Measurements were made over the shaded plane covering the thickness and diameter of the disc. For the radial/axial component the table moves in the horizontal plane (X, Y) with the height, Z, fixed. For the hoop/axial case the sample moves in the table (Y, Z) plane. The sample was rotated by 90° between the two sets of measurements, so that the measurements were made at the same physical locations in the disc. The two sets of axial strains agreed to within 100×10^{-6}.

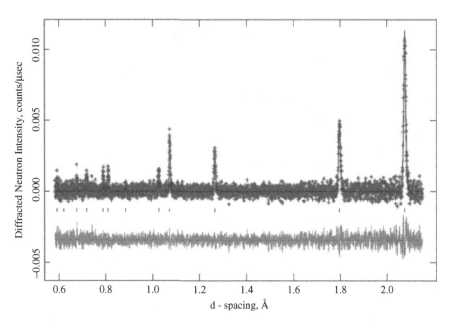

Figure 8.8 Upper line: Time-of-flight spectrum, intensity versus lattice spacing, for the measurement of strain in the indented 316 stainless steel plate. The crosses indicate the data and the line is the Rietveld refinement fit. The tick-marks indicate the positions of the face-centered cubic peaks. Lower line: difference between the data and Rietveld refinement fit (from [34]). Reproduced with permission from [34], Copyright 2009 ASME

Table 8.4 Wavelengths and velocities and time elapses for the range of d-spacings covered in Figure 8.8

λ (Å)	v (msec^{-1})	$T_0 + T_1$ (μ sec)
0.85	4662.2	6971.0
2.97	1332.0	24399.4

Reference measurements were made on $5 \times 5 \times 5$ mm^3 coupons cut from an annealed sample taking care that the gage volume was always within the reference coupons. The strains were calculated from the lattice parameters obtained by Rietveld fits to the complete time-of-flight diffraction spectra for the intact sample and the references. It was assumed that the principal axes were radial, r, hoop, θ and axial, a, and the stresses were calculated for σ_r as:

$$\sigma_r = \frac{E}{(1+\nu)(1-2\nu)}((1-\nu)\varepsilon_r + \nu\varepsilon_\theta + \nu\varepsilon_z) \qquad (8.27)$$

where E and ν are the macroscopic elastic constants for 316 stainless steel, taken to be 208 GPa and 0.33. The results for the hoop, axial and radial stresses at the mid-thickness of the disc across a diameter are shown in Figure 8.10, taken from the comprehensive results

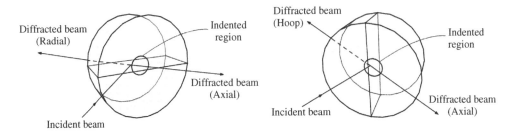

Figure 8.9 Set-up of the indented plate to measure the radial/axial strains and the hoop/ axial strains

Table 8.5 Root mean square differences in stress (MPa) between pairs of measurements [34] on the indented disc. Parentheses indicate the corresponding strain differences in units of 10^{-6}

	Contour 2	Neutron diffraction	FE model
Contour 1	20 MPa (104 με)	27 MPa (140 με)	32 MPa (166 με)
Contour 2		28 MPa (145 με)	33 MPa (171 με)
Neutron diffraction			33 MPa (171 με)

presented by Pagliaro et al. [34]. There is quantitative agreement between the experimental measurements and both these match the finite element calculations as shown in Table 8.5.

Interestingly, poor agreement between the initial finite element model and the measurements in the indented region prompted a sophisticated improvement in the model to take account of plasticity changes upon un-loading.

8.8.2 Residual Stress in a Three-pass Bead-in-slot Weld

The results described here were made under the auspices of the European network on Neutron Techniques Standardisation for Structural Integrity (NeT) on a sample designed and manufactured to allow prediction and measurement of residual stresses in a three-pass bead-in-slot weld in an AISI 316LN stainless steel plate, Figure 8.11. The experiments and modeling were reported by Muransky et al. [37]. The prediction of stresses in a weld is difficult because of the complexity of the welding process. To provide data for this prediction, the sample was instrumented to record the time-temperature history during welding. The sample was also measured by the spiral slit technique [38] employing high-energy synchrotron X-rays that permitted a mapping of the longitudinal and transverse stresses over the complete plate on the assumption that the plate normal stress was zero.

The experiments were carried out with the KOWARI instrument [17] at the Australian Nuclear Science and Technology Organisation at Lucas Heights, Australia. The instrument employs a focusing bent silicon single crystal monochromator on a 70 m bent thermal neutron guide to bring neutrons from the reactor to a low background guide hall. The neutron wavelength was 1.52 Å, which gives diffraction from the {113} planes of face-centered

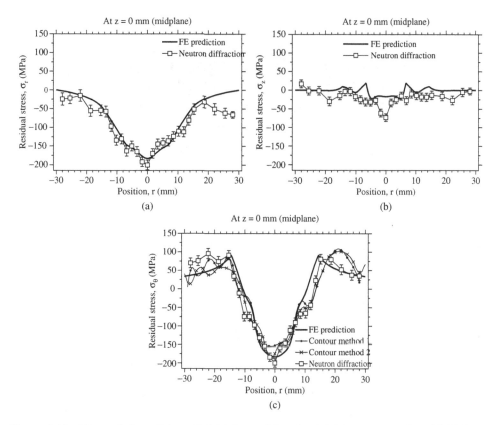

Figure 8.10 The variation of the radial (a) hoop, (b) and axial (c) stresses at the mid-thickness (z = 0) of the indented plate derived from neutron diffraction measurements (open squares) and the finite element model by the solid curves. The hoop stresses derived from the contour measurements are also shown in [34]. Reproduced with permission from [34], Copyright 2009 ASME

cubic 316LN at 92.8°. The incident beam was defined by a slit in absorbing Cd which was 2 mm wide and 2 mm high situated 30 mm from the center of rotation of the instrument. The diffracted beam was defined by a slit 2 mm wide and 10 mm high, situated 30 mm from the center of rotation. This provided a $2 \times 2 \times 2\,mm^3$ gage volume, but with enhanced vertical divergence on the diffracted side, and hence higher count rate, without loss of spatial resolution. The detector was a cross-wired position sensitive detector spanning ±7.5°. After collection of the longitudinal strain data with the longitudinal direction along the scattering vector, the sample was re-oriented in turn to measure the transverse and plate normal strains.

Three reference samples were measured in each of the three principal directions, one in the parent metal, one in the region of the root pass and one in the final third pass. There was no difference between the three directions for the first two reference samples and their lattice parameters were equal to within the assigned uncertainty. This indicated that intergranular effects were small, as is usually the case for the {113} reflection. In the third

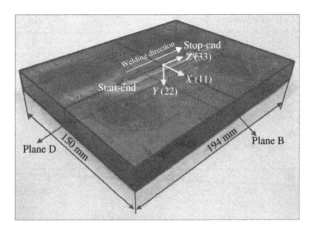

Figure 8.11 The three-pass bead-in-slot weld in the 316LN stainless steel plate. Reproduced with permission from [37], Copyright 2012 Elsevier

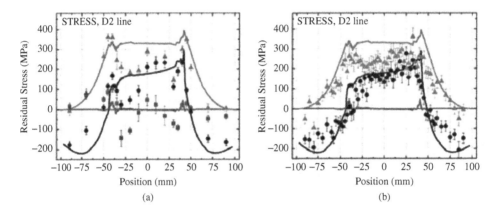

Figure 8.12 Measurements of the longitudinal (triangles), transverse (points) and normal (squares) residual stresses made 2 mm below the surface of the plate along the longitudinal direction beneath the weld. (a) Neutron diffraction measurements and (b) synchrotron X-ray measurements using the spiral-slit method. The solid curves show the model calculations. Reproduced with permission from [37], Copyright 2012 Elsevier

reference sample there were no differences between the three directions but the lattice parameter differed from the first two reference samples. For this reason a single value of reference lattice spacing was adopted for all but measurements in the top-most pass in the weld metal. One of the problem areas in austenitic materials is large grain size, especially in relatively slow cooling weldments. To improve the statistical sampling of the {113} grains the intact sample and the reference samples were rotated continuously by ±5° on either side of the intended measurement direction, about the vertical axis.

The strains were calculated following Equation (8.3) and the stresses calculated using Equation (8.8) with values of the diffraction elastic constants, E_{113} and v_{113} of

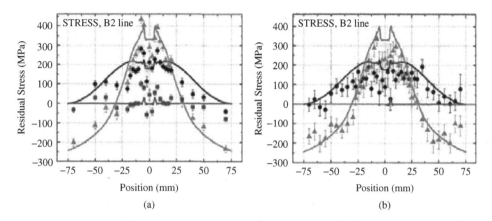

Figure 8.13 Measurements of the Longitudinal (triangles), transverse (points) and normal (squares) residual stresses made 2 mm below the surface of the plate transverse to the weld through the weld center. (a) Neutron diffraction measurements and (b) synchrotron X-ray measurements using the spiral-slit method. The solid curves show the model calculations. Reproduced with permission from [37], Copyright 2012 Elsevier

183.6 GPa and 0.306 respectively. Stresses were mapped on planes parallel to (plane D) and perpendicular to (plane B) the weld, and line scans from these planes 2 mm below the top surface are shown in Figures 8.12 and 8.13 for longitudinal, transverse and normal components. There is good agreement between the neutron and synchrotron results and the finite element model gives an excellent account of the measured stresses.

Acknowledgments

The author wishes to acknowledge the contribution of his former colleagues at Chalk River, namely J. H. Root, S. R. McEwen, R. A. Holt and R. R. Hosbons to the understanding of stress measurements with neutrons and M. A. Bourke and D. W. Brown for initiating him into the methods of time-of-flight diffraction. He also wishes to acknowledge a useful correspondence with G. S. Schajer, M. B. Prime and P. J. Withers.

References

[1] Krawitz, A. D. (2001) *Introduction to Diffraction in Materials Science and Engineering*. Wiley: New York.
[2] Reimers, W., Pyzalla, A. R., Schreyer, A., Clemens, H., eds. (2008) *Neutrons and Synchrotron Radiation in Engineering Materials Science*. Wiley-VCH: Weinheim.
[3] Noyan, I. C., Cohen, J. B. (1987) "Residual Stress: Measurement by Diffraction and Interpretation." In *Materials Research and Engineering*, eds. Ischner, B. I., Grant, N. J., Springer-Verlag: New York.
[4] Hutchings, M. T., Withers, P. J., Holden, T. M., Lorentzen, T. (2005) *Introduction to the Characterization of Residual Stress by Neutron Diffraction*. Taylor and Francis: Boca Raton.
[5] Squires, G. L. (1994) *Introduction to the Theory of Thermal Neutron Scattering*. Dover, Mineola, New York.
[6] ISO (2005) Non-destructive Testing-standard Test Method for Determining Residual Stresses by Neutron Diffraction. Technical Specification ISO/TS 21432.

[7] Kröner, E. (1958) Berechnung der elastischen Konstanten des Vielkristalls aus den Konstanten des Einkristalls. *Zeitschrift für Physik A: Hadrons and Nuclei* 151(4):504–518.

[8] Turner, P. A., Tomé, C. N. (1994) "Study of Residual Stresses in Zircaloy-2 with Rod Texture." *Acta Metallurgica et Materialia* 42(12):4143–4153.

[9] Clausen, B., Lorentzen, T., Leffers, T. (1998) "Self-consistent Modelling of the Plastic Deformation of FCC Polycrystals and its Implications for Diffraction Measurements of Internal Stresses." *Acta Materialia.* 46(9):3087–3098.

[10] Priesmayer, H. G., Stalder, M., Vogel, S., Meggers, K., Trela, W. (1997) Proc. International Conference on Neutrons and Research in Industry, *SPIE Proceedings Series* 2867:164–167.

[11] Santisteban, J. R., Edwards, L., Steuwer, A., Withers, P. J. (2001) "Time-of-flight Neutron Transmission Diffraction." *Journal of Applied Crystallography* 34(3):289–297.

[12] Tremsin, A. S., McPhate, J. B., Vallerga, J. V., Siegmund, O. H. W., Feller, W. B., Lehman, E., Dawson, M. (2011) "Improved Efficiency of High Resolution Thermal and Cold Neutron Imaging." *Nuclear Instruments and Methods in Physics Research A* 628(1):415–418.

[13] Sears, V. F. (1992) "Neutron Scattering Lengths and Cross Sections." *Neutron News* 3(3):29–37. A.-J. Dianoux and G. Lander (2003) Neutron Data Booklet Institute Laue-Langevin, Grenoble.

[14] Christ, J., Springer, T. (1962) "The Development of a Neutron Guide at the FRM Reactor." *Nukleonik* 4:23–25.

[15] Wagner, V., Mikula, P., Lukas, P. (1993) "A Doubly Bent Silicon Monochromator." *Nuclear Instruments and Methods in Physics Research A* 338(1):53–59.

[16] Withers, P. J., Johnson, M. W., Wright, J. S. (2000) "Neutron Strain Scanning with a Radially Collimated Diffracted Beam." *Physica B* 292(3–4):273–285.

[17] Kirstein, O., Brule, A., Nguyen, H., Luzin, V. (2008) "KOWARI–The Residual Stress Spectrometer for Engineering Applications at OPAL." *Materials Science Forum* 571–572:213–217.

[18] Prince, J. (1983) "The Effect of Finite Detector Slit Height on Peak Positions and Shapes in Powder Diffraction." *Journal of Applied Crystallography* 16(5):508–511.

[19] Johnson, M. W., Daymond, M. R. (2002) "An Optimum Design for a Neutron Diffractometer for Measuring Engineering Stresses." *Journal of Applied Crystallography* 35(1):49–57.

[20] Edwards, L. E., Withers, P. J., Daymond, M. R. (2000) ENGIN-X: a neutron stress diffractometer for the 21st century. In Proc. 6th Int. Conf. on Residual Stress, Oxford 2000, IOM Communications, Ltd., London 2:1116–1123.

[21] James, J. A., Edwards, L. (2007) "Application of Robot Kinematics Methods to the Simulation and Control of Neutron Beam Line Positioning Systems." *Nuclear Instruments and Methods in Physics Research. A* 571(3):709–718.

[22] Carpenter, J. M., Robinson, R. A., Taylor, A. D., Picton, D. J. (1985) "Measurement and Fitting of Spectrum and Pulse Shapes of a Liquid Methane Moderator at IPNS." *Nuclear Instruments and Methods in Physics Research A* 234(3):542–551.

[23] Rietveld, H. M. (1969) "A Profile Refinement Method for Nuclear and Magnetic Structures." *Journal of Applied Crystallography* 2(2):65–71.

[24] Larsen, A. C., Von Dreele, R. B. (1994) General Structure Analysis System, GSAS, LAUR-80-748, Los Alamos National Laboratory, NM, USA.

[25] Carr, D. G., Ripley, M. J., Holden, T. M., Brown, D. W., Vogel, S. C. (2004) "Residual Stress Measurements in a Zircaloy-4 Weld by Neutron Diffraction." *Acta Materialia* 52(14):4083–4091.

[26] Webster, P. J., Mills, G., Wang, X. D., Kang, W. P., Holden, T. M. (1996) "Impediments to Efficient Through-surface Strain Scanning." *Journal of Neutron Research* 3(4):223–240.

[27] Wang, X. L., Spooner, S., Hubbard, C. R. (1998) "Theory of the Peak Shift Anomaly Due to Partial Burial of the Sampling Volume in Neutron Diffraction Residual Stress Measurements." *Journal of Applied Crystallography* 31(1):52–59.

[28] Lorentzen, T. (1997) "Numerical Analysis of Instrument Resolution Effects on Strain Measurements by Diffraction Near Surfaces and Interfaces." *Journal of Neutron Research* 5(3):167–180.

[29] Wang, X-L., Wang, Y. D., Richardson, J. W. (2002) "Experimental Error Caused by Sample Displacement in Time-of-flight Neutron Diffractometry." *Journal of Applied Crystallography* 35(5):533–537.

[30] Holden, T. M., Hosbons, R. R., Root, J. H., Ibrahim, E. F. (1989) Neutron diffraction measurements of strain and texture in welded Zr2.5wt.% Nb tube. *Materials Research Symposium*, Materials Research Society, Pittsburgh, 142:59–64.

[31] Pang, J. W. L., Holden, T. M., Mason, T. E., Panza-Giosa, R. (1997) "Texture and Residual Strain in an A17050 Billet." *Physica B* 241(1–4):1267–1269.

[32] Clausen, B., Lorentzen, T., Bourke, M. A. M., Daymond, M. R. (1999) "Lattice Strain Evolution During Tensile Loading of Stainless Steel." *Materials Science and Engineering A* 259(1):17–24.

[33] Holden, T. M., Holt, R. A., Dolling, G., Powell, B. M., Winegar, J. E. (1988) "Characterization of Residual Stress in Bent INCOLOY-800 Tubing by Neutron Diffraction." *Metallurgical Transactions A* 19(9):2207–2214.

[34] Pagliaro, P., Prime, M. B., Clausen, B., Lovato, M. L., Zuccarello, B. (2009) "Known Residual Stress Specimens Using Opposed Indentation." *Journal of Engineering Materials and Technology*, 131(3):031002 1:10.

[35] Prime, M. B. (2001) "Cross Sectional Mapping of Residual Stresses by Measuring the Surface Contour After a Cut." *Journal of Engineering Materials and Technology*, 123:162–168.

[36] Bourke, M. A. M., Dunand, D. C., Üstündag, E. (2002) "SMARTS – A Spectrometer for Strain Measurement in Engineering Materials." *Applied Physics A* 75(S1):1707–1709.

[37] Muransky, O., Smith, M. C., Bendeich, P. J., Holden, T. M., Luzin, V., Martins, R. V., Edwards, L. (2012) "Comprehensive Numerical Analysis of a Three-pass Bead-in-slot Weld and its Critical Validation Using Neutron and Synchrotron Diffraction." *International Journal of Solids and Structures*, 49(9):1045–1062.

[38] Martins, R. V., Honkimäki, V. (2003) "Depth Resolved Strain and Phase Mapping of Dissimilar Friction Stir Welds Using High Energy Synchrotron Radiation." *Textures and Microstructures* 35(3–4):145–152.

9

Magnetic Methods

David J. Buttle
MAPS Technology Ltd., GE Oil & Gas, Oxford, UK

9.1 Principles

9.1.1 Introduction

Ferromagnetic metals such as iron, nickel and cobalt are naturally magnetic and can be magnetized by exposure to a magnetic field, for example, from an electromagnet coil. Figure 9.1 shows the typical non-linear relationship between the induced magnetism (the "B" field, measured in units of Tesla) and the applied magnetic field strength (the "H" field, measured in units of Amperes/metre). The diagram illustrates the characteristic S-shaped curves, with saturation of the magnetic induction at high positive and negative applied field strengths, and hysteresis in the response as the magnetic field strength is cycled.

In addition to being non-linear, the magnetic response is not smooth, as can be seen in the small inset in Figure 9.1. The induced magnetism responds to the applied magnetic field in small irregular jumps corresponding to magnetic changes in small local regions of the material. These magnetic jumps can induce voltage pulses across a sensor coil, known as magnetic Barkhausen noise, and some of them create a random emission of low-level acoustic noise within the material, called Acoustic Barkhausen Emission. The presence of residual stress influences the way in which the local magnetic changes occur, affecting in a repeatable way the amount and character of the noise produced. Thus, the measurement and analysis of the Barkhausen noise can provide an indication of the stresses present in a ferromagnetic material. This chapter describes various practical measurement and analytical techniques that can be used to quantify residual stresses from magnetic-based measurements.

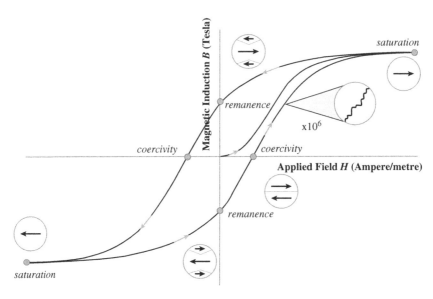

Figure 9.1 The hysteresis loop showing how magnetic induction varies with applied field strength, starting from a demagnetized condition (graph origin) and cycling the applied field (larger inset shows magnetic induction proceeds by small variable steps, smaller insets show simple domain distributions ignoring closure domains and rotation). Reproduced by permission of MAPS Technology Ltd

9.1.2 Ferromagnetism

Materials that have a net magnetic moment due to unpaired electrons and furthermore whose orbitals interact sufficiently with those of lattice neighbors to align the moments are said to be ferromagnetic [1,2]. This alignment of atomic moments leads to a very high spontaneous magnetization of the material, for example, along the [1,0,0] axes in b.c.c. iron or the [1,1,1] axes in f.c.c. nickel. These are referred to as the magnetic 'easy' axes, which arise directly from the different separation of the atomic moments along the different crystallographic axes. Ferromagnetic elements include iron, nickel, cobalt and gadolinium, and materials most relevant to stress measurement in industry include ferritic steels and some nickel alloys.

9.1.3 Magnetostriction

Magnetic materials change very slightly in dimensions during magnetization/demagnetization in a process called magnetostriction. This effect occurs because of the distortion of the material crystal lattice by the magnetization. In iron, the strain is positive along the direction of magnetization and negative in orthogonal directions, while nickel shows opposite responses. Materials have magnetostriction constants (= strain at magnetic saturation) for "easy" and "hard" axes $\sim\pm 10\,\mu\varepsilon$. An inverse effect occurs where applied strains and therefore stresses influence material magnetization response. It is this effect that provides the basis for the magnetic methods for measuring residual stresses.

9.1.4 Magnetostatic and Magneto-elastic Energy

At the crystal lattice level, ferromagnetic materials are permanently fully magnetized. However, in the bulk state they generally do not appear to be permanently magnetized because of the existence of magnetic domains. These are regions smaller than the crystal grains within which the atomic moments are aligned with one of the easy axes. These domains are separated by domain walls (DWs), where the magnetic moment direction moves smoothly from one easy axis to another. The DWs are thin ($\sim 10^2$ atoms thick). The mixture of many small magnetic domains with varying magnetization directions reduces the aggregate magnetic effect, tending to zero when the alignment mixture is spatially isotropic. Two energy types can be identified: the magnetostatic energy associated with the strength of the bulk magnetic field, and the magnetic anisotropy energy associated with the atomic moments in the DWs being aligned away from the easy axes. Reducing the size of the magnetic domains reduces the magnetostatic energy but increases the anisotropy energy so there is a balance, as Figure 9.2 illustrates.

At low fields, magnetization is the process whereby the domains are preferentially aligned so that their combination produces a net effect at the bulk level. When there is internal tensile stress, the total elastic energy can be reduced when, for positive magnetostriction, the magnetic domains are pointing along the stress axis (this magneto-elastic energy partially offsetting the conventional elastic energy), or for internal compression, away from the stress axis (Figure 9.3 upper part). Applying a magnetic field will cause magnetic domains already aligned with the field to grow at the expense of other domains, but the net volume fraction of domains that align will depend upon the initial domain distribution. For a field aligned with a tensile stress, the magnetization will be greater than for a compressive stress and hence the magnetic permeability is greater (see Figure 9.3

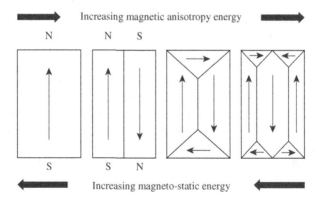

Figure 9.2 Magnetic domains schematic demonstrating balance between magneto-static and anisotropy energies for b.c.c. iron within a single crystal. Reproduced by permission of MAPS Technology Ltd

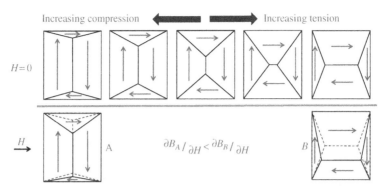

Figure 9.3 Schematic showing partial magnetic domain alignment due to tensile and compressive stresses (upper half) and applied field (two cases only in lower half) in single crystal b.c.c. iron. (This simple 2D domain representation is inadequate for a 3D material.) Reproduced by permission of MAPS Technology Ltd

lower part). Thus the stress driven changes in magnetic domain distribution mean that magnetic properties are functions of the stress tensor and, as a result, changes in the magnetic signature can be used to deduce the underlying state of stress. Several examples of analytical models for the tensor stress dependence of magnetic properties have been published [3–5].

9.1.5 The Hysteresis Loop

The bulk magnetic properties of ferromagnetic materials are normally represented as a plot of the magnetic induction, B, against applied field strength, H, and is known as the hysteresis or B-H loop (Figure 9.1). A number of properties can be derived from the B-H loop as well as the maximum permeability $(B/H)_{max}$ and differential permeability (dB/dH). When fields below that needed to saturate the material are used, the coercivity (the magnetic field strength required to reduce the magnetization to zero, see Figure 9.1) and remnance (the magnetization remaining after the removal of the magnetic field) parameters are better named as coercive force and remnant induction, these being smaller.

During the cyclic magnetization process, reduction from saturation is characterized by DW nucleation and rotation away from the applied field direction towards the nearest easy magnetic axis. As the field drops below the "knees" of the hysteresis loop the DW movement is characterized by large volumes of the material undergoing reversible and irreversible 180° movement resulting in increased permeability. The magnetization stages are reversed as the applied field is increased towards material saturation in the opposite direction. For further information on basic magnetic properties, please refer to the books listed in Section 9.11.

9.1.6 An Introduction to Magnetic Measurement Methods

Magnetoelastic methods for stress measurement have the great advantage of being rapid, entirely non-destructive and easily adapted for *in-situ* measurements on components and industrial plant using relatively small portable units and well-designed probes. A large number of magnetic and electromagnetic methods are used to characterize materials, material degradation processes and internal stresses [6], and these may be loosely classified into methods that use macroscopic magnetic properties, such as hysteresis loop and related parameters, where 'macroscopic' refers to properties averaged over longer times and involve the movement of DWs in many grains; and those that use micro-magnetic properties such as Magnetic Barkhausen Emission (MBE) and Acoustic Barkhausen Emission (ABE), where the signals are sensed from individual or few magnetic DW movements over a shorter time during the magnetization cycle. There are many different methods used to monitor these magnetic properties, the term 'monitor' being used as practical approaches often measure signals that depend only indirectly upon the magnetic properties of interest. Their complex dependence on material composition, thermo-mechanical treatment and microstructural condition as well as the large number of magnetic *parameters* that can be measured using different field strengths and frequencies, and the different designs of probes that can be used for practical application, have limited the widespread use for evaluation of engineering stresses in an unambiguous and robust way. It is not the purpose here to review all these approaches, particularly as currently very few have matured to a sufficient degree for general field application and are in widespread use. So this guide is limited to just three techniques, the multi-parameter MAPS (Magnetic Anisotropy and Permeability System) method, and the MBE and ABE methods.

For other methods refer to Section 9.11 for the magnetic anisotropy system "MAS" or "Sigmatron" [7], stress-induced magnetic anisotropy [8], directional-effective permeability [9], non-linear harmonic analysis [10] and magnetically-induced velocity change [11].

9.2 Magnetic Barkhausen Noise (MBN) and Acoustic Barkhausen Emission (ABE)

9.2.1 Introduction

Magnetic Barkhausen Noise (MBN), also known as Barkhausen Emission, is one of the earliest practical magnetic measurement techniques [12,13]. The MBN signal (noise) constitutes electromagnetic pulses (10^2–10^6 Hz bandwidth) measured across a sensor coil whose voltage is the time derivative of the magnetic induction. The pulses arise from irreversible DW movement in the nearby test material when under an applied varying magnetic field (see inset Figure 9.1). Since the domain distribution depends upon the internal stress, the MBN signal will also be stress sensitive. The irreversible DW movements arise from microstructural second phases, inclusions and dislocations, all of which introduce discontinuities in magnetic properties within the grains thereby causing DW pinning, and consequently MBN is particularly sensitive to the microstructure of the material.

The MBN signal can be measured as a pulse height distribution or an rms noise level as a function of applied field. Figure 9.5(a) gives an example of the latter. Both features provide several parameters such as maximum pulse size, number of pulses above a threshold, number of peaks, peak height and position (coercive force), peak area, total MBN energy (integrated squared voltage) per cycle. All of these may be combined with empirical models or calibration to characterize material properties and stress.

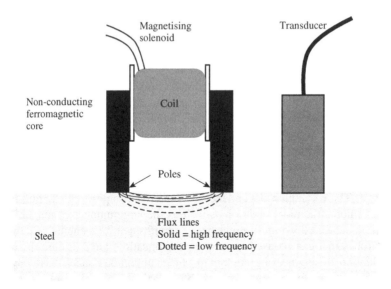

Figure 9.4 ABE is generated by 90° DWs movement in steel in the region directly below the magnetizing solenoid and is detected with an ultrasonic transducer. For MBN the sensor coil is positioned between the poles near the sample surface (not shown). Reproduced by permission of MAPS Technology Ltd

In steel, magnetic domains are separated by either 180° or the less numerous 90° DWs, whose movements both contribute to the MBN signal. However, for 90° DW movement, the local material swept out by the DW causes a sudden change in magnetostrictive strain and hence generates an ultrasonic shear wave. This Acoustic Barkhausen Emission (ABE), sometimes known as Magneto-acoustic Emission (MAE), is normally detected using piezo-electric transducers coupled to the component surface, monitoring from 10 kHz to 1 MHz (Figure 9.4). ABE signals are relatively weak, requiring low-noise high-gain electronics. Like MBN, ABE can be monitored as an rms noise as a function of applied field, Figure 9.5(b), the highest 90° DW movement being observed around the knees of the hysteresis loop.

9.2.2 Measurement Depth and Spatial Resolution

Practical measurements are usually limited to the surface of a component because the magnetic field penetration into the bulk is limited by eddy current screening. For MBN and ABE, if the probe active diameter is large compared with the field penetration, then a simple exponential formula can be used for the magnitude of the applied field with depth, z, below the material surface, $H(z)$, which defines a skin depth δ as:

$$H(z) = H_0 e^{-z\sqrt{\pi \mu_0 \mu_r f/\rho}} = H_0 e^{-z/\delta} \qquad (9.1)$$

where ρ is the electrical resistivity, μ_r the relative permeability and f the magnetic field frequency. When the applied field amplitude is large ($\gg 100\,\mathrm{Am}^{-1}$ in steel) the

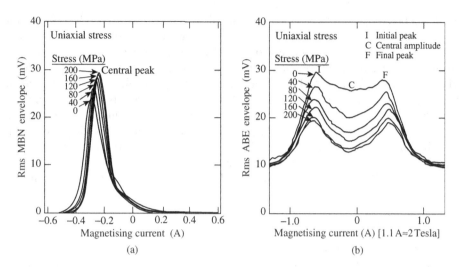

Figure 9.5 (a) MBN rms and (b) ABE rms signal envelopes as a function of magnetic field (magnetizing current) for a range of applied uniaxial stresses in a mild steel. Reproduced with permission from [16]

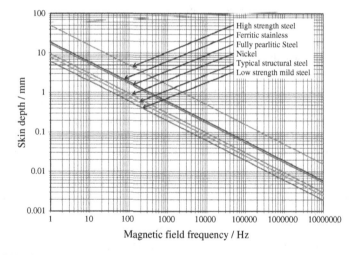

Figure 9.6 Skin depths as a function of applied magnetic field frequency in typical ferromagnetic industrial materials. Reproduced by permission of MAPS Technology Ltd

permeability cannot be assumed to be constant (Figure 9.1). Furthermore, material and stress gradients in the sample will introduce a permeability depth profile ($\mu_r(z)$). Figure 9.6 gives examples of typical field penetration for various materials, assuming the probe size is large compared with the penetration depth.

For MBN the measurement penetration is limited by the frequency content of the detected signal, for example, 0.5 to 100 kHz, and not that of the energizing field like ABE. This makes MBN more of a near surface measurement technique. Since the MBN signal has a broad bandwidth and the energizing field attenuates with depth, it is difficult to have any precision as to the depth of MBN measurement other than tens to a few hundred microns. However, using a low energizing field frequency and a lower signal frequency measurement bandwidth, information from up to 1 mm depth has been extracted [14,15].

The spatial resolution of ABE is determined by the size of the magnetizing solenoid, typically 20 × 20 mm up to 100 × 100 mm. A relatively large core is needed to apply large amplitude magnetic fields. High energizing frequency MBN measurement employs much lower applied field strengths, so the magnetizing solenoid can be smaller, for example, ∼10 × 10 mm. Low frequency MBN utilizes larger cores like ABE. The spatial resolution for MBN is set by the size of the ferrite core mounted within the pick-up coil, typically ∼1 mm, and so MBN has the highest potential spatial resolution of the three techniques.

9.2.3 Measurement

MBN measurement can be broadly separated into two categories, high and low frequency. Low frequency measurements employ magnetic field frequencies below 1 Hz with a signal detection bandwidth typically 0.5 to 100 kHz. This method allows full MBN profiles as in Figure 9.5(a) and so uses much higher amplitude magnetic fields. High frequency measurements employ field frequencies above 10 Hz and a detection bandwidth of 2 to 1000 kHz, the lower limit being well above harmonics of the drive frequency.

The MBN or ABE probe is driven from a bipolar power supply fed from a function generator that provides a triangular or sinusoidal drive waveform at the desired frequency. The MBN sensor signal is amplified (∼60 dB in soft materials, ∼72 dB in hard steels for low frequency MBN) and band-pass filtered in the required range before being acquired electronically. This is done using a PC-based system with suitable DAQ software. It extracts the rms MBN profile (Figure 9.5(a)) and other parameters such as tangential magnetic field together with a signal for the drive current. This latter signal may either be a direct input from the function generator (if the drive frequency is only a few Hz, or the voltage measured across a small sense resistor in series with the probe. Typically MBN profile measurements are repeated 1 to >10 times (the higher number being used for high frequency MBN) and averaged for better signal accuracy. The ABE transducer signal is amplified using a charge amplifier and then further amplified (typically 60 dB) and band-pass filtered before again being acquired by a PC-based system to derive rms profiles as a function of drive current (Figure 9.5(b)).

One method used to enhance and average ABE noise profiles is to superimpose a small amplitude higher frequency on the drive waveform. The frequency of this modulation also sets the measurement depth. For example the main drive waveform may be between 0.01 and 1 Hz and the drive modulation waveform between 10 and 100 Hz. The amplitude of the modulation waveform is set to be large enough for good signal to noise but small enough to prevent too much integration of the profile as a function of drive current. This method requires no further signal averaging.

Figure 9.5 also shows an example of how the MBN and ABE profiles change with applied uniaxial stress for a mild steel. Profile measurements made at different angles in the presence of a biaxial stress field enable the principal stress axes to be found, see Figure 9.7.

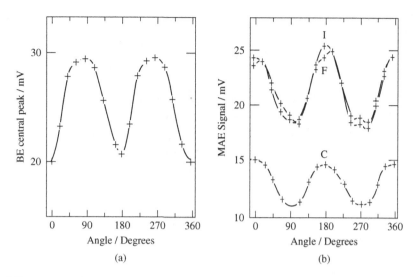

Figure 9.7 MBN (BE) peak and MAE (ABE) peak parameters (I,F & C Figure 9.5(b)) from profiles measured with the probe field at various angles to a biaxial stress state of 40 MPa at 0° and 120 MPa at 90° in a mild steel. Reproduced with permission from [16]

9.2.4 Measurement Probes and Positioning

The MBN or ABE electromagnet core should be a soft magnetic material with low losses, such as laminated silicon iron, permundar or ferrite. The latter is best suited to low field strength and high frequencies as used by the high frequency MBN method. As the lift-off strongly influences the flux density induced in the sample, the poles should be in close contact with the steel test piece. The drive current and wire turns should be optimized for achieving the required field strength.

For low frequency MBN, the sensor should consist of a high number of turns of fine wire, this requirement being relaxed somewhat when using high frequency MBN. To increase sensitivity the coil can be ferrite cored. The sensor must not be too close to the electromagnet poles (≥ 5 mm), particularly if ferrite cored. Figures 9.8 and 9.9 show typical MBN probes for low and high frequency measurement respectively, showing examples for specific geometries. The ABE transducer is typically a 1 to 5 MHz damped piezoelectric with a contact area between 5×5 mm and 20×20 mm, the size has no bearing on the measurement spatial resolution.

9.2.5 Calibration

Micromagnetic methods such as MBN and ABE are strongly influenced by the microstructure of the material. This makes it essential to have a calibration on the specific material of interest for quantitative stress measurement. Also, particularly for the parameters used in MBN, the signal response is dependent upon the particular probe design too, so it is advisable that a calibration is made with the same sample geometry and probe design. Ideally, it is best to examine the magnetic response to different property variations in the

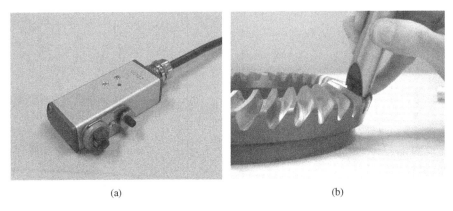

Figure 9.8 High frequency MBN devices for measurement on (a) flat surface and (b) gear flanks. Courtesy of Stresstech, Finland

Figure 9.9 Low frequency MBN devices for measurement on (a) flat surface and (b) smaller tooth-width gear flanks. Images courtesy of Design Unit, University of Newcastle

material prior to stress. Also, for MBN, these investigations may naturally begin using the low frequency method with an applied field range close to saturation. The field range can then be reduced to a minimum where enough information can still be extracted to solve the problem.

Most studies have concentrated only on measuring the response to varying uniaxial stress in, for example, a cantilever or four-point bend test. Some limited biaxial work has been done for MBN [16] however, although the author knows of no applications using MBN for biaxial stress measurement. Most work has been concerned with quality control on components where combinations of material and stress changes arising from various forms of defect are assessed in a semi-quantitative manner. Figure 9.10 shows an example of the variation in peak MBN with applied biaxial tensile stresses from 0 to 200 MPa in a mild steel. The graph shows an increase in peak MBN for parallel stress and decrease for orthogonal stress.

Figure 9.11 shows an example of the biaxial response of the ABE I and C profile parameters. Note that the ABE signal magnitude decreases for stress of either sign. ABE has

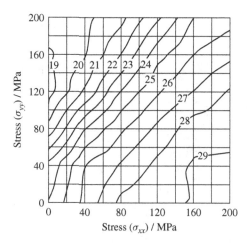

Figure 9.10 MBN peak parameter for magnetic field in x as a function of biaxial tensile stress in mild steel. Reproduced with permission from [16]

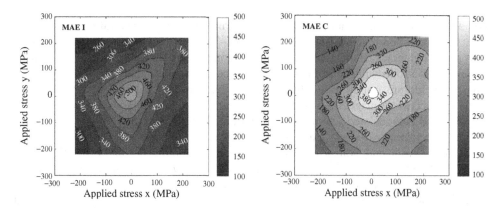

Figure 9.11 ABE (MAE) I & C Parameters with field in x as a function of applied biaxial stress in S355 steel. Reproduced by permission of MAPS Technology Ltd

been applied to quantitative biaxial stress measurement in the laboratory to assess welding stresses, see Section 9.9.1.

9.3 The MAPS Technique

9.3.1 Introduction

MAPS (Magnetic Anisotropy and Permeability System) was specifically developed for absolute biaxial stress measurement in the surface plane of industrial plant and components. It incorporates several sensors and analysis methods into a single unit, enabling absolute biaxial stress levels to be determined in a wide range of industrial materials by

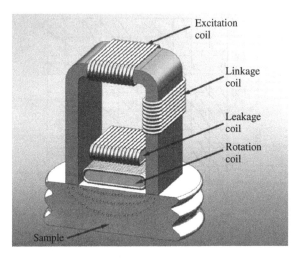

Figure 9.12 Schematic of a one of the probe coil arrangements used to measure local magnetic properties with MAPS. Reproduced by permission of MAPS Technology Ltd

providing sufficient information to partially discriminate microstructural changes and to allow full correction for poor surface quality, lift-off variation and geometric influences.

To carry out practical measurement on components a C-core electromagnet is used to apply a controlled magnetic field. In the probe configuration shown in Figure 9.12, magnetic fields are sensed using flux linkage, flux leakage and flux rotation sensors. Single or successive magnetic field frequencies are used and voltages from the sensors amplified and demodulated into amplitude and phase components.

Usually the principal stress axes are not known and so for a standard measurement the induced sensor coil voltages are monitored while rotating the probe on its vertical axis through 360° at uniform angular increments (Figure 9.13). Measurements are made using a very stable constant current amplifier driving the electromagnet at one or multiple specified frequencies. The absolute amplitude of the induced flux density in the component is much lower than saturation to ensure good stress sensitivity.

Several parameters are measured, each giving their own principal values and axes at one or more applied field frequencies. The standard analysis algorithms deliver parameters that are almost entirely invariant to lift-off variation (Section 9.6.4) but depend upon a combination of the electrical and magnetic properties such as:

- Material Delta Value for the linkage sensor (PMD)
- Material Delta Value for the flux leakage sensor (FMD)
- Stress-induced Magnetic Anisotropy from the rotation sensor (SMA).

These parameters describe a $\cos 2\theta$ variation with probe rotation (Figure 9.13). The principal magnetic parameter amplitudes for PMD and FMD are then given by the two extreme values and the stress axes by the angles of these extremes. The SMA parameter is sensitive to the in-plane shear stress and its extremes occur at the shear axes, only its peak-to-peak amplitude being important (not both peak values). In addition, the analysis

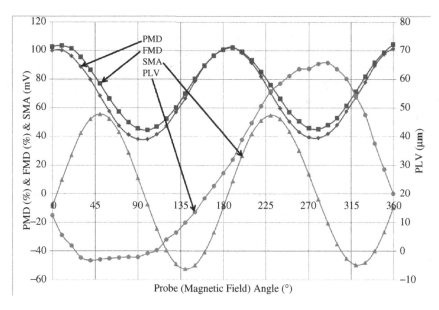

Figure 9.13 Typical magnetic parameters derived from MAPS sensors for a measurement using a 70 Hz magnetic field at various angles to a 366 MPa tension applied at 0° in 220 grade rail steel rectangular bar. Reproduced by permission of MAPS Technology Ltd

can provide parameters that are insensitive to material properties (including the stress) to give a relative measure of the probe lift-off (an average distance from the probe face plate to the component surface). Figure 9.13 includes this parameter for the linkage sensor (PLV) showing that the probe was tilted with a ≈60 µm variation during rotation.

The accuracy of the principal stress axes, $\delta\theta$, depends upon the number of angular measurements, $N(>2)$, and for PMD and FMD (*MD*) with evenly spaced angles is given by:

$$\delta\theta = \sqrt{\frac{2}{N}} \frac{\delta MD}{|MD1 - MD2|} \frac{180}{\pi} \tag{9.2}$$

where δMD is the accuracy of the cosine fit and $MD1$ and $MD2$ the extreme values. This ignores probe placement accuracy on the component and in many cases yields $\delta\theta \ll 1°$.

9.3.2 Measurement Depth and Spatial Resolution

For MAPS, provided the energizing solenoid is large compared to the skin depth, Equation (9.1), the measurement depth is the skin depth (like ABE). The measurement penetration can be adjusted by varying the drive and detected frequency. This proves to be a useful tool for investigating stress depth profiles, see Sections 9.9.3 and 9.9.4. Measurement frequencies from 3 Hz to 3 kHz (standard digiMAPS) or from 500 Hz to 10 MHz (high frequency MAPS) can be used thus allowing a wide range of measurement penetrations (Figure 9.6). When working on a thin sample the maximum penetration depth should not exceed half the component thickness unless a correction is made for the flux distortion due to the proximity of the sample back surface.

The measurement resolution in the plane of the surface is linearly related, for a linkage sensor, to the size of the electromagnet core, and for a leakage sensor, the size of the sensor. Specifically for a probe whose core has a footprint of $a \times b$ that contains a leakage sensor with dimension $c \times d$, the spatial resolutions (full width half max.) are approximately:

$$r_{\text{linkage}} \approx 2\delta + 3\sqrt{a^2 + b^2}/8 \quad \text{and} \tag{9.3}$$

$$r_{\text{leakage}} \approx \sqrt{c^2 + d^2}/2 \tag{9.4}$$

where it is assumed the probe is rotated through 360° for measurement. For the leakage coil a secondary influence from the size of the core has been neglected. For typical MAPS probes, linkage sensor resolutions from 2 mm to 30 mm are available. For a linkage sensor the measurement penetration must always be smaller than the spatial resolution. However, because the leakage sensor is decoupled from the energizing electromagnet, this limit is weakened.

9.3.3 MAPS Measurement

Figure 9.14 illustrates the digiMAPS measurement system. It is controlled via USB by a PC that hosts the proprietary control and analysis software. The unit contains a highly linear constant current amplifier to drive the probe electromagnet with a sinusoidal current at the specified frequencies and field strengths. Signals from each of the three sensors, linkage, leakage and flux rotation, are amplified ensuring no phase distortion before being digitized and demodulated into in-phase and quadrature components with respect to the

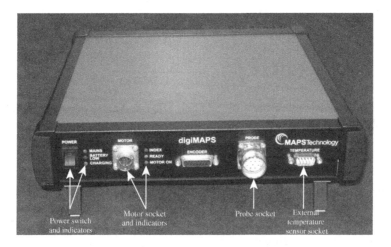

Figure 9.14 digiMAPS biaxial stress measurement unit (3–3000 Hz). Reproduced by permission of MAPS Technology Ltd

drive current phase and stored. The sensor readings are processed using software analysis algorithms to extract the different magnetic parameters as described previously.

9.3.4 Measurement Probes and Positioning

A suitable measurement probe should be chosen based on the requirements of the spatial resolution, measurement depth, typical lift-off levels to be encountered and the required stress accuracy. There is a trade-off between these factors, where larger probes operate over a higher lift-off range and offer better performance at high penetration (low frequency) but they have poorer spatial resolution. Conversely if shallow penetration and/or high spatial resolution are required a small probe is needed and the lift-off variation will need to be minimized.

MAPS probes are available with spatial resolutions from 2 mm up to 50+ mm. During measurements the probe may be rotated either manually or mounted in a motor drive unit controlled by the computer. Figure 9.15 shows a range of probes for 2 to 17 mm spatial resolution and some of the motor assemblies. Measurements should be made using a motor unit when possible, where the angular positioning of the probe is controlled by the operator through software. However if access or space is too small or the sample geometry does not allow it, it may be necessary to manually position and manipulate the probe. In that case the probe must be held in a suitable guide block to hold its axis normal to the sample surface and allow the probe to be rotated even when the surface is curved. Typically the acceptable alignment accuracy can be taken as normal ±0.5°, with

Figure 9.15 Examples of some MAPS probes (front) and motor units (back) and fittings. Reproduced with permission from [17]

a corresponding error of ±5 MPa when measuring to a depth of 1 mm in a typical mild steel. Angular increments are usually relatively small for motorized rotation, 10, 15 or 22.5°, and larger, 22.5, 45 or 90° for manual rotation, the number of angles depending upon required principal axes accuracy and time available (Equation (9.2)).

9.3.5 Calibration

To convert measurements into stress it is necessary to have a suitable calibration for the material grade. Both the magnetic properties corresponding to zero stress, either single parameter values for homogeneous materials or pairs of parameters and axes for sufficiently textured materials, and the variation of magnetic properties with stress (sensitivity) must be measured. A common way to do this is with calibration samples fitted with strain gages and loaded in a suitable testing machine. A calibration sample must be representative of the material component(s) to be inspected being the same steel grade and ideally having experienced the same thermo-mechanical treatment. Also, if shaped probes are required for the inspection, then the calibration sample should have a similar local geometry. The calibration sample need not have similar internal stresses; indeed it is best if it contains very little residual stress since these residual stresses need to be determined. There are several approaches to setting up a material biaxial stress calibration depending upon the level of accuracy desired and the effort and the available material.

1. **Experimental biaxial method**: This approach is the most elaborate, requiring a special cruciform sample to be manufactured, strain gaged and tested in a suitable four-axis testing machine. A range of biaxial loads (generating elastic strains over the biaxial plane as complete as possible) are applied and the various MAPS parameters measured. This method is usually the most accurate.
2. **Theoretical (thermodynamic) method**: Thermodynamic models, neglecting hysteresis, are used together with the MAPS measurements at known stress levels (at least six biaxial states) to evaluate theoretical biaxial maps. Typically one or more uniaxial tests are performed. If more than one, additional tests make use of a different stress state, for example, residual, orthogonal to the applied stress direction.
3. **Experimental bi-uniaxial method with complex scaling**: If a biaxial calibration for a similar material is already available then these data can be compared to uniaxial data made on the test material to generate a modified biaxial calibration. Typically, if the material is textured, a uniaxial test is carried out using samples cut along the two orthogonal material axes. The existing biaxial data are then scaled in two dimensions to fit the uniaxial test data.
4. **Mixed map method**: As biaxial calibrations for a range of materials now exist, it is possible to generate all intermediate variants by combining combinations of the existing maps such that the resulting mixed map agrees with data measured on the test material at a few known stresses.

Figure 9.16 shows how the PMD parameter from probe rotations varies for different levels of applied uniaxial stress in an S355 mild steel rectangular bar (as may be used for methods 2–4 above). Note how the PMD orthogonal to the applied stress also varies. The relationship between magnetic properties and stress is a tensor one and so, although

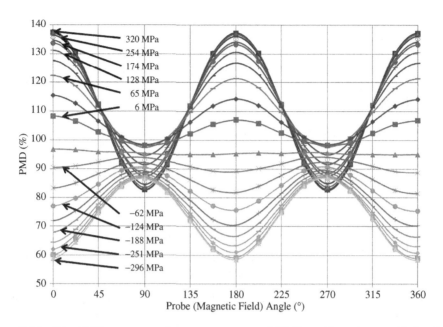

Figure 9.16 The PMD parameter for measurement at 2950 Hz with various applied uniaxial stresses between −296 and +320 MPa at 0° in S355 grade mild steel rectangular bar (only every other trace is labelled). Reproduced by permission of MAPS Technology Ltd

the parameter in the direction of the varying stress often changes the most, the magnetic properties in all axes are affected. A plot of the PMD measured in the direction of the applied uniaxial stress is shown in Figure 9.17 that also shows insensitivity to variations in probe lift-off.

A common way of applying uniaxial stress is by bending. However allowance must be made for the measurement penetration below the surface. As Figure 9.18 demonstrates, successively lower measurement frequencies "sample" deeper into the steel and so are influenced by successively lower average stress levels.

A careful look at Figure 9.16 shows that the smallest PMD variation with probe angle is not at minimum applied stress, actually it is at −28 MPa. Similarly, in Figure 9.18, the maximum PMD is observed at 2144 Hz not 2950 Hz. Both of these are due to small levels of residual stress in the calibration sample, the latter being highest and compressive at the surface. These residual stresses must be determined and added to the applied stresses in order to derive the correct calibration. Usually the measurement data can be used together with models that describe the underlying symmetry of the calibration maps on the biaxial stress plane to evaluate these residuals.

Figures 9.19 and 9.20 show examples of biaxial calibrations generated using methods 1 and 2. To extract stress levels and errors at each measurement location each magnetic parameter value defines a contour line on the biaxial plane with a width equal to the standard error of the measurement. The region where all parameter lines cross (PMD 1 and PMD 2 contour lines in calibration example of Figure 9.19) then defines the unique biaxial stress state together with the standard error.

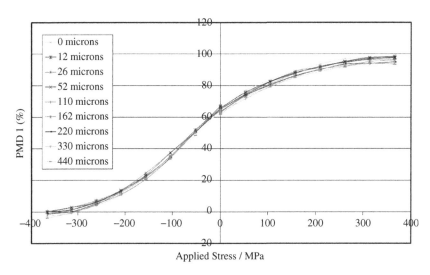

Figure 9.17 PMD 1 Principal parameter measured at 70 Hz as a function of applied uniaxial stress in a 220 grade rail steel bar using a 10 mm probe with a 150 μm face plate. Additional stand-off varied from 0 to 440 μm. Reproduced by permission of MAPS Technology Ltd

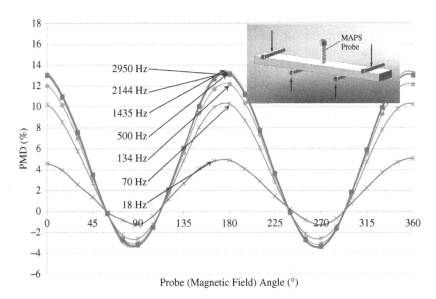

Figure 9.18 The PMD parameter for measurement at various frequencies for an applied uniaxial stress 320 MPa at 0° in S355 grade mild steel rectangular bar under four-point loading (inset). Reproduced by permission of MAPS Technology Ltd

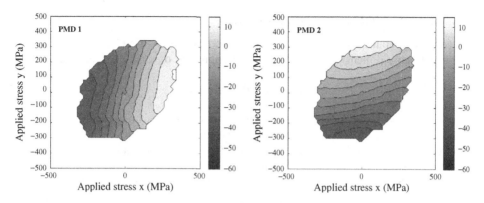

Figure 9.19 Example of biaxial calibrations for PMD 1 and PMD 2 parameters generated using the first method. 220 grade rail steel. Reproduced by permission of MAPS Technology Ltd

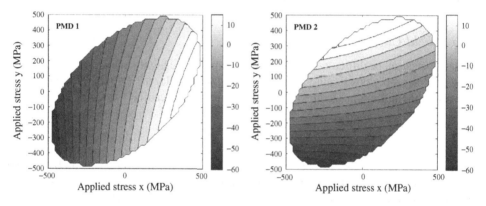

Figure 9.20 Example of biaxial calibrations for PMD 1 and PMD 2 parameters generated using the second method. 220 grade rail steel. Reproduced by permission of MAPS Technology Ltd

9.4 Access and Geometry

9.4.1 Space

To place the probe onto the surface and make any necessary adjustments there must be sufficient access on the component. In practice, the operator must be able to place the probe and their hand into position while being able to see that the probe is placed, operating properly and, if appropriate, that the probe can be held in position firmly. If the surface area is flat or uniformly curved over a region smaller than the motor assembly footprint, then consideration should be made for taking the measurements by hand using a small non-metallic guide block.

9.4.2 *Edges, Abutments and Small Samples*

Both edges and abutments will cause signal distortion if they are close to the probe. For a probe with core dimension $a \times b$ the closest approach, d, to an edge is given by:

$$d = \alpha\delta + 0.5\sqrt{a^2 + b^2} \qquad (9.5)$$

where the factor α is set depending upon the desired final measurement accuracy, a value of 2 being typical, and 3 to reduce the distortion to a negligible level.

Samples large enough to satisfy Equation (9.5) may nevertheless allow alternative eddy current paths from the energizing magnetic field and result in signal distortion. For example a probe placed with the field direction along the longitudinal axis of a narrow thin bar will result in some induced currents flowing around the back face of the bar as well as the normal flow around the poles. Therefore a probe rotation in the unstressed case will exhibit an apparent magnetic anisotropy. In this case the usual stress sensitivity calibration must be on the same geometry or an adequate theoretical model must be used to describe the distortion of a calibration taken on a larger sample.

9.4.3 *Weld Caps*

Measurements can be difficult on a weld cap because of the highly uneven surface, and this can affect the proper positioning of the probes. If acceptable, the weld cap can be dressed flush. Then the normal procedures for surface preparation can be used, see Section 9.5.

9.4.4 *Stranded Wires*

When the sample has dimensional aspects smaller than the measurement probe, several new factors influence the measurement and can make conventional analysis methods fail. A common example is stranded steel rope (also flexible risers Section 9.9.5). In these cases the geometry may limit or prevent conventional eddy current flow around the probe poles, thereby reducing sensitivity to stress. Also, independent movement of the strands under the probe as well as uncontrolled electrical shorts between them contribute additional variations to the measured signals. Analysis methods have been recently developed for MAPS that allow separation of lift-off, electrical inter-connection variation and wire movements from stress.

9.5 Surface Condition and Coatings

It is generally advisable not to attempt to prepare the surface of industrial components using hand tools or machining unless it has been abused or is in some way not typical of the in-situ condition. The surface stresses may be part of the information to be determined. Calibration samples are an exception where the material is ideally uniform with low or known residual stress levels.

Since the measurement penetration is usually relatively high, the MAPS and ABE methods are less surface-critical than many other surface stress measurement techniques. Conversely the surface condition (from oxides, microstructure and stress gradients) will have a stronger effect for the smaller measurement penetration of MBN and high frequency MAPS. Manufacturing processes such as machining, grinding and shot-peening can result in severe thermal damage, plasticity and changes in the residual stress state at the surface. For high frequency measurement, surface preparation using chemical and/or electro-chemical methods is preferred over mechanical methods. Generally, low frequency measurement should be used in cases where the variations in surface properties can be ignored and subsurface properties are considered important. Where some level of surface preparation cannot be avoided a detailed guide for hand or machined preparation, dealing with rust and loose scale, surface roughness and pitting and cast surfaces can be found at [17].

MAPS measurements can be made through coatings such as paint or conductive cladding such as non-magnetic stainless steel. However, the uncertainty in stress will increase depending upon the coating thickness, see Section 9.6.4. High frequency MBN measurements are very sensitive to near surface properties, and the coating will strongly affect the field penetration and hence the MBN signal detection in this case. If the coating needs to be removed the normal procedures for surface preparation should be used. For the low frequency MBN measurement, a thin coating would not significantly affect the MBN signal level.

Thick cladding will strongly affect the magnetic field penetration, requiring large probes. For example the MAPS-FR probes with their 85×60 mm footprint are designed to routinely measure through up to 30 mm of sheath (Section 9.9.5).

9.6 Issues of Accuracy and Reliability

Primary factors affecting the accuracy of magnetic stress measurement techniques are:

- The ultimate precision and stability of the equipment (Section 9.6.5).
- Ferromagnetic issues such as material magnetic and loading history (Section 9.6.1).
- Calibration issues such as the intrinsic magneto-elastic response of the material and microstructure variation (Section 9.6.2).
- Operational factors such as probe design, measurement field frequency and probe lift-off (Sections 9.6.3 and 9.6.4).
- Application aspects such as surface condition (Section 9.5) and geometry (Section 9.4) and the presence of electric currents (Section 9.6.6).

The following sections expand on many of these factors.

9.6.1 Magnetic and Stress History

9.6.1.1 Origins

The magnetic and stress history of the component influences measurements due to the magnetic hysteresis properties of the material. Both prior applied magnetic fields (leaving the material at a remanent point) and prior applied stress (leaving domains partially aligned to the prior maximum stress axis) alter the permeability. For prior applied magnetic fields, it is not usually the remanent field that is the problem, unless it moves the material around the hysteresis loop significantly. Rather, it is the partial alignment of the magnetic domains. The size of the problem depends on three factors:

1. The likely size of prior applied magnetic fields or stresses. This is often unknown.
2. The magnetic hardness (coercivity) of the material. For relatively soft magnetic materials such as low carbon steels these problems should be small, typically causing uncertainties of the order of a few MPa. For harder materials such as fully pearlitic or martensitic steels the arising uncertainties become large, typically a few tens of MPa (>100 MPa in some very high strength steels).
3. The amplitude of the MAPS or MBN applied magnetic fields. Both MAPS and high frequency MBN use fields significantly below saturation but the high fields used by the low frequency MBN method erase the prior domain states thus removing the problem. For intermediate field strengths the hysteresis problem depends upon the ratio of the coercive force from the applied field and material coercivity.

In cases where these errors are too large, a local demagnetization procedure should be used prior to each measurement.

9.6.1.2 Demagnetization and Magnetic Conditioning

Demagnetization, the removal of remnant fields from the component, is not the same as randomizing the magnetic domain distribution. A demagnetization is achieved by applying a high amplitude low frequency alternating field that decays to zero over a number of cycles. However the magnetic domains will then be partially aligned parallel and antiparallel to the applied field axis introducing a magnetic anisotropy. It is difficult to fully randomize the domains in all three axes, but in-plane isotropic methods can be used. There are various ways of achieving this *magnetic conditioning*. Figure 9.21, which is an example of a uniaxial stress calibration carried out with and without magnetic conditioning, shows the increased accuracy obtained. It is important to use magnetic conditioners that randomize magnetic domains over a volume, including in the depth direction, at least as large as that interrogated by the measurement.

9.6.2 Materials and Microstructure

9.6.2.1 Alloy Content

The magnetic properties of materials are affected by the chemical composition and structure. At a fundamental level the alloy content will affect whether the material is ferromagnetic. Thus alloys that are close to 100% iron or nickel with unpaired d-electrons are strongly ferromagnetic, but alloys with intermediate concentrations of iron or nickel with other elements may be either ferromagnetic or not, depending on the filling on the d-band. The alloy composition also determines whether the structure is f.c.c. or b.c.c., which in turn affects the magnetic properties. In steels, the carbon content is important mainly for determining magnetic softness or hardness because the carbon tends to influence the DW movement whether in solution (causing crystal lattice strain) or precipitate form (pinning the DWs). Thus the sensitivity of a magnetic parameter to stress varies with alloy composition.

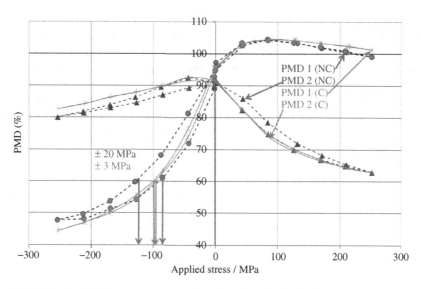

Figure 9.21 Principal PMD parameters as a function of applied uniaxial stress measured on a mild steel with no magnetic conditioning (NC) and with conditioning (C). Reproduced by permission of MAPS Technology Ltd

Low strength steels containing reasonably large grains of ferrite have good stress-sensitivity (nickel alloys and superalloys can be better). Cast iron and low alloy steels have lower sensitivities but are usually acceptable as are high-strength bearing steels. Higher carbon steels and partially magnetic stainless steels can also give acceptable results. Weakly magnetic stainless materials can be inspected but the risk of a poor result is higher.

9.6.2.2 Grain Alignment

The grain boundaries in polycrystalline material are regions of magnetic discontinuity due to the grain's different crystallographic orientation, and so domains do not cross grain boundaries. In addition, the resulting magnetic free poles at the boundaries generate demagnetizing fields and hence influence local DW movement. The grain size and any grain alignment, such as in rolled plate, may be expected to influence the mean free path of DW displacement and hence the magnetization process. However, when the domains are very much smaller than the grains, as in most steels excluding low strength steel sheet, these effects may be small.

9.6.2.3 Phases and Material Property Variability

Domain walls are pinned by various microstructural features introduced during fabrication, including grain boundaries, phase boundaries, precipitates, inclusions and dislocations; the pinning depending upon the nature of the phases present. If only one of the phases is magnetic or there is a large difference in properties, then the phase boundary will strongly pin the walls, raise the coercivity and cause the hysteresis loop to broaden and the material to harden magnetically, for example, in pearlite with alternating ferrite and cementite (iron carbide) laths compared to ferrite grain structure. In the case of precipitates or inclusions, it depends on their magnetic properties; if they are non-magnetic, as is often the case in steels, then they pin the DWs strongly, again raising the coercivity [18,19]. Thus the sensitivity of a magnetic measurement to stress can change during heat treatments such as annealing and ageing, and may vary from specimen to specimen of nominally the same material. An important example arises near welds where the heat-affected zone will contain significant variations in microstructure just where knowledge of residual stress levels may be desirable.

Another important example is case hardening of steel to produce a martensitic surface layer. Martensite has a fine lath structure with dissolved carbon, and this structure strongly pins DWs relatively increasing MBN. However, because of its different crystal structure, martensite is heavily strained, so that there is an internally generated residual stress that will change the local permeability. Other materials that can be problematic are: cast iron (property variations), duplex steel, and weakly magnetic stainless steels with ferritic content variation. For most magnetic techniques it is not generally possible to fully separate the microstructural and stress effects.

9.6.2.4 Plastic Strain

Thermo-mechanical treatment, in-service creep and fatigue loading will plastically deform the material, in which case it will contain networks of dislocations and different substructures. Also bending beyond the elastic limit will generate a depth dependent combination of plastic and elastic strains, which will remain unless the material is annealed. Dislocations will pin the magnetic DWs, although the effect may not be great unless there has been heavy deformation. High dislocation density, such as that of quenched steel, will give a low level of MBN due to restricted DW movement and reduced magnetic permeability ([20,21]). A simple magnetic measurement will be sensitive to both stress and plastic strain.

9.6.3 Magnetic Field Variability

Magnetic properties depend upon the applied field strength (Section 9.1.1) and so methods that use low fields (MAPS and low applied field MBN) rely on controlling that field strength (in the sample) within some tolerance. Thus even with good lift-off compensation (Section 9.6.4) a variation in lift-off will modify the observed stress sensitivity slightly. The impact on absolute accuracy depends upon the material and working magnetic field strength.

9.6.4 Probe Stand-off and Tilt

MAPS measurements are automatically compensated against uncontrolled variations in lift-off, although increasing lift-off results in increased measurement uncertainty due to the decreasing signal to noise ratio. The non-linear relationship between MAPS PMD parameter and stress results in different stress states yielding different errors. For a typical pipeline steel and averaging the response over ±80% of the elastic range, Figure 9.22

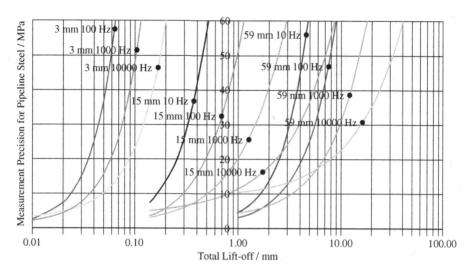

Figure 9.22 How the measurement accuracy depends upon the probe size, measurement frequency and lift-off (pipeline steel example). Reproduced by permission of MAPS Technology Ltd

shows how probe size, measurement frequency and probe lift-off affect the accuracy of the equipment. So, although accuracy degrades with standoff, larger probes deliver better accuracy at a particular standoff and higher frequencies are more accurately measured provided they are below the frequency limit of the probe and the hardware can drive the probe at that frequency and required current.

9.6.5 Temperature

Electrical resistivity and magnetic permeability both change with temperature. Also the electronic hardware and probe may exhibit some temperature sensitivity.

9.6.5.1 Electrical and Magnetic Properties

The eddy current density induced in the steel below the probe depends upon the electrical resistivity. The electrical resistivity has a positive linear temperature coefficient (typically $\sim 5 \times 10^{-3}/°C$) so as the temperature changes, both the magnetic field penetration (Equation 9.1) and the induced eddy currents in the surface of the steel will change. As the temperature increases the measurement penetration will also increase causing a slight drop in the magnetic flux density in the steel surface.

Ferromagnetic magnetization initially drops only very slowly with absolute temperature until the Curie temperature is approached (770°C in iron, 358°C in nickel), when the decline becomes much more rapid. The magnetic permeability typically initially increases with temperature before reaching a peak and dropping dramatically to zero at the Curie temperature. Thus for a constant current probe and a fixed standoff, the flux density in the surface of the steel should increase slowly with temperature for ambient temperatures.

9.6.5.2 Measurement Probe Response

Temperature sensitivity of the measurement probe arises from thermal expansion of certain parts and sensors, and, if the probe is in a metal can, changes in resistivity of the can. These are normally compensated using a temperature sensor in the probe together with temperature calibration coefficients.

9.6.5.3 Electronics Stability

The precision and stability of the equipment is beyond the scope of this book. For MAPS, the hardware must be considered both in terms of the ability to deliver an accurate and stable current to the probe drive coil and in the measurement of both the amplitude and phase of each sensor signal with respect to the coil current with minimal drift and sufficient resolution. It is good practice to leave equipment turned on for a short period prior to measurement to allow the electronics to reach thermal equilibrium. The performance of the hardware contributes to achieving the minimum random error for stress measurement.

9.6.5.4 Temperature Summary

The different physical factors influence measurement in opposing directions making accurate prediction difficult. For mild steel, comparisons of the changes in experimental measurements with the typical stress sensitivity indicate, that at low stresses, a $1\,^{\circ}\mathrm{C}$ change in temperature of the test component will equate to a typical equi-biaxial error in stress of $0.7\,\mathrm{MPa}$. Therefore provided temperature differences are kept to a few degrees, these sources of error should be small. For large temperature changes, accurate work or work with an insensitive steel will be necessary to apply temperature compensation coefficients.

9.6.6 Electric Currents

Some consideration should be given as to whether there are likely to be electric currents flowing in the test component (e.g., traction power supply return currents or signalling currents in rails). Such currents may not only directly induce voltages in the sensors but, if they are large enough, induce significant magnetic fields in the component to reduce the magnetic permeability. Direct induction across the sensor coils may not be a problem if the frequencies are well away from the measurement frequencies (allowing for bandwidth) but changes in magnetic properties must be dealt with either by measuring only when the currents are low enough or by directly calibrating their effect on the measurement.

9.7 Examples of Measurement Accuracy

Table 9.1 shows typical examples of expected or realized accuracy for single frequency MAPS measurements, where pragmatic choices have been made for the calibration method and measurement procedures employed in industrial applications. Random errors assume a single frequency biaxial measurement, except Case 5.

When multi-frequency measurements are deconvolved to obtain stress depth profiles, the random errors are unevenly distributed with reduced errors for the near surface stress

Table 9.1 MAPS Measurement accuracies for several applications using different probes and frequencies on various components. Case 5 used multi-frequency depth deconvolution

Case	Measurement Conditions				
	Probe Active Footprint Size (mm) [Spatial Resolution]	Measurement Frequency (Hz) [skin depth (mm)]	Typical Stand-off (mm)	Measurement Angles (°)	Magnetic Conditioning (Yes/No)
1	10 × 12 [7] 10 × 12 [10]	1000 [0.4] 35 [2]	0.3	0,22.5,45...360	No
2	3 × 3 [2.6] 3 × 3 [4]	3000 [0.5] 400 [1.3]	0.05	0,90	No
3	15 × 13 [9] 60 × 60 [70]	3000 [1] 10 [19]	0.5	0,45,90...360	Yes
4	10 × 12 [6] 10 × 12 [14]	3000 [0.26] 17 [4.3]	0.3	0,22.5,45...360	No
5	45 × 45 [28]	1000 to 9 [0.4–4]	2	0,45,90...270	Yes
6	60 × 42 [24]	70,140,280 [N/A]	16	90	No
7	3 × 3 [1.6]	150,000 [0.026]	contact	0,90	No

Case	Application				Accuracy (± Std. Deviation)	
	Industry	Component	Steel Grade	Geometry	Calibration Type	Random (MPa) [%yield]
1	Oil & Gas	Plate	S355	Flat	Experimental biaxial	9 [2.5] 30 [9]
2	Oil & Gas	Tensile wire riser re-inforcement	As case 6	Locally Flat	Theoretical method	10 [2.5] 25 [1.0]
3	Aerospace	Landing Gear	300 M	Flat	Experimental biaxial	25 [1.3] 10 [0.7]
4	Oil & Gas	Pipeline	X52	Radius +300 mm	Mixed map	5 [1.4] 27 [7.6]
5	Railroad	Rail web	136 RE	Radius ~−500 mm	Multiple rails Axial Loading	5 [1]
6	Oil & Gas	Riser	UNS G10600 ASTM A29	Multi-strands 12 × 7 mm section	Riser Loading	30 [3]
7	Aerospace	Bearings	M50	Double Radii +150 & −12 mm	Experimental biaxial	15 [1]

components and increased ones at the larger depths. However components can exhibit higher variability in stress at the surface, a sub-surface measurement often being more useful depending upon the application. Also systematic errors can arise from the calibration, which without care and effort can be larger than the random ones.

9.8 Example Measurement Approaches for MAPS

Before commencing measurements, several practical aspects should be considered. First, it is helpful to understand how the data are to be used and what stress distributions may be expected in the component. For example, are absolute stresses required or only bending moments or load assessments? This information will influence the requirements for an optimum calibration and also where the measurements should be taken, the choice of suitable probes and what measurement characteristics should be used. Depending upon the surface condition or if a surface treatment exists on the component, a careful choice of penetration depth for the measurements needs to be made, with a decision whether to use a stress profiling option during data collection. In some cases it will be necessary first to prepare the surface.

The number and location of measurements will depend primarily on the requirements of the test program and on the component geometry, location of welds, features, and so on, and may be limited in some cases by the need for access and the shape of the component surface and proximity of edges and abutments. The probe manipulation procedure (static non-rotating, manual rotate by hand, motor automatic rotation, automatic translation etc.) should be planned carefully depending upon access and geometry and an appropriate method used to accurately place the probe orthogonal to each measurement position.

9.8.1 Pipes and Small Positive and Negative Radii Curvatures

Many important applications involve measurement on curved component surfaces and pipework. The use of flat probes on curved surfaces is possible but an estimate of the optimum size must be made, larger probes being more tolerant of the stand-off, but smaller ones yielding less stand-off. Performance data like Figure 9.22 should be combined with simple average standoff calculations, for both axial and hoop orientations, to decide the best size. There are limits to this and a shaped core probe set can alternatively be used. These cannot be rotated and so using a shaped 0° and 90° probe pair allows measurements in the geometry axes, or a third 45° probe can be added, allowing principal stresses to be measured.

9.8.2 Rapid Measurement from Vehicles

Probe rotation is not practical and so a short train of static probes, each aligned in one direction, is used. In order to avoid magneto-dynamic distortion, the measurement frequency must be higher than the ratio of the vehicle velocity and probe size, thus limiting the maximum measurement penetration. Conventional measurements also have limited spatial resolution, as large probes are typically used to operate over 2 to 5 mm lift-off variation.

9.8.3 Dealing with 'Poor' Surfaces in the Field

Rough and pitted surfaces with loose oxides should be cleaned with a wire brush and multi-frequency measurements used together with depth deconvolution to extract sub-surface properties for the component. Surface measurements are unreliable.

9.9 Example Applications with ABE and MAPS

9.9.1 Residual Stress in a Welded Plate

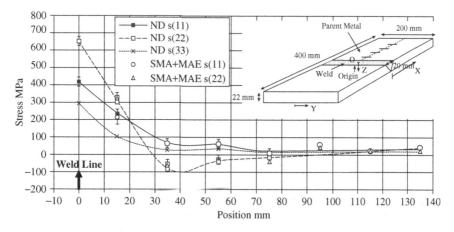

Figure 9.23 Stress measurements made on a butt welded S355 steel plate (inset) using neutron diffraction and a magnetic technique combination (MAE(ABE) & SMA). Reproduced from [9]

9.9.2 Residual Stress Evolution During Fatigue in Rails

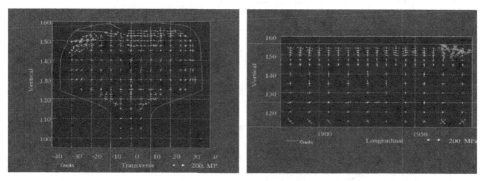

Figure 9.24 MAPS biaxial stress levels measured over the transverse and longitudinal cut faces of worn steel rails showing a ~10 mm deep compressive layer introduced by traffic. Each biaxial measurement is represented by a pair of orthogonal arrows aligned with the principal axes with a length proportional to the stress magnitude (refer to scale bar). Red arrows pointing out indicate tensile and green arrows pointing in compressive stress levels. Note how gage corner cracks disrupt the compressive stress layer with significant tensile stress. These tensile stresses evolve prior to the onset of cracking. Reproduced with permission from [22]. Copyright 2004 British Institute of Non-Destructive Testing

9.9.3 Depth Profiling in Laser Peened Spring Steel

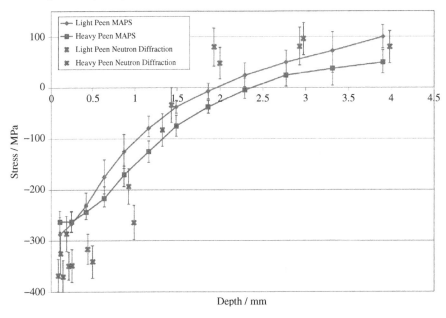

Figure 9.25 Example of depth deconvolved multi-frequency MAPS data compared with neutron diffraction data measured on two laser shock-peened spring steel samples with permission from NPL project MPP8.5 on "Advanced Techniques for Residual Stress Measurement", which is part of the programme on Measurements for the Processability and Performance of Materials (MPP), funded by the Engineering Industries Directorate of the UK Department of Trade and Industry

9.9.4 Profiling and Mapping in Ring and Plug Test Sample

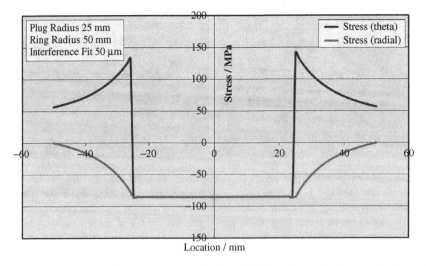

Figure 9.26 Predicted radial and theta stress levels within the ring and plug sample for an interference fit of 50 μm following heating ring and cooling plug to fit

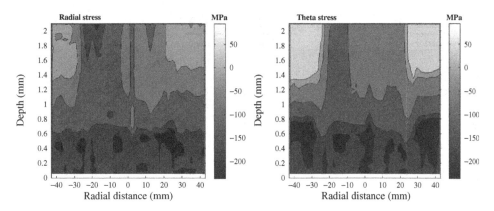

Figure 9.27 Radial and theta stress components 'depth deconvolved' from multi-frequency MAPS measurements on the shot peened ring and plug sample surface. Expected radial and theta stress patterns "emerge" below the high surface compressive peening stresses. Reproduced by permission with permission from NPL project MPP8.5 on "Advanced Techniques for Residual Stress Measurement", which is part of the programme on Measurements for the Processability and Performance of Materials (MPP), funded by the Engineering Industries Directorate of the UK Department of Trade and Industry

9.9.5 Measuring Multi-stranded Structure for Wire Integrity

Figure 9.28 Full MAPS-FR R151 inspection tool (bottom left) mounted on a flexible riser (example section top left) in Dynamic Testing Rig (right) monitoring for broken armor wires. Broken wires carry reduced stress over 10 to 20 m from their break

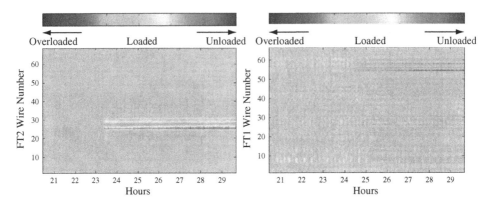

Figure 9.29 Stress levels monitored in outer (left plot FT2) and inner armor wires (right plot FT1) showing two or three outer wire breaks after ~23 hours and increased stresses on some inner wires a short time after outer wire failure. MAPS-FR tool is ~8 m from breaks [23]. Reproduced with permission from [23]. Copyright 2012 International Society of Offshore and Polar Engineers

9.10 Summary and Conclusions

Magnetic techniques offer a practical approach to stress measurement *in-situ* on industrial components and structures that are not accessible by conventional methods. They are relatively fast, making it possible to make detailed maps of the stress distribution over surfaces. Three methods have been reviewed in this chapter, MBN, ABE and MAPS. The first two are based on micro-magnetic processes and the third on the macroscopic magnetic properties. All three techniques are influenced by the internal stress state via magnetostriction.

The speed and ease of application in the field of the various methods are tempered by the indirect relationship between the measured parameters and the engineering stress state. This makes calibration the most important issue to solve for a successful deployment. For MAPS, this procedure equates to a calibration for material grade, but for MBN it must also include the different test geometries and different probes as their response differs. For MAPS, although more generic, the calibration is difficult because of the complicated tensor relationship between magnetic properties and stress. This is also true for MBN but quantitative biaxial stress measurement has not been attempted. Instead MBN is typically used for qualitative assessment of residual stress variation with simpler uniaxial calibration work.

The surface of the component to be inspected needs careful consideration. The following actions are helpful:

- Avoid surface waviness or high near-surface stresses introduced by grinders or machining.
- Remove any loose rust.
- For MBN consider preparing the surface using chemical or electro-chemical methods.
- For MAPS and ABE always use the largest probe compliant with the required spatial resolution when working on flat surfaces.

- If the material is magnetically hard consider a local demagnetization or "magnetic conditioning" procedure prior to measurement.
- Always ensure the probe is supported accurately orthogonal to the surface.

It is important to consider what stress information is required, absolute, bending and so on, because it allows tailoring of the calibration. Also if the surface is damaged or treated in some way and bulk stresses are required then use multiple measurement frequencies with depth deconvolution. Single "spot" measurements should be avoided if short scans can be used.

MBN is classified into two types, high frequency rapid measurement and lower frequency noise profile measurement, the latter giving more information and greater depth penetration but requiring the use of larger probes.

The absolute biaxial stress measurement of MAPS is unique to magnetic methods, each measurement yielding principal biaxial stress levels and axes in the plane of the surface with a sampling depth depending upon the applied magnetic field frequency. The stress orthogonal to the surface plane is currently assumed to be zero.

References

[1] Bozorth, R. M. (1993) *Ferromagnetism*, Wiley: New York.
[2] Bleaney, B. I., Bleaney, B. (1993) *Electricity and Magnetism*, 3ed, Oxford University Press: Oxford.
[3] Hauser, H. (1994) "Energetic Model of Ferromagnetic Hysteresis," *Journal of Applied Physics* 75(5):2584–2597.
[4] Sablik, M. J., Riley, L. A., Burkhardt, G. L., Kwan, H., Cannell, P. Y., Watts, K. T., Langman, R. A. (1994) "Micromagnetic Model for Biaxial Stress Effects on Magnetic Properties." *Journal Magnetism and Magnetic Materials* 132:131–148.
[5] Buttle, D. J., Scruby, C. B. (2001) *Residual Stresses: Measurement using Magnetoelastic Effects, The Encyclopaedia of Materials: Science and Technology*, Pergamon Press: Elsevier Science.
[6] Jiles, D. (1998) *Introduction to Magnetism and Magnetic Materials*, Chapman & Hall / CRC, Boca Raton.
[7] Abuku, S. (1977) "Magnetic Studies of Residual Stress in Iron and Steel Induced by Uniaxial Deformation," *Japanese Journal of Applied Physics* 16:1161–1170.
[8] Langman, R. A. (1981) "Measurement of the Mechanical Stress in Mild Steel by Means of Rotation of Magnetic Field Strength," *NDT Int*, Part 1 Oct, 1981 255–262, Part 2 Apr 1982 91–97.
[9] Allen, A. J., Buttle, D. J., Dalzell, W., Hutchings, M. T. (2000) Residual Stress in Butt Weldments of 50D Steel Measured by Neutron Diffraction and Magnetic Techniques, Proc. 6th Int. Conf. Residual Stresses (ICRS6), Oxford, UK, 923–931.
[10] Schneider, E. (1998) Nondestructive Analysis of Stress States in Components using Micromagnetic and Ultrasonic Techniques – An Overview, Proc 7th ECNDT, Copenhagen, Denmark, 3(11).
[11] Kwan, H. (1986) "A Non-destructive Measurement of Residual Bulk Stresses in Welded Steel Specimens by Use of Magnetically Induced Velocity Changes for Ultrasonic Waves." *Materials Evaluation* 44:1560–1566.
[12] Pasley, R. L. (1970) "Barkhausen Effect – An Indication of Stress," *Materials Evaluation* 28(7):157–161.
[13] Tiitto, S. (1977) "On the Influence of Microstructure on Magnetisation Transitions in Steel," *Acta Polytechnica Scandinavica, Applied Physics* 119:1–80.
[14] Moorthy, V., Shaw, B. A., Mountford, P., Hopkins, P. (2005) "Magnetic Barkhausen Emission Technique for Evaluation of Residual Stress Alteration by Grinding in Case-carburised En36 Steel," *Acta Materialia* 53:4997–5006.
[15] Moorthy, V., Shaw, B. A. (2008) "Magnetic Barkhausen Emission Measurements for Evaluation of Material Properties in Gears," *Nondestructive Testing and Evaluation* 23(4):317–348.
[16] Buttle, D. H., Dalzell, W., Scruby, C. B., Langman, R. A. (1989) Comparison of three magnetic techniques for biaxial stress measurement. Proc. of Rev of Progress in Quantitative NDE, Bowdoin College, Brunswick, Maine, USA, July 23–28, Plenum Publishing.

[17] Buttle, D. J., Moorthy, V., Shaw, B. (2006) Determination of Residual Stresses by Magnetic Methods, Measurement Good Practice Guide No. 88, National physical Laboratory, Teddington, UK.
[18] Buttle, D. J., Scruby, C. B., Jakubovics, J. P., Briggs, G. A. D. (1987) The Measurement of Stress in Steels of Varying Microstructure by Magnetoacoustic and Barkhausen Emission, Proc. Royal Society of London A414:469–497.
[19] Moorthy, V., Vaidyanathan, S., Jayakumar, T., Baldev, R. (1998) "On the Influence of Tempered Microstructures on Magnetic Barkhausen Emission in Ferritic Steels," *Philosophical Magazine* A77(6):1499–1514.
[20] Maker, J. M., Tanner, B. K. (1998) "The In-situ Measurement of the Effect of Plastic Deformation on the Magnetic Properties of Steel: Part I Hysteresis Loops and Magnetisation." *Journal of Magnetism and Magnetic Materials* 184:193–208.
[21] Moorthy, V., Choudhary, B. K., Vaidyanathan, S., Jayakumar, T., Bhanu Sankara Rao, K., Baldev, R. (1999) "An Assessment of Low Cycle Fatigue Damage Using Magnetic Barkhausen Emission in 9Cr-1Mo Ferritic Steel," *International Journal of Fatigue* 21:263–269.
[22] Buttle, D. J., Dalzell, W., Thayer, P. J. (2004) "Early Warnings of the Onset of Rolling Contact Fatigue by Inspecting the Residual Stress Environment of the Railhead," *Insight* 46(6):344–348.
[23] McCarthy, J. C., Buttle, D. J. (2012) MAPS-FR Structural Integrity Monitoring of Flexible Risers, 22nd International Ocean and Polar Engineering Conference, Rhodes, Greece.

10

Ultrasonics

Don E. Bray
Don E. Bray, Inc., Texas, USA

10.1 Principles of Ultrasonic Stress Measurement

The presence of stress within a material slightly alters the speed of acoustic waves traveling within the material. This is called the acoustoelastic effect. Thus, accurate measurement of the acoustic wave speed can provide an evaluation of the stresses present. Such measurements are typically done using ultrasonic waves and provide a practical method for evaluating residual stresses.

Ultrasonic stress measurement is accomplished using one of two common probe arrangements: the critically refracted longitudinal L_{CR} wave and the shear wave in birefringence mode. Parameters that are important in the stress measurement are the direction of the stress field being investigated and the particle motion and propagation path of the ultrasonic wave. The L_{CR} wave travels parallel to the surface of the specimen shown in Figure 10.1(a) and has particle motion also parallel to the surface. This velocity is designated V_{11}, where the first subscript designates the direction of travel and the second the particle motion. This wave has the maximum sensitivity to stress, as reported by Egle and Bray [1]. For the shear wave propagating across the thickness of the block, Figure 10.1(b), the greatest sensitivity to stress is for the wave with particle motion parallel to the stress field, V_{21}. The L_{CR} wave, traveling in the 11 direction is more sensitive to stress than the shear wave travelling across the field. The least sensitivity is for the particle motion perpendicular to the stress field, V_{13} and V_{23}.

Shear wave birefringence uses two contact shear probes acting across the thickness of the part. Shear waves are polarized so that the particle motion is perpendicular to the direction of propagation. With this method, the velocity (or the measured arrival times) associated with V_{23} and V_{21} can be compared. For homogeneous isotropic material, the stress effect would be shown by the travel time for V_{21} and the material zero stress travel time shown by V_{23}. Real materials are seldom isotropic and homogeneous, but

---

Practical Residual Stress Measurement Methods, First Edition. Edited by Gary S. Schajer.
© 2013 John Wiley & Sons, Ltd. Published 2013 by John Wiley & Sons, Ltd.

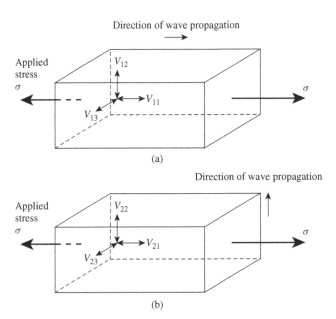

Figure 10.1 Particle motions and direction of wave travel for plane waves (a) L_{CR} wave and (b) shear wave. Reproduced with permission from [20], Copyright 1997 Taylor and Francis

nevertheless, this arrangement has been used successfully. It is essential that the two measurement surfaces are parallel. Temperature affects the wave speed and calculation of the stress must be adjusted accordingly, as will be discussed in a later section.

The critically refracted longitudinal wave L_{CR} is typically excited just underneath the surface in plates and bars, at approximately the first critical angle as defined by Snell's Law. Snell's Law states that the incident and refraction angles at an interface are governed by a constant that is the ratio of the incident and the refracted wave speeds of particular waves. Figure 10.2 illustrates a typical arrangement where an incident longitudinal wave (T) at speed C_1 in material A at angle θ strikes the interface with material B. The first critical angle is where the refracted longitudinal wave is at 90°, as shown, and travels parallel to the surface. Typically, polymer (PMMA) wedges, or immersion probes, are used in this arrangement. In material B, which has a longitudinal wave speed, C_1' greater than C_1 in material A, the longitudinal wave will be excited at an angle θ_1' greater than the incident angle θ. The critical angle occurs when the refracted angle θ_1' is at 90° in material B, the test piece. The L_{CR} longitudinal wave travels at a bulk longitudinal wave speed in the test piece, parallel to the surface, and is received by probes at R_1 and R_2 that are inclined at an angle equal to the first critical angle of the transmitter wedge. As shown, a shear wave also is excited in the test material. This is usually of no concern since firstly it is travelling at a speed slower that the longitudinal wave and secondly it also reflects across the material surfaces. The assumed depth of penetration of the L_{CR} is also shown.

Figure 10.4 shows a practical example of the dual probe setup schematically illustrated in Figure 10.3. It is mounted on a steel plate with the transmitter T at the left and receivers

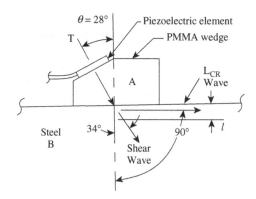

Figure 10.2 Single L_{CR} probe showing PMMA wedge. Courtesy of Don E. Bray, Inc.

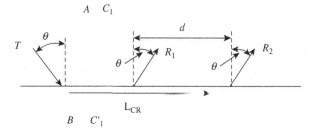

Figure 10.3 Dual receiver probe arrangement for L_{CR} stress measurement. See text for explanation of symbols. Courtesy of Don E. Bray, Inc.

Figure 10.4 Dual receiver probe with rotating wedges for ultrasonic stress measurement. Courtesy of Don E. Bray, Inc.

R_1 and R_2 at the right. Commercial 12 mm square transducers are used in this example. The distance d from R_1 to R_2 is 50.8 mm here. This distance can be adjusted to fit the circumstances of the measurement. The wedges in this design are able to pivot to better match the surface of the part, so enabling the hydraulic piston to apply sufficient force against the hole in the top to minimize the couplant thickness. This is a necessity for precise travel time measurements for stress measurement. The distance from T to R_1 is not critical.

A couplant is needed for impedance (ρC) matching between two dissimilar materials, where ρ is the material density and C is the wave speed. Values for these properties are available in ultrasonic reference sources. Typically a PMMA (low impedance) material is used for the wedge, while the material being inspected might be a high impedance metal or other material. Fluids such as motor oil, grease and glycerine-based gels as well as water are frequently used for couplants. A hydraulic piston may be used to place a large measured force on the probe assembly, and therefore the interface. A typical piston unit might be a Ram-Pac model number RC-5-LP-.5S. Hydraulic force can be generated with a hand pump such as Power Team Model No. P55. By using a piston that has a 1 square inch area, a pressure gage will directly show the force in pounds on the piston and the probe assembly. So, 300 psi gives 300 lbs (1.33 kN) force on the interface and typically gives satisfactory results. It is also possible to use air-actuated systems.

Waves propagating as bulk longitudinal waves, but near the surface, are correctly labeled as L_{CR} waves [2]. Other names have been used, however, such as surface skimming longitudinal waves (SSLW) [3–5]. The essential characteristic of these waves is that they travel just below the surface at bulk wave speeds and generally are free from any effect of surface conditions such as scratches, rust, and so on.

Where change in the travel time in the travel path from R_1 to R_2 in Figure 10.3 is the indicator of stress related velocity change, the full path of the wave will include t_w, the time in the PMMA wedges, t_c, the time in the couplant as well as in the travel time in the material being investigated. Clearly, the travel time in the wedge and couplant is crucial for accurate measurement. This is usually established by measurement in a known stress free zone of the same material, for example, the parent metal far away from the weld metal in a welded structure.

When describing the acoustoelastic effect, the stress change $\Delta \sigma$ can be determined from the velocity change ΔV using:

$$\Delta \sigma_1 = \frac{E\,(dV_{11}/V_{11})}{L_{11}} = \frac{E}{L_{11} \times t_o}\,dt \qquad (10.1)$$

where E is Young's modulus, L is the acoustoelastic coefficient, which must be determined experimentally for the material, and t_o is the travel-time in stress-free conditions. The measured travel-time change Δt indicates the stress change. Equation (10.1) can be simplified further and represented as:

$$K_{11} = (\Delta V/V)/\Delta \sigma \qquad (10.2)$$

where the elastic modulus is now incorporated in the constant, K_{11}.

Acoustic (or shear wave) birefringence is a description adapted from optics where an energy packet launched into a material assumes different characteristics due to the influence of the material. Shear waves have particle motion perpendicular to the direction of travel, often referenced as SH and SV waves for shear horizontal and shear vertical, related to the particle motion. In homogeneous isotropic materials, SH and SV waves have the same velocity. When excited in material with texture or a stress field, one can observe the wave breaking up into components dictated by the anomalous field, that is, particle motion aligned with the principal axes of the stress field will have maxima and minima. This feature allows measurement of stresses.

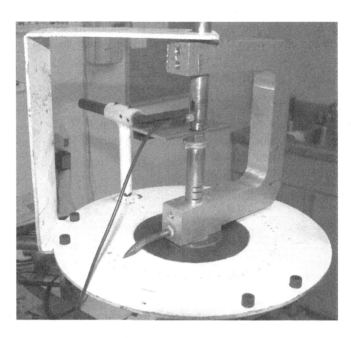

Figure 10.5 Shear wave probe set up for birefringence investigation. Courtesy of Don E. Bray, Inc.

In a homogeneous, isotropic plate, a shear wave is launched across the thickness of the plate using a normally incident contact probe, as shown in Figure 10.5. Here the shear wave probe is contained in the upper vertical tubular member of the apparatus and the plate is held fixed to the rear. Particle motion polarization is typically along the direction of the cable connection to the probe and as the probe is rotated on the surface, the particle motion moves with the rotating probe. For the arrangement shown in Figure 10.5, shear waves are propagated across the thickness of the specimen with the angle of polarization θ relative to a reference angle R. The angles can be read from the scale marked on the base plate. R defines the orientation of the coordinate system relative to some characteristic of the plate or bar, for example, the rolling direction or the longitudinal direction, or simply a direction line marked on the plate or bar. Typically, the operator will observe and record both the arrival time of the wave giving the velocity, and the orientation angle of the arrival of interest (e.g., the fastest wave). Following that, the probe will be rotated typically 90° to find the opposite component and the arrival time and angle recorded. This technique requires a very viscous couplant and care must be taken to ensure that the couplant has stabilized at the interface to eliminate error.

Practitioners using shear wave birefringence may first scan the velocity field in a part and identify the orientation of the fastest (earliest arriving) wave and then that of the slowest wave. They may arbitrarily associate R, for example, with the fastest arrival and then note the relationship of R to a landmark or feature in the part being inspected. Next, they will measure the angle (θ) for the slowest arrival. This gives a map of the stress field. There are other features of the stress field that may be important, for example, the sharpness of the fastest and slowest peaks and troughs.

The stress difference in the two directions is calculated using Equations (10.3) and (10.4), where the birefringence is determined using Equation (10.4).

$$\sigma_\theta - \sigma_R = \frac{B - B_o}{C_A} \qquad (10.3)$$

where:
σ_θ = stress in the direction θ, in the cylindrical system $R \times \theta$
σ_R = stress in the direction R, in the cylindrical system $R \times \theta$
C_A = acoustolastic constant for the material
B_o = birefringence, unstressed state

The birefringence can be calculated using:

$$B = 2\,\frac{t_R - t_\theta}{t_R + t_\theta} \qquad (10.4)$$

where:
t_θ = time-of-flight in the direction θ
t_R = time-of-flight in the direction R

Since the velocity is also affected by texture, this technique is also very useful for texture analysis.

10.2 History

Ultrasonic measurement of stresses has a long history. Among the first publications on the subject, the work of Noronha, Chapman and Wert [6], Hsu [7] and Noronha and Wert [8] made significant advances in achieving a workable system for non-destructive stress measurement. Clotfelter and Risch [9] showed practical results for application to aircraft grade aluminum. Following this, Egle and Bray [1,10] proceeded to demonstrate that ultrasonic wave speed change could be successfully applied in the field for measuring stress changes in railroad rails. Subsequently, Brokowski and Deputat [11] presented a workable system for use in rail rolling plants. Further examples were presented for application to welded plates and pressure vessels by Leon-Salamanca and Bray [12] and Bray and Junghans [13] and Bray [14].

While shear wave birefringence measurements require accurately parallel surfaces on both sides of the item being inspected, this is not a serious limitation for many shapes. For example, the rims of railroad wheels meet this criterion, as reported by Schramm [15]. Santos et al. have applied the technique to stress measurement in rolled plates [16]. Material anisotropy can have a serious effect on the observed travel times and users of this technique must be aware of this occurrence.

As these developments occurred, capabilities of the technique as well as deficiencies became clearer. While the physics of wave speed change and stress was well established and confidence was built in developing a workable technique, it was clear early on that the need to reliably measure very small travel time changes with conventional instrumentation would press existing electronics. Also, it was clear that variations in the acoustoelastic coefficient within otherwise similar materials were significant, making absolute stress measurements challenging.

10.3 Sources of Uncertainty in Travel-time Measurements

10.3.1 Surface Roughness

Because the L_{CR} wave travels beneath the surface, the L_{CR} method for stress measurement is largely unaffected by surface roughness arising from manufacturing, for example, mill scale and casting oxidation. Experience has shown that cleaning a surface with a wire brush and a rag is usually a sufficient preparation. However, large gouges and other major surface irregularities can adversely affect probe placement and couplant thickness.

10.3.2 Couplant

Any variation in couplant thickness can have a serious effect on the accuracy and repeatability of ultrasonic stress measurements. Since the couplant fills in vacancies in a rough surface and since all rough surfaces are non-uniform, practitioners must be keenly aware of variations in couplant affects. Where a typical glycerin or oil-based material is used, with a velocity of 1700 m/s, a variation of 0.01 mm in couplant thickness could result in a 6 ns travel-time deviation at each interface, which is comparable to the stress induced travel-time changes. In repeatability tests, experience shows that the error is greatest for the first application and is lower after that. This is likely due to some wetting phenomena, but was not investigated further.

10.3.3 Material Variations

Material variations such as grain size and orientation (texture), as occurring in ordinary rolling and cooling, seriously affect travel-times obtained for L_{CR} stress measurement. Santos et al. [17] studied the effect of rolling direction in API 5L X70 pipe steels on ultrasonic stress measurement. Comparing the L_{11} values obtained for samples longitudinal and transverse to the axis of the pipe, they found an almost 30% variation for non-stress-relieved samples, the largest value occurring in the longitudinal direction. Walaczek et al. [18] reported variations in the acoustoelastic coefficient (L_{11}) for weld metal and heat affected zone in P460 HLE and P265 steel. Following that work, Buenos et al. analyzed the effect of the mean grain size on time of flight measurements for L_{CR} waves in ASTM A36 steel [19]. As the grain size increased, they found a corresponding increase in time of flight. For the L_{CR} technique, probe placement assuring uniform material conditions has been found to have a significant effect on repeatability, as will be discussed in a following section.

10.3.4 Temperature

While temperature has a significant effect on wave speed in materials, it can be factored into a process so that reliable data may still be obtained. The effect is more significant in PMMA probe material than in metals. There are two ways of reconciling the temperature effect, first by taking data in constant temperature conditions. This is often reasonable for stress measurement in shops and labs that are temperature controlled. In the absence of these conditions, though, the temperature of the wedges may be monitored with thermocouples or other device. Of course, values taken in an immersion technique should be monitored through taking the temperature of the fluid.

Bray and Stanley [20] report work from Egle and Bray [21] on wave speed variations with temperature in steel and PMMA, materials typically encountered in ultrasonic stress measurement. For longitudinal waves in steel traveling parallel to the rolling direction of the forged steel, the wave speed variation was found to be

$$C_1 = C_1^0 - (dC/dT) \Delta T \quad (10.5)$$

where
C_1^0 = longitudinal wave speed at a reference temperature
dC/dT = speed change constant
ΔT = temperature change in °C.

For a nominal longitudinal wave speed of 5900 m/s at 25 °C (77 °F) the results are

$$C_1 = 5900 \text{ m/s} - 0.55 \, (T - 25) \, ^\circ\text{C} \quad (10.6)$$

Shear wave speed changes in the same material are given by

$$C_2 = 3228 \text{ m/s} - 0.38 \, (T - 25) \, ^\circ\text{C} \quad (10.7)$$

Longitudinal waves in PMMA at a nominal speed of 2690 m/s at 25 °C were found to vary according to the following

$$C_1 = 2690 \text{ m/s} - 2.3 \, (T - 25) \, ^\circ\text{C} \quad (10.8)$$

Comparing the expected wave speed changes over a typical temperature range of 2.8 °C to 47.2° (37 °F to 117 °F) shows that both the longitudinal wave speed and shear wave speed in the steel varies by just less than one-half of one percent. On the other hand, longitudinal wave speeds in PMMA vary by just less than 4% over the same temperature range.

10.4 Instrumentation

Ultrasonic stress measurement has greatly benefited from improved instrumentation capability, namely in allowing arrival time measurements with increased resolution. Early work in this area involved frequency counters and signal generators, in addition to oscilloscopes. Researchers persevered with those early instruments and advanced rapidly to the era of digital oscilloscopes, instrumentation software for desktop and laptop computers, and finally very high resolution in commercial ultrasonic flaw detectors.

10.5 Methods for Collecting Travel-time

Application of either the shear wave birefringence or L_{CR} techniques requires knowledge of the acoustoelastic coefficient, L or B. These values have been established by numerous researchers. Table 10.1 lists acoustoelastic coefficients for various materials and Table 10.2 describes the capabilities of various travel-time collection systems. The minimum resolution of the instrumentation must be able to obtain reliably the accurate arrival times needed for stress measurement. Given this, a longer travel path in the test piece will deliver better overall resolution. The compromise is that the stress measured will be that which occurs over an average length, no smaller than the probe separation. Repeatability for the probe can be established by repeated removal and replacement tests. In general, the values of L for similar materials reported by different researchers are consistent.

Variations of probe and instrumentation arrangements not listed in Table 10.2 include Bray and Leon-Salamanca [36] with a T $R_1 - R_2$ probe with 215.9 mm separation between R_1 and R_2. The probe was clamped with vice-grip pliers to collect zero-force travel-times on the neutral axis of short rail samples. Santos et al. [37] used a rigid T – R probe connected by an aluminum bar to map the travel-time field on API 5L X65 steel. Pathak et al. used an L_{CR} probe in T – R arrangement with 109 mm probe separation and flexible plate connection to map stresses in the rim of a turbine disk [38]. Bray et al. [39] used a short, rigid probe to map travel-times on a compressor rotor. Here the probe was made from a single block of PMMA and the L_{CR} travel distance was 57 mm. The probe was clamped using a circular screw type clamp and the pressure was increased until the signal height was at a predetermined level. Using a calibration sample, the repeatability of the probe was shown to be 2 ns with this technique. A specially contoured L_{CR} probe was used to map laser shock peening induced residual stresses on turbine blades [40]. These probes were each made from a block of PMMA with a nominal travel distance of less than 25 mm. Probe frequency was 20 MHz. The shock peening patterns were successfully mapped.

10.5.1 Fixed Probes with Viscous Couplant

To achieve good repeatability and accuracy, several researchers have performed tests with probes attached to a rigid frame and coupled to the test piece through a viscous couplant. While good distance control is achieved, the analysis must accommodate expansion and contraction of the probe with temperature change as well as errors induced by differences in couplant thickness.

Figure 10.6 shows an example of a simple rigid probe held down by spring clamps with an L_{CR} travel distance $(R_2 - R_1)$ of 125 mm and repeatability of 3 ns. This arrangement showed variations in the plates of approximately 35 MPa (5 ksi) [35]. Here, the transducers were square air-backed plates glued to the wedges.

10.5.2 Fixed Probes with Immersion

Using a fixed, rigid probe in an immersion bath with a fluid as a couplant has several advantages because it can easily be moved to new locations, thereby enabling faster and cheaper scanning. Errors induced by surface irregularities would be the same as with viscous couplant. Belahcene and Lu [32] described work using the immersion technique to scan the stress field in butt welds, as will be described in a later section.

Table 10.1(a) Acoustoelastic constants (L_{ij}) for longitudinal and shear waves in engineering materials Bray [22]. Courtesy of Don E. Bray, Inc.

Material	Load	L_{21}	L_{23}	L_{22}	L_{11}	L_{12}
Aluminum [23]	Compressive	−2.0	+0.6			
Aluminum [7]	Compression	−2.1	+0.57			
Aluminum [24]	Tension – RD		+0.46	+0.68	−2.7	
Aluminum [24]	Tension – TD			+0.93	−3.1	
Aluminum 6061 [25]	Tension – RD				−3.39	
Aluminum 7050 [25]	Tension RD				−2.9	
Aluminum 7175 [25]	Tension RD				−2.87	
Aluminum 7175 [25]	Tension TD				−2.93	
Aluminum 5052 [26]	Tension RD				−2.34	
Aluminum 5052 [26]	Tension TD				−2.46	
Aluminum 5086 [24]	Tension RD				−2.7	
Aluminum 5086 [24]	Tension TD				−3.1	
Aluminum 6056 [27]					−3.83*	
Aluminum 7198 T9 [28]					−3.77*	
Ductile Cast Iron [29]						
As-cast	Compressive				−2.15	
Annealed	Compressive				−3.89	
Normalized	Compressive				−3.92	
Q & T	Compressive				−2.98	
Rail Steel [1](1080)	Tension	−1.5	+0.09	+0.27	−2.38	−0.15
	Compressive				−2.45	
Cold rolled Steel Bar [30]	Tension				−2.38	
4140 Steel [31]	Tension (2.25 MHz)				−2.2	
	Tension (5 MHz)				−2.36	
Steel P460 HLE [32]	Tension				−2.82*	
Steel P460 HLE [32]	Tension				−3.38*	
Steel P265 [18]	Tension				−2.66*	
Steel P265 [18]	Tension				−2.96*	
S355 steel RD [32]	Tension				−2.52	
S355 steel [32]	Tension				−2.2	
316L Stainless Steel [32]	Tension	−1.5	−1.2	∼0	−2.1	
Clear acrylic, aircraft grade [33]	Tension				−2.14	
Polyethylene – cross-linked natural [34]	Tension				−0.85	
Polyethylene – cross-linked black [34]	Tension				−1.2	

*Reported values for L_{11} were obtained by multiplying by the appropriate Young's modulus.

Table 10.1(b) Acoustoelastic constants (LR_{ij}) for Rayleigh waves in engineering materials [24]

Material	Load	LR_{13}	LR_{23}
Aluminum	Tension – RD	−1.1	+0.5
Aluminum	Tension – TD	−0.48	+0.5
316L Stainless Steel	Tension		

Table 10.2 Travel-time variations with various L_{CR} applications

Project Description	Time measurement instrument	Instrument Precision (ns) t_P	Probe type connection	Probe d mm	Probe t_o μs	t_P/t_o %	Δt ns	Probe Repeatability @ σ = 0 Δt/t_o %	ksi	Mpa
Kaiser aluminum plates [35]	LeCroy 400MHz/2GHz	0.5	T, $R_1 - R_2$ Rigid	125	19.8	0.0025	3	0.01	0.3	2.1
Steel tension bar [41]	Gage 265 and StressMap	0.1	T – R flexible strip	201	34	0.0003	3	0.008	1.0	6.9
Steel tension bar [41]	Panametrics Epoch III	7	T – R flexible strip	201	34	0.02	3	0.008	1.0	6.9
Steel calibration frame [42]	Gage 265 and StressMap	0.1	T – R flexible strip	201	34	0.0003	3	0.008	1.0	6.9
Steel calibration frame [42]	Gage 265 and StressMap	0.1	Dual receiver – probes rotate	50.8	8.5	0.0011	2.4	0.028	3.4	23
Aluminum Bar – 3 receivers	Gage 265 and StressMap	0.1	T – R_1 Dual axis rotation	101.6	16.1	0.0006	2	0.01	0.3	2.1
Aluminum Bar – 3 receivers	Gage 265 and StressMap	0.1	$R_1 - R_2$ Dual axis rotation	241.05	38.14	0.00026	2	0.005	0.15	1
Aluminum Bar – 3 receivers	Gage 265 and StressMap	0.1	$R_1 - R_3$ Dual axis rotation	342.6	54.2	0.00018	2	0.0037	0.11	0.8
Aluminum Bar – 3 receivers	Gage 265 and StressMap	0.1	$R_2 - R_3$ Dual axis rotation	101.6	16.1	0.0006	2	0.01	0.3	2.1

Figure 10.6 Rigid, fixed probe used for residual stress measurement in aluminum. Courtesy of Don E. Bray, Inc.

10.5.3 Fixed Probes with Pressurization

Adding measurable force to the probe interface and thereby squeezing the couplant at the interface to a thinner film is a way of reducing error. Use of devices such as mechanical vice-grip pliers or screw presses can achieve this goal. The person applying the pliers can feel the amount of force placed on the interface. An improved system could be used where a pressure gage in a hydraulic system could monitor the amount of pressure on the interface.

10.5.4 Contact with Freely Rotating Probes

The minimum couplant error is achieved with a pressurized probe and wedges freely rotating about the entry point, thereby enabling them to conform to the surface contours. Figure 10.7 shows one version of such a probe where 2.4 ns repeatability was achieved with a probe spacing of 50.8 mm. This is far more compact than the longer rigid-frame model, and easier to use in smaller spaces, although the pressurization system requires some external hardware. Here the pressure is applied mid-point between the two receivers. This probe arrangement has performed well in a number of setups, ranging from plates to pressure vessels.

10.6 System Uncertainties in Stress Measurement

Considering the combined roles of the material variations, couplant and instrumentation as well as environmental variations, the question arises of what is a reasonable expectation of measurement accuracy of residual stresses with ultrasound. Table 10.2 shows the combined contribution for these factors. The variables are the time resolution of the instrumentation and the probe arrangement. Longer, dual probe arrangements give better travel-time and stress resolution. In addition, rotating probes and smooth polished surfaces minimize the couplant error, and also improve stress resolution. Dual receivers also reduce error due to temperature changes because both probes should be the same temperature. Confined spaces, however, sometimes do not allow a dual probe setup. In the early tests on railroad rail, the technique used was pulse overlap, where a timing pulse was overlaid on the oscilloscope screen with the ultrasonic pulse [1,10]. Digital oscilloscopes such as the

Figure 10.7 Dual receiver rotating probe arrangement under hydraulic pressure. Courtesy of Don E. Bray, Inc.

LeCroy 400 MHz improved the precision (t_p) to 0.5 ns over a travel path of 125 mm. In the aluminum this gave a nominal travel time between probes (t_o) of 19.8 µs. Therefore, the user could expect a minimum accuracy due to instrument and equipment (t_p/t_o) of 0.0025 percent. Repeatability for the rigid probe used here was 3 ns, or 0.01 percent. Digitizing boards and appropriate software for desktop and laptop computers improve the convenience of data handling, with the sacrifice of some resolution. Using the Gage 265 board with a probe having a flexible connection on steel with a travel path 201 mm gave a t_p/t_o of 0.0003 percent. For this system the repeatability was 3 ns or 0.008 percent.

Several notable characteristics are shown in Table 10.2. Instrumentation resolution, as shown at the left, varies from 0.1 ns for the digital oscilloscope to 7 ns for the commercial flaw detector. Experiments described by Santos et al. [41] and Bray [42] for the steel tension bar and the calibration frame showed an ability to resolve reasonable stress changes. There are obvious advantages for each, depending on the demands of the test. The maximum resolution is with the long, three-receiver dual axis probe where repeatability tests showed a travel time resolution of 2 ns giving a predicted resolution 0.8 MPa (0.11 ksi) for the longer probe separation (342.6 mm). In a tension testing machine, this probe showed an ability to resolve 3.4 MPa (500 ksi) in aluminum. The probe performed satisfactorily in the intended factory setting with hydraulic pressure applied to the top of the frame.

In the experimental results of Santos et al. [41] and Bray [14] a much wider error occurred at low stress levels. At higher stresses, the relationship of stress and travel-time was almost linear. The cause of this effect was not studied further.

10.7 Typical Applications

10.7.1 Weld Stresses

An early application of ultrasonic stress measurement was for stresses in welds and the region surrounding the weld seam for both hot rolled and cold rolled plates [12]. Two

19 mm thick 762 mm long and 254 mm wide carbon steel plates were welded for testing. The single sender and two-receiver probe had the receivers connected to a thin flexible strip, enabling some vertical rotation. The probes were clamped to the plate using vise-grip pliers. Probe frequencies were 2.25 MHz and the separation of the two receivers was 219 mm. The plates were stress-relieved after testing. Both the hot rolled and cold rolled plates showed travel time peaks at the centerline, which would be expected for the tensile stresses there. Following thermal stress relief, the peak travel times disappeared. A second experiment using a patch-welded plate showed and expected travel time profile as the probe was advanced toward the patch at the center [13].

Immersion of the L_{CR} probes in water enables more precise and smoother control of the probes relative to the specimens being evaluated and therefore enables scans of large areas and the generation of plots for weld stresses, as described by Sajauskas [3]. Belahcene and Lu [32] reported results using an immersion scanning system, where they plotted stress profiles of welds in S355 steels. Five frequencies were used (2.25, 3.2, 5, 6.6 and 10 MHz), to enable investigation of the depth of penetration of the L_{CR} wave. The results showed at 2.77 mm penetration for the 2.25 MHz probe and less than 1 mm for 10 MHz. Other frequencies showed expected penetration between the two limits.

Santos, Andrino, Bray and Trevisan [37] reported welding induced stresses for API 5L X65 steel using a sender-receiver type probe and a viscous couplant with hydraulic pressure at the interface. Stresses in the centerline of the weld showed values of 1500 to 2000 MPa before stress relief and generally 1000 to 1500 MPa after stress relief. Using a 20 mm thick, 395 mm long by 400 mm wide plate as the specimen, they moved their probe to either side of the weld line in increments at distances of 130 mm and 360 mm from the plate edge. They had one sender and two receivers. Their results clearly showed peak longitudinal stresses of 600 MPa at the centerline of the weld, falling off symmetrically to zero stress at approximately 40 mm from the centerline. Interestingly, the 2.25 MHz and 10 MHz probe results showed about the same behavior.

Gachi et al. [28] reported results using the L_{CR} to measure stresses in friction stir welds. This work clearly showed the small difference in the compressive stresses in the base material and the tensile stresses in the weld.

Using a L_{CR} probe on the outer surface of a nominal 304 mm diameter welded steel pressure vessel, Bray plotted the stress pattern adjacent to the weld [14]. The results were shown to be very near to results obtained by other researchers using a hole drilling technique on a similar welded pressure vessel. It should be noted here that a precise spacer block was used to assure position consistency in the placement of the probe with the result being a 2 ns repeatability.

10.7.2 Measure Stresses in Pressure Vessels and Other Structures

To apply ultrasonic residual stress measurement in industrial applications, there needs to be developed additional knowledge of the basic material properties for engineering materials as well as an easily accessible library of acoustoelastic coefficients (L_{11}) for these materials. Gonulal, Aras and Ozhan [43] have provided some guidance for the effect of material properties and reports Santos et al. [17] and Walaczak et al. [18] describe the effect of material properties on stress measurement. This being done, the engineering knowledge of the expected stress field will need to be established so that deviations from the expected may be confidently noted.

10.7.3 Stresses in Ductile Cast Iron

Ductile cast iron structures may be complex in shape and susceptible to warpage and breakage due to unfavorable residual stresses. Residual stresses may be evaluated in these materials using the L_{CR} technique, as discussed by Srinivasan et al. [29,44,45]. With knowledge of these conditions, the foundry can adjust the process to reduce the residual stresses.

10.7.4 Evaluate Stress Induced by Peening

Compressive residual stresses induced by several techniques in parts subject to fatigue have been known for many years to reduce the onset of fatigue failure. Both shot peening, using high impact particles, and shock peening, using lasers, have been evaluated with this technique. [40,47] Ultrasonic analysis of titanium parts and actual aircraft turbine blades has shown an association of the L_{CR} results and the residual stresses due to shock peening. Since texture is also affected by the shock or shot peening, and texture affects the speed, work remains to develop a process useful to industry. The ease and convenience of collecting data with the L_{CR} ultrasonic method was well demonstrated.

10.7.5 Measuring Stress Gradient

Since material properties often vary with depth, there is a need for a technique to evaluate the stress gradient. The effective penetration depth of the L_{CR} wave has been demonstrated by several researchers, including Sajauskas and Bray and Tang [3,46], to be approximately equal to one wavelength. Varying frequency, and therefore the wavelength, leads to the possibility of evaluating the gradient. The ability of the technique to do this is limited by the fact that it interrogates an average from the surface to the wavelength depth.

10.7.6 Detecting Reversible Hydrogen Attack

In many chemical and petroleum operations there is serious risk of the small hydrogen atom creeping between the grain boundaries, creating a stress buildup at the surface and initiating a crack. If undetected, this crack could ultimately lead to failure. Bray and Griffin [48,49] present experimental data showing that both the velocity and the spectrum of the L_{CR} wave are affected by hydrogen in 4140 steel. While the velocity changes are very small, they are measurable.

10.8 Challenges and Opportunities for Future Application

10.8.1 Personnel Qualifications

The education and experience required to conduct successful ultrasonic stress measurement go substantially beyond the needs for basic ultrasonic NDE. Required knowledge includes advanced ultrasonics, materials properties, engineering structures and instrumentation. As the techniques evolve, there will undoubtedly be more use of automated systems. This must be approached with caution, however, until an adequate database is built so that engineers and technicians can have a better grasp on expected and unexpected responses.

10.8.2 Establish Acoustoelastic Coefficients (L_{11}) for Wider Range of Materials

While the list furnished in Table 10.1 has grown through the years, there is a much larger range of materials with different grain metallurgy and heat treatments that need to be studied and classified and the results stored in a manner so that users and researchers may have access to this data bank.

10.8.3 Develop Automated Integrated Data Collecting and Analyzing System

Effective application of any non-destructive technique for materials evaluation is well known to evolve into a repetitive task prone to operator mistakes. Certainly ultrasonic stress measurement is in this group. This drives the demand for automated systems where manipulation of the probes and decisions are made using automated techniques.

10.8.4 Develop Calibration Standard

A calibration standard that would enable performance evaluations for probe properties would be very useful for further development of L_{CR} probes. Bray [42] describes the results obtained for two ultrasonic probes using the frame described by Kypa [50]. The frame was a 3-bar frame with a middle bar that had been cooled in a bath of liquid nitrogen to enable it to be inserted in the frame. Expansion upon heating should have resulted in compressive stresses in the middle link and tension in the outer two. This did occur and, a useful stress pattern was created. However, unanticipated manufacturing difficulties caused the design to give a twisting stress pattern in the outer links. This difficulty prevented the full achievement of an accurate calibration standard, and this remains a goal for future effort.

10.8.5 Opportunities for L_{CR} Applications in Engineering Structures

Design engineers have long been aware that the unknowns in a stress field demand the use of factors of safety. These range from just over 1 for some aircraft applications to over 10 for elevators. While this approach added confidence in the safe performance, it also added to the weight, complexity and cost of a machine, structure or pressure vessel. It was a necessary factor to assure safe performance. Certainly developers of industry inspection and design codes, for example, [51] and [52], would welcome progress in research in the various non-destructive methods for residual stress measurement so that more efficient designs and performance standards may be adopted.

References

[1] Egle, D. M., Bray, D. E. (1976) "Measurement of the Acoustoelastic and Third-Order Elastic Constants for Rail Steel," *Journal of the Acoustical Society of America* 60(3):741–744.

[2] ASTM (2010) Standard Terminology for Nondestructive Examinations, E1316-10a, *American Society for Testing and Materials*, West Conshohocken, PA, USA.

[3] Sajauskas, S. (2004) *Longitudinal Surface Acoustic Waves*, Kaunas University of. Technology, Lithuania.

[4] Langenberg, K. J., Fellinger, P., Marklein, R. (1990) "On the Nature of the So-called Subsurface Longitudinal and/or the Surface Longitudinal Creeping Wave," *Research in Nondestructive Evaluation*, 2(2):59–81.

[5] Schneider, E. (1997) "Ultrasonic Techniques" 4:522–563. In: Hauk V (ed) *Structural and Residual Stress Analysis by Nondestructive Methods*, Elsevier: Amsterdam.

[6] Noronha, J., Chapman, J., Wert, J. (1973) "Residual Stress Measurement and Analysis Using Ultrasonic Techniques," *Journal of Testing and Evaluation*, 1(3):209–214.

[7] Hsu, N. N. (1975) Shear Wave Birefringence, Proceedings of a Workshop on Nondestructive Evaluation of Residual Stress, Nondestructive Testing Information and Analysis Center (NTIAC), NTIAC-76-2:173–178.

[8] Noronha, P. J., Weil, J. (1975) "An Ultrasonic Technique for Measurement of Residual Stresses," *Journal of Testing and Evaluation* 3(2):147–152.

[9] Clotfelter, M., Risch, E. (1974) Ultrasonic Measurement of Stress in railroad wheels and in long lengths of welded rail, NASA Technical Memorandum NASA TM X-64863.

[10] Egle, D. M., Bray, D. E. (1979) "Ultrasonic Measurement of Longitudinal Rail Stresses," *Materials Evaluation* 378(4):41–46, 55.

[11] Brokowski, A., Deputat, J. (1985) *Ultrasonic Measurements of Residual Stresses in Rails*. Proc 11th World Conference on Nondestructive Testing, American Society for Nondestructive Testing, Columbus, Ohio 592–598.

[12] Leon-Salamanca, T., Bray, D. E. (1995) "Residual Stress Measurements in Steel Plates and Welds Using Critically Refracted (LCR) Waves," *Research in Nondestructive Evaluation*, 7(4):169–184.

[13] Bray, D. E., Junghans, P. G. (1995) Applications of the LCR Ultrasonic Technique for Evaluation of Post-Weld Heat Treatment in Steel Plates, NDT&E International, 28(4):235–242.

[14] Bray, D. E. (2002) "Ultrasonic Stress Measurement in Pressure Vessels, Piping and Welds," *Journal of Pressure Vessel Technology* 124(3):326–335.

[15] Schramm, R. E., Clark, A. V., Mitrakovic, D. V., Schaps, S. R. (1991) "Ultrasonic Measurement of Residual Stress in Railroad Wheels," *Proceedings of Practical Application of Residual Stress Technology*, (3):61–67.

[16] Santos, A. A., Kypa, J., Bray, D. E. (1999) Determination of Stresses in Plates Using Ultrasonic Shear Waves, Proceedings of the XV COBEM Congresso, Brazil.

[17] Santos, A., Bray, D. E., Caetano, S., Andrino, M., Trevisan, R. (2004) Evaluation of the Rolling Direction Effect in the Acoustoelastic Properties for API 5L X70 Steel used in Pipelines, ASME Conf Proc.

[18] Walaczek, H., Hoblos, J., Bourse, G., Robin, C., Bouteille, P., Lieurade, H. (2004) Ultrasonic stress measurement in welded component by using LCR waves: analysis of the microstructure effect, Recent Advances in Nondestructive Evaluation Techniques for Material Science and Industry, ASME/JSME Pressure Vessels and Piping Conference, PVP-Vol 484, San Diego, CA. PVP2004-2829:147–152.

[19] Buenos, A., Santos, A., Pereira, P., Santos, C. (2012) Effect Of Mean Grain Size In The Time Of Flight For L_{CR} Waves, Proc. ASME 2012 International Mechanical Engineering. Congress and Exposition, IMECE2012, Houston, TX.

[20] Bray, D. E., Stanley, R. (1997) *Nondestructive Evaluation*, CRC, Boca Raton, FL.

[21] Egle, D., Bray, D. (1978) Nondestructive Measurement of Longitudinal Rail Stresses: Application of the Acoustoelastic Effect to Rail Stress Measurement, Federal Railroad Administration, Washington, DF. C., Report No. FRA-ORD-77-341, PB-281164m NTIS, Springfield, VA.
[22] Bray, D. E. (2012) http://brayengr.com/
[23] Becker, F. L. (1973) *Ultrasonic Determination of Residual Stress*, Battelle-Pacific Northwest Laboratories, Richland, WA.
[24] Tanala, E., Bourse, G., Fremiot, M., De Belleval, J. F. (1995) "Determination of Near Surface Residual Stress on Welded Joints using Ultrasonic Methods," *NDT&E International* 28(2):83–88.
[25] Bray, D. E., Ainsworth, P. (2010) Acoustoelastic Coefficients for Commercially Produced 6XXX and 7XXX Series Aerospace Aluminum Plate, http://brayengr.com/Acoustoelastic_Coeffi_4149_Rev2B.pdf
[26] Andrino, M. H., Santos, Jr, A. A., Caetano, S. F., Gonçalves, R. (2005) Determination Of 5052 Aluminum Alloy Acoustoelastic Coefficients, Proc. COBEM 2005 18th Int. Congress of Mech. Engr, ABCM November 6–11, 2005, Ouro Preto, MG Brazil.
[27] Ya, M., Marquette, P., Belahcene, F., Lu, J. (2004) "Residual Stresses in Laser Welded Aluminum Plate by Use of Ultrasonic and Optical Methods," *Materials Science and Engineering A* 382:257–264.
[28] Gachi, S., Belahcene, F., Boubenider, F. (2009) "Residual Stresses in AA7108 Aluminum Alloy Sheets Joined by Friction Stir Welding," *Nondestructive Testing and Evaluation*, 24(3):301–309.
[29] Srinivasan, M. N., Chundu, S. N., Bray, D. E., Alagarsamy, A. (1992) "Ultrasonic Technique for Residual Stress Measurement in Ductile Iron Continuous Cast Round Bars," *Journal of Testing and Evaluation* 20(5):331–335.
[30] Santos, A., Bray, D. (2000) "Ultrasonic Stress Measurement Using PC Based and Commercial Flaw Detectors," *Review of Scientific Instruments* 71(9):3464–3469.
[31] Tang, W., Bray, D. E. (1996) Stress and Yielding Studies Using Critically Refracted Longitudinal Waves, NDE Engineering Codes and Standards and Material Characterization, Proc 1996 ASME PVP Conf, Montreal, PQ, PVP-Vol.322, NDE-Vol. 15, Cook, Sr, J. F., Cowfer, C. D., Monahan, C. C. (eds) *The American Society of Mechanical Engineers*, New York, 41–48.
[32] Belahcene, F., Lu, J. (2002) "Determination of Residual Stress Using Critically Refracted Longitudinal Waves and Immersion Mode," *Journal of Strain Analysis* 37(1):13–20.
[33] Shaikh, N. (1992) "Transducer and Technique for Ultrasonic Nondestructive Evaluation of Structural Plates," Review in Progress in Quantitative Nondestructive Evaluation, Vol. 11, D. O. Thompson and D. E. Chimenti, Eds. Plenum Press: New York, 1992, 1831–1835.
[34] Bray, D. E., Vela, J. M., Al-Zubi, R. (2005) "Stress and Temperature Effects on Ultrasonic Properties in Cross Linked and High Density Polyethylene," *Journal of Pressure Vessel Technology*, 127(3):220–225.
[35] Bray, D. E., Kim, S-J., Fernandes, M. (1999) Ultrasonic Evaluation of Residual Stresses in Rolled Aluminum Plates, Proc Ninth International Symposium on Nondestructive Characterization of Materials, Green RE (ed), Sydney Australia, American Institute of Physics, Melville, NY, 443–448.
[36] Bray, D. E., and Leon-Salamanca, T. (1985) "Zero-force Travel-time Parameters for Ultrasonic Head Waves in Railroad Rail," *Materials and Evaluation*, 43:854–858,863.
[37] Santos, A., Andrino, M., Bray, D., Trevisan, R. (2008) "Evaluation of Stresses Generated by Welding in API 5L X65 Steel using Acoustoelasticity," *Materials Evaluation* 66(8):858–864.
[38] Pathak, N., Bray, D. E., Srinivasan, M. N. (1992) Detection of Stress in a Turbine Using the LCR Ultrasonic Technique, Serviceability of a Petroleum, Process and Power Equipment, PVP-Vol.239/MPC-Vol.33, Proceedings 1992 ASME Pressure Vessels Piping Conference, New Orleans, LA, .
[39] Bray, D. E., Tang, W., Grewal, D. (1996) Ultrasonic Stress Evaluation in a Turbine/Compressor Rotor, Review of Progress in Quantitative NDE, Brunswick College, Brunswick, ME, 1691–1697.
[40] Bray, D., Suh, U., Hough, C. (2001) Ultrasonic Evaluation of the Effects of Treatment and High Cycle Fatigue in Aircraft Engine Turbine Blades, Rev Prog in QNDE, Brunswick, Maine, Vol. 21B, American Institute of Physics, 615, 1643–1650.
[41] Santos, A., Bray, D. E. (2000) "Ultrasonic Stress Measurement Using PC Based and Commercial Flaw Detectors," *Review of Scientific Instruments* 71(9):3464–3469.
[42] Bray, D. E. (2005) Performance Comparison of Two LCR Probes Using a Reference Frame, Review of Progress in Quantitative Nondestructive Evaluation, Brunswick, Maine, 31 July – 5 August 2005, Vol. 25B, American Institute of Physics, 820:1442–1449.
[43] Gonalal, M., Aras, H., Ozhan, N. (2011) "The Correlation Between Ultrasonic Testing and Metallurgy," *Materials Evaluation* 69(10):1139–1142.

[44] Srinivasan, M., Bray, D. E., Junghans, P., Alagarsamy, A. (1991) "Critically Refracted Longitudinal Waves Technique: A New Tool for the Measurement of Residual Stresses in Castings," *AFS (American Foundrymen Society) Transactions*, 91(157):265–267.

[45] Srinivasan, M. N., Chundu, S. N., Bray, D. E., Alagarsamy, A. (1992) "Detection of Stress in Ductile Iron Bars Using Critically Refracted Longitudinal Wave Techniques," *AFS (American Foundry Society) Transactions* 92(114):309–312.

[46] Bray, D. E., Tang, W. (2001) "Evaluation of Stress Gradients in Steel Plates and Bars with the L_{CR} Ultrasonic Wave," *Nuclear Engineering and Design*, 207:231–240.

[47] Belahcene, F., Lu, J., Thomas, F., Zhou, X. (2004) Determination of compressive residual stresses using critically refracted longitudinal (LCR) waves, Recent Advances in Nondestructive Evaluation Techniques for Material Science and Industry, ASME/JSME Pressure Vessels and Piping Conference, PVP-Vol 484, San Diego, CA. PVP2004–2827, 137–146.

[48] Bray DE, Griffin RB (2011) Monitoring Hydrogen Release from 4140 Steel using the LCR Ultrasonic Technique, Nondestructive Testing of Materials and Structures, O. Büyüköztürk et al. eds. NDTMS-2011, 15–18 May 2011, Istanbul, Turkey, Session 5D, RILEM Bookseries 689–694.

[49] Bray, D. E., Griffin, R. B. (2010) Ultrasonic Characterization of Hydrogen Induced Stress in 4140 Steel, International Offshore Pipeline Forum, Houston, TX. http://www.asme-ipti.org/iopf2010/

[50] Kypa, J. (1999) *Computer Aided Analysis for Residual Stress Measurement using Ultrasonic Techniques*, MS Thesis, Department of Mechanical Engineering, Texas A&M University.

[51] Abou-Hanna, J. J., Marriott, D., McGreevy, T. (2012) *Update and Improve Subsection NH – Simplified Elastic and Inelastic Design Analysis Methods*, ASME Standards Technology, New York, NY.

[52] Ball, D. L. (2008) "The Influence of Residual Stress on the Design of Aircraft Primary Structure," *Journal of ASTM International*, 5(4):1–18.

11
Optical Methods

Drew V. Nelson
Stanford University, Stanford, California, USA

Optical methods provide an important and increasingly used approach to residual stress measurement. The main attractions of optical techniques are that they are non-contacting and provide full-field data. The non-contacting aspect avoids the need for time-consuming attachment of strain gages or other measurement devices, and the full-field aspect provides a rich source of data from which sophisticated residual stress evaluations can be made. Several different optical methods are available, each with its particular features. Depending on the optical configuration used, it is possible to measure both in-plane and out-of-plane displacements and also derivatives of those displacements, further augmenting the range and sophistication of the possible residual stress evaluations. The major optical techniques in common use for measuring residual stresses are holographic interferometry/ESPI, Moiré interferometry and digital image correlation. This chapter describes the main features of these techniques and gives examples of practical applications of each. Less frequently applied residual stress measurement techniques using optical methods are summarized. Since this chapter considers a variety of optical approaches for determining residual stresses, a substantial number of references are included, which provide details of implementation too numerous to cover here.

11.1 Holographic and Electronic Speckle Interferometric Methods
11.1.1 Holographic Interferometry and ESPI Overview

This section summarizes basic principles of holographic interferometry and electronic speckle pattern interferometry (ESPI) to provide a background description of their use in determining residual stresses. Examples of commercially available systems for implementation of the methods will also be cited. At the outset, it should be noted that methods

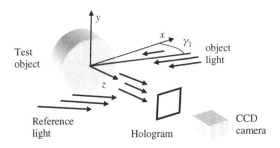

Figure 11.1 Holographic interferometry set-up

for determining residual stresses using holographic interferometry can also generally be applied with ESPI instead, and vice versa.

Holographic interferometry [1,2] provides quantitative information about small surface displacements (typically from about ten nanometers to ten microns [3]). It is usually carried out by illuminating a region of interest on a diffusely reflecting object with coherent light, generally from a low-power laser. Light is scattered from the object towards a location where a hologram is to be recorded, as depicted in Figure 11.1. The hologram location is also illuminated by reference light from the same laser source. The reference light and object light combine to produce an interference pattern (and corresponding intensity distribution) at the hologram location. In analog holography, the interference pattern is recorded rapidly by an automated camera in a photosensitive material. Holograms can be erased in milliseconds and the recording material re-used many times [4]. If viewed through a microscope, a hologram may appear as tiny dark and light features of no particular significance. Information stored in a hologram can be retrieved by re-illuminating it with reference light. The hologram acts as a complicated diffraction grating, producing a replica image of the test object as viewed through the hologram, a phenomenon known as reconstruction. (The image can be seen even if the test object is removed.) If the region of interest is re-illuminated with both reference and object light, as is typically done in real-time holographic interferometry, the reconstructed image of the region coincides with the light directly scattered from the test object. Now suppose that the illuminated region of the object displaces, with displacements typically varying in magnitude and direction over the region. The path lengths of light rays scattered from the region to the hologram will change. This alters the phase of the scattered light because phase is proportional to path length change. Superposition with the original scattered object light causes optical interference fringes to be seen on the object, as viewed through the hologram. Figure 11.2 shows an example of such fringes. Each fringe corresponds to a phase change $\Delta\varphi$ of 2π relative to a neighboring fringe. The change in phase can be related to the surface displacements. For instance, in Figure 11.2, $\Delta\varphi = (2\pi/\lambda)\,d$, where λ is the wavelength of laser light, d is the in-plane, fringe-to-fringe displacement, and angle γ_1 in Figure 11.1 approaches zero.

Digital holographic interferometry [1,5,6] may be implemented using a set-up similar to that in Figure 11.1 except that the interference pattern contained in a hologram is recorded electronically, such as via a charge coupled device (CCD) [or a CMOS] image sensor, and stored digitally. After surface displacements occur, a second digital hologram is recorded and stored. The "before" and "after" digital holograms can be reconstructed

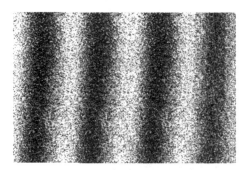

Figure 11.2 Holographic interference fringes associated with in-plane displacements in a thin plate, stretched uniformly

numerically (such as by a Fresnel transform) and phase information as well as interference fringes recovered. Surface displacements can be obtained from the phase information.

Electronic Speckle Pattern Interferometry (ESPI) [7–9], also known as digital speckle pattern interferometry, can be carried out using a set-up similar to that shown in Figure 11.1, except with a CCD array at the hologram location. Object and reference light superpose on the CCD array to create a speckle interferogram that can be stored digitally. A fringe pattern resulting from surface deformations can be found by digital subtraction of the "before" and "after" interferograms.

The intensity distribution of an optical interference fringe pattern from holographic interferometry or ESPI is related to phase change by:

$$I(x, y) = a + b \cos [\Delta\varphi(x, y)] \quad (11.1)$$

Determination of $\Delta\varphi(x, y)$ for situations in which knowledge of the expected displacement field is available, as for the fringes in Figure 11.2, can be relatively straightforward. In general, though, $\Delta\varphi(x, y)$ cannot be found unambiguously from a single pattern of $I(x, y)$ obtained by analog holography or ESPI. A common technique for determining $\Delta\varphi(x, y)$ is phase shifting [10], in which a set of intensity patterns is recorded, each pattern having a different known phase shift α_i for use in Equation (11.2). To create different values of α_i, a piezoelectrically actuated mirror is often used to alter the path length of reference light by a fraction of the wavelength of the laser light being used.

$$I_i(x, y) = a + b \cos [\Delta\varphi(x, y) + \alpha_i] \quad (11.2)$$

Equation (11.2) is solved for $\Delta\varphi(x, y)$, producing an arctangent result for $I_i(x, y)$. The nature of that function generates 2π jumps when phase change reaches π or $-\pi$, creating what is known as a "wrapped" or "modulo 2π" phase distribution. An unwrapping algorithm [10] is applied to provide a continuous $\Delta\varphi(x, y)$ distribution that can be used to compute corresponding surface displacements. This "unwrapped" phase is particularly useful because it allows direct full-field evaluations of residual stresses [31,33]. In digital holographic interferometry, numerical reconstruction of a hologram yields a wrapped phase distribution without the need for phase stepping hardware and software.

For both holographic and ESPI methods, substantial mechanical stability must exist during recording of a hologram or interferogram (and during phase shifting if used).

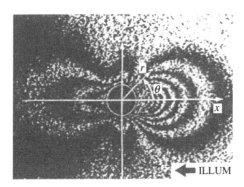

Figure 11.3 Typical fringe pattern from holographic-hole drilling for uniaxial stress in the direction of illumination (The unsymmetrical pattern results from an optical effect created by illumination from one side of the hole and the mixture of in-plane and out-of-plane displacements generated by the release of stress by the hole.)

During recording, the path lengths of reference and object light should not change by more than a small fraction of the wavelength of light being used [11].

Compact, portable systems for general-purpose deformation measurements by analog holographic interferometry have been developed [12,13] and sold in recent years by Tavex America and Optrion. ESPI systems are available from companies such as American Stress Technologies, Dantec Dynamics, Steinbrichler, GOM and Optonor. Digital holographic interferometry systems have also been developed [5,14,15]. Selection of the type of interferometric system will depend on the specific measurements needs and budget of a prospective user, as well as the state of technological development of the different systems.

A number of approaches have been investigated for determining residual stresses with holographic interferometry or ESPI. The following section provides an overview of those approaches.

11.1.2 Hole Drilling

Interest in applying hole drilling with optical methods rather than strain rosettes stems from advantages such as elimination of the time and costs associated with installation of a rosette and milling guide and an ability to be applied to regions where installation would be infeasible. The use of optical methods in conjunction with hole drilling has been under development since the 1980s [16–20] and in recent years a transition from laboratory R&D use to applications in manufacturing and in the field has been occurring.

The hole drilling method can be applied using holographic interferometry or ESPI, among other optical methods. The displacements resulting from release of residual stresses by hole drilling create an interference fringe pattern such as that in Figure 11.3 obtained with holographic interferometry. The displacements for a blind hole drilled to a certain depth into residual stresses constant with depth can be expressed as [21]:

$$\begin{Bmatrix} u_r \\ u_\theta \\ u_z \end{Bmatrix} = \begin{bmatrix} A + B\cos 2\theta & A - B\cos 2\theta & 2B\sin 2\theta \\ C\sin 2\theta & -C\sin 2\theta & -2C\cos 2\theta \\ F + G\cos 2\theta & F - G\cos 2\theta & 2G\sin 2\theta \end{bmatrix} \begin{Bmatrix} \sigma_x \\ \sigma_y \\ \tau_{xy} \end{Bmatrix} \quad (11.3)$$

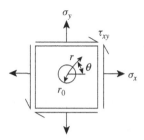

Figure 11.4 Hole drilling terminology

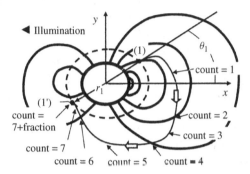

Figure 11.5 Illustration of fringe counting path (which should not cross the same fringe twice)

where u_r, u_θ are in-plane displacements in the radial (r) and tangential directions (θ) seen in Figure 11.4, u_z is out-of-plane displacement, σ_x, σ_y and τ_{xy} are residual stress components, $A = r_o(1+\nu)a/2E$, $B = r_o b/2E$, $C = r_o c/2E$, $F = r_o f/2E$, $G = 4\nu r_o g/2E$, r_o is hole radius, E = modulus of elasticity, ν = Poisson's ratio, and a, b, c, f and g are non-dimensional coefficients available [21] as functions of radial location normalized by hole radius (r/r_o) and hole depth normalized by hole diameter. The displacements cause changes in the path length of light reflected from the region around a hole, which, in turn causes phase changes and a fringe pattern that can be processed to obtain residual stresses. Different approaches have been developed to perform that processing. Choice of an approach will depend on user preferences and resources available.

One approach makes use of a fringe pattern obtained by analog or digital holographic interferometry or ESPI without the need for phase shifting and unwrapping. Referring to Figure 11.5, a radial location in a pattern, denoted by point (1) and angle θ_1, is selected. Then the number of light and dark fringes crossed while following an arbitrary path to a diametrically opposite point (1') is counted. Using the same radius as for point (1), similar counts can be made for two other starting points with different angles θ_2 and θ_3. The counts n_1, n_2, and n_3 can then be entered in the following relation [21] to find residual stress components $\sigma_x, \sigma_y, \tau_{xy}$ (for stresses uniform over the depth of the hole)

$$\begin{Bmatrix} \sigma_x \\ \sigma_y \\ \tau_{xy} \end{Bmatrix} = \pi \begin{bmatrix} C_{11} & C_{12} & C_{13} \\ C_{21} & C_{22} & C_{23} \\ C_{31} & C_{32} & C_{33} \end{bmatrix}^{-1} \begin{Bmatrix} n_1 \\ n_2 \\ n_3 \end{Bmatrix} \qquad (11.4)$$

with

$$C_{i1} = K_x [\cos \theta_i (A + B \cos \theta_i) - C \sin \theta_i \sin 2\theta_i]$$

$$C_{i2} = K_x [\cos \theta_i (A - B \cos \theta_i) + C \sin \theta_i \sin 2\theta_i]$$

$$C_{i3} = 2 K_x [B \cos \theta_i \sin 2\theta_i + C \cos 2\theta_i]$$

$$K_x = (2\pi/\lambda) \cos \gamma_1 \tag{11.5}$$

where the x-direction is taken in the direction of illumination as in Figures 11.1 and 11.3, γ_1 is the angle shown in Figure 11.1, λ is the wavelength of laser light used, and constants A, B and C are defined with Equation (11.3) (The fringe counting procedure cancels the effect of out-of-plane displacements and thus constants F and G in Equation (11.3) are not needed in Equations (11.4) and (11.5))

The sign of stress can be determined by a simple procedure described in [21]. A number of other fringe counting methods are also available [e.g., 22–25]. Alternatively, a fringe pattern can be processed by methods that take advantage of the additional information in an entire pattern, e.g., [26,31,33].

Many users may prefer hole drilling with a phase shifting capability. Most current systems with that capability are based on ESPI, but phase shifting can be used with other interferometric methods as well. Phase shifting and associated software provide the advantage of a detailed map of phase change $\Delta\varphi(x, y)$ as an input to computational models for determining residual stresses from hole drilling [e.g., 27–29]. An automated, single beam ESPI hole drilling system and a full-field method for converting resulting optical data to residual stresses are available [30–32]. Methods that enable residual stresses to be determined from ESPI-hole drilling data, even in the presence of possible rigid body motions, have also been developed [33,34]. To determine the variation of residual stresses with depth, incremental drilling versions of the hole drilling method are available for use with ESPI [35] or holographic interferometry [36]. Figure 11.6 shows an example of a fringe pattern found by hole drilling with ESPI.

ESPI systems have been developed specifically for determining residual stresses by hole drilling. Figure 11.7 shows some equipment from StressTech and Dantec Dynamics. Several different compact portable ESPI hole-drilling systems have also been developed for use in field environments [39–41]. Figure 11.8 shows an example of an ESPI hole drilling system consisting of a hole drilling module, optical components and a base that can be magnetically clamped to a test object. It is designed to be sensitive to radial in-plane displacements through use of a special diffractive optical element [42], and it makes use of a laser diode as a light source, an advantage for compactness, portability and cost.

Optical methods such as holographic interferometry and ESPI can also be applied with machining of slots [44,45] as a means of releasing residual stresses instead of circular holes. Also, to extend the depth to which hole drilling can be applied to find residual stresses, data can be gathered by optical means from drilling a small hole, then milling a larger hole over the small hole to create a fresh surface for capture of fringes produced when a second hole is drilled in the bottom or the larger hole, and so forth [46].

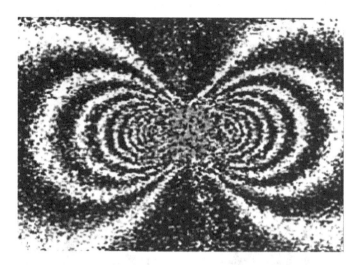

Figure 11.6 A fringe pattern found by ESPI-hole drilling with stress and illumination in the horizontal direction. Reproduced with permission from [25], Copyright 1998 The Optical Society of America

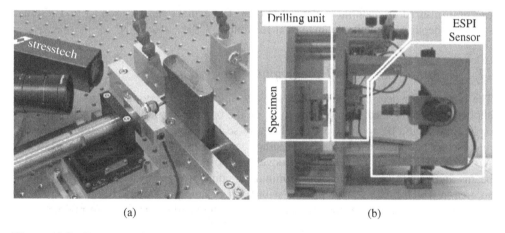

Figure 11.7 Examples of equipment developed for hole drilling with ESPI: (a) Reproduced with permission from [37]. Copyright 2009 SEM, (b) Reproduced with permission from [38]

11.1.3 Deflection

Residual stresses can be determined from deflections that occur as layers of material containing the stresses are removed incrementally. Deflections can be measured by optical means. For instance, holographic interferometry or ESPI can be applied to monitor the defection of cantilever beams or sheets as material is removed by chemical etching [47,48], producing fringes similar to those seen in Figure 11.2, except that the fringes correspond to out-of-plane displacements (with each fringe representing a displacement equaling half the wavelength of laser light used). Alternatively, deflections

Figure 11.8 Compact portable system for hole drilling residual stress determination. Reproduced with permission from [43], Copyright 2009 SPIE

can be measured as material is deposited [49]. Deflections can be used to find residual stress vs. depth with available analytical relations.

11.1.4 Micro-ESPI and Holographic Interferometry

The methods for determining residual stresses described in previous sections have been applied to "macro-sized" components (i.e., on the order of centimeters or larger). In principle, they should be applicable at smaller size scales. For example, a digital holographic microscope has been used to study residual stresses in micromechanical devices [50]. A micro-ESPI system [51] has been developed to measure the deflection of thin film beams and should be capable of determining residual stresses by layer removal. This is an area in which further developments with ESPI or holographic interferometry could be forthcoming, including the possible application of hole drilling in this size regime.

11.2 Moiré Interferometry

11.2.1 Moiré Interferometry Overview

Moiré interferometry is a technique for determining surface displacements with high sensitivity [52]. Figure 11.9 shows a four-beam Moiré interferometry setup. A cross-line grating is applied to a region of interest, which must be flat in virtually all cases. The various references cited in this section give details on installation of gratings. Laser light $B1$ and $B2$ illuminate the grating, causing light to be diffracted in the z-direction. Displacement of the surface and grating causes the diffracted light to interfere in the image plane of a recording device, producing a fringe pattern that can be related to in-plane x-displacements U_x. Similarly, beams $B3$ and $B4$ form a fringe pattern related to

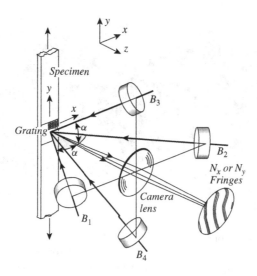

Figure 11.9 Schematic of an experimental set-up for Moiré interferometry. Reproduced with permission from [53], Copyright 2001 Sage

the y-displacements U_y. The x-direction pattern is recorded with light B3 and B4 blocked, and the y-pattern with light B1 and B2 blocked. If the grating has a typical frequency of 1200 lines/mm, the displacement between fringes like those depicted in Figure 11.9 is approximately 0.4 µm [53]. Mechanical stability of a Moiré setup comparable to that mentioned in Section 11.1.1 for ESPI and holographic interferometry is needed.

A compact, general purpose Moiré interferometer was manufactured and sold by IBM in the 1990s (based on R&D at Virginia Tech), and later by Photomechanics Inc. (Vestal, NY). Other research groups worldwide have developed their own Moiré interferometry setups. A variety of methods have been developed for determining residual stresses with the use of Moiré interferometry. The following sections summarize several such developments.

11.2.2 Hole Drilling

Residual stresses can be determined using displacement data obtained from fringe patterns and computational relations between displacements and residual stresses similar in nature to those used for the strain rosette implementation of hole drilling. A number of such relations [54–57] have been developed, and one that is representative [56] is summarized below as an example. The release of residual stresses by introduction of a blind hole drilled to certain depth will cause a pattern of Moiré fringes such as that in Figure 11.10.

The radial displacements from hole drilling can be expressed from Equation (11.3) as

$$u_r(r, \theta) = A(\sigma_x + \sigma_y) + B[(\sigma_x - \sigma_y)\cos 2\theta + 2\tau_{xy}\sin 2\theta] \qquad (11.6)$$

where A and B are computed from non-dimensional coefficients that have been developed by finite element analyses [56], and from modulus of elasticity, Poisson's ratio, the ratios of hole depth-to-diameter and radial position-to-hole radius. U_x and U_y are displacements

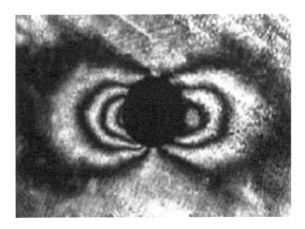

Figure 11.10 Moiré interferometric fringe pattern (U_x- displacement field with illumination in the horizontal x-direction). The pattern is nearly symmetric from one side of a hole to another because the formation of fringes is a function of in-plane displacements in this case. Reproduced with permission from [58], Copyright 2004 Elsevier

in x and y directions that are related to fringe orders N_x and N_y by

$$U_x = (1/2f_s)\ N_x \qquad\qquad U_y = (1/2f_s)\ N_y \qquad (11.7)$$

where f_s is the grating frequency.

Radial displacements can be expressed in terms of U_x and U_y displacements by

$$u_r(r, \theta_i) = U_x(x_i, y_i) \cos\theta_i + U_y(x_i, y_i) \sin\theta_i$$
$$x_i^2 + y_i^2 = r^2, \qquad \theta_i = \tan^{-1}(y_i/x_i) \qquad (11.8)$$

where θ_i is an angle such as that shown in Figure 11.11. Combining Equations (11.6) to (11.8) yields a relation between fringe orders determined at three values of θ_i (e.g., 0°, 45° and 90°) and residual stresses:

$$[N_x(x_i, y_i)\ N_y(x_i, y_i)] \begin{bmatrix} \cos\theta_i \\ \sin\theta_i \end{bmatrix}$$
$$= 2f_s[A + B\cos 2\theta_i \quad A - B\cos 2\theta_i \quad 2B\sin 2\theta_i] \begin{bmatrix} \sigma_x \\ \sigma_y \\ \tau_{xy} \end{bmatrix} \qquad (11.9)$$

Fringe orders N_x and N_y can be found for a given θ_i and radial position (e.g., $r/r_o = 1.2$). In Figure 11.11, a zero order fringe is assigned to the region remote from the hole. Increasing fringe orders are assigned approaching the hole, as seen by values such as 0.5, 1.0, 1.5. The sign of stress can be found by temporarily perturbing a fringe pattern using a method described in [56]. The Moiré-hole drilling method can be used to quantify how residual stresses vary with depth using incremental drilling and a computational approach [56] similar to that used for hole drilling with strain rosettes.

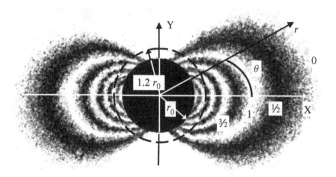

Figure 11.11 Moiré-hole drilling fringes showing radial location $(r/r_o) = 1.2$ and whole and half-order fringes N on one side of the hole

Systems to implement the Moiré hole-drilling method have been developed on a customized basis. For example, [57] describes a compact interferometer with an air turbine drilling device and specimen held in place by a C-clamp. The Moiré hole-drilling method can also be applied to determine how residual strains vary through the thickness of fiber reinforced composite laminate specimens [59,60]. Computational methods are available to determine residual stresses from holes drilled though the thickness in orthotropic materials [61,62] and layer-by-layer in laminates from holes drilled incrementally [63].

The Moiré-hole drilling method offers the advantages of high sensitivity to small deformations, the existence of a computational methodology for blind holes and incremental drilling, equipment available for implementation, and numerous successful applications. Drawbacks include the need to apply a grating on a smooth, flat surface and vibration isolation sufficient to enable interferometric fringe patterns to form and be recorded.

11.2.3 Other Approaches

Moiré interferometry can be used in conjunction with sectioning to determine residual stresses. As an example, Figure 11.12 shows Moiré fringes resulting from deep slots machined to relax residual stresses in a railway rail. A methodology for finding residual stresses from Moiré fringes caused by sectioning can be found, for example, in [65].

Measurements of curvature and out-of-plane displacements can be performed with the shadow Moiré method [66], in which the shadow of a grating is cast on a surface. For instance, residual stresses resulting from processes such as curing of composite laminates [67] or deposition of material [68] can be determined using that method, with implementation details given in the references just cited. Other approaches for applying Moiré interferometry to find residual stresses, such as the cure referencing method for composites, are summarized in [69].

11.2.4 Micro-Moiré

Moiré interferometry can be perfomed on smaller size scale with optical microscope techniques [53] or within a scanning electron microscope using gratings created by electron

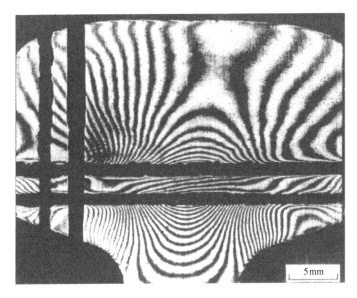

Figure 11.12 Moiré fringe pattern from sectioning. Reproduced with permission from [64], Copyright 1997 Elsevier

beam lithography [70] or focused ion beam milling. For example, the Moiré technique has been used to measure strains produced in a MEMS cantilever as residual stresses in a layer of SiO_2 were released by chemical etching [71]. Other methods of releasing residual stresses such as hole drilling, slotting, and so on could potentially be used with micro-Moiré techniques.

11.3 Digital Image Correlation

11.3.1 Digital Image Correlation Overview

In macroscopic 3D digital image correlation (DIC) [72,73], a surface region is illuminated with ordinary white light and viewed with a pair of high-resolution CCD cameras, as illustrated in Figure 11.13 (a single camera can be used for 2D measurements). A reference image of the surface is recorded and digitized prior to deformation. Sub-sets of an image (typically a square of about 10 to 20 pixels on each side) are then defined. By comparing the locations of corresponding subsets in images taken by two cameras, 3D coordinates of the surface can be determined. After deformation, 3D displacements can be found by tracking the movement of corresponding sub-sets between the reference image and the image after deformation. DIC is not an interferometric technique, and so has much less strict mechanical stability requirements. Cameras are often mounted on a tripod.

General-purpose digital image correlation cameras and software for implementation are available from providers such as GOM (Aramis), Correlated Solutions, Dantec, ISI-SYS and LaVision. Important practical considerations involved in applying DIC are discussed

Figure 11.13 Digital image correlation set-up for measurement of tyre deformations showing two cameras and ordinary lamps. Reproduced with permission from [74], Copyright 2007 Wiley

in a series of bi-monthly articles in the journal *Experimental Techniques* beginning in January 2012 [75].

11.3.2 Hole Drilling

Residual stresses can be measured by hole drilling by first capturing a reference image of an area to be drilled. The region should have surface features suitable for correlation, or have features added such as by spraying tiny black dots on the surface. Next a hole is drilled to release residual stresses. Then a second image is captured and processed by DIC system software to determine displacements resulting from release of stresses. From Equation (11.3), one approach [76] for obtaining the residual stresses is to use radial displacements measured for a given radial position (r/r_o) and a number of different angles θ_i. Residual stress components σ_x, σ_y and τ_{xy} can be expressed in terms of radial displacements u_{ri} from by

$$u_{ri} = (A + B \cos 2\theta_i)\, \sigma_x + (A - B \cos 2\theta_i)\, \sigma_y + (2B \sin 2\theta_i)\, \tau_{xy} \qquad (11.10)$$

Equation 11.10 may be written as $Ks = u$, where K is an n x 3 matrix containing values of the three terms within parentheses in Equation (11.10) for n values of θ_i, s is a 3×1 vector containing the stresses components to be determined, and u is an $n \times 1$ vector with the measured radial displacements. A least-squares solution for the stress components can then be computed from

$$K^T K s = K^T u \qquad (11.11)$$

where superscript T denotes a transpose. The approach just described is applicable to a hole drilled to a depth over which stresses are approximately constant. A similar approach

[77] has been developed that also allows, in principle, the determination of how stresses may be varying across the surface region in which a hole is drilled. Another approach [78] for determining residual stresses by DIC-hole drilling performs image correlation with an optimization procedure that takes advantage of analytical knowledge of an expected hole drilling displacement field for an assumed set of residual stress components (σ_x, σ_y, τ_{xy}). This is in contrast to the approach represented by Equation (11.11) in which displacements are measured with a general-purpose image correlation system, and then used to solve for stresses. As with hole drilling performed with holographic interferometry, ESPI and Moiré interferometry, an incremental version of hole drilling has also been developed [79] to enable stresses vs. depth to be determined.

The DIC-hole drilling approach has the advantages of ordinary white light illumination, a more direct determination of displacements than by interferometric methods, applicability to rough or curved surfaces, commercially available systems to implement DIC, no need for interferometric-level mechanical stability, and an ability to correct for rigid body motions. In addition, DIC gives displacement data in multiple orthogonal directions. In tests conducted by the author and colleagues, it was possible to record a set of images by DIC, then remove the test object for hole drilling at another location, then return the specimen to approximately the same initial location for recording of a set of "post-drilling" images. Drawbacks of the approach are the need to "calibrate" cameras prior to use in a given test and approximately a factor of 10 less sensitivity to deformations than ESPI or similar interferometric techniques [73]. In spite of the lower sensitivity, it appears possible to obtain reasonably accurate results with DIC-hole drilling.

11.3.3 Micro/Nano-DIC Slotting, Hole Drilling and Ring Coring

Digital image correlation can also be used to find residual stresses at micro-scale sizes through use of images obtained in a scanning electron microscope (SEM). Residual stresses are relaxed by material removal, typically accomplished through focused ion beam (FIB) milling within an SEM chamber. Many of the DIC system providers mentioned earlier in this section have software that can also be used to process SEM images.

Slotting by FIB can be used to find residual stresses [80–83]. After an SEM image of small region is captured, a slot is made, and the resulting displacement field determined by DIC. In-plane displacements normal to the slot, u_x, can be related to residual stress σ in the same direction by an analytical expression [81] such as

$$u_x = (\sigma/E)\{[2\,(x^2+a^2-y^2)^{1/2} - (1+\nu)(\,x^2/\,(x^2+a^2-y^2)^{1/2})] \\ \times (|x|/x) - (1-\nu)\,x\} \tag{11.12}$$

where E is modulus of elasticity, ν is Poisson's ratio, $2a$ is slot length, and x, y are coordinates with an origin at the center of a slot. To determine a profile of residual stress with depth, incremental slotting can be applied with a suitable computational methodology [84].

Hole drilling may be used instead of, or in addition to, slotting to find residual stresses in microscale structures. Consider the case of a thin micro-machined membrane with equibiaxial residual stress σ. Suppose a through-hole is created by FIB, creating a displacement

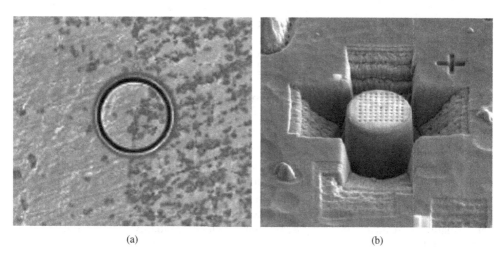

Figure 11.14 (a) SEM image of ring core crossing a grain boundary. Reproduced with permission from [88]. Copyright 2009 IOP and (b) a ring core showing a pattern of a grid of FIB milled dots used to facilitate image correlation. Reproduced with permission from [89], Copyright 2010 Elsevier

field by stress relaxation. Stress can be found from the in-plane displacements using [85]:

$$u_x(r,\theta) = [(\sigma R^2)/E\, r](1+v)\cos\theta \qquad u_y(r,\theta) = [(\sigma R^2)/E\, r](1+v)\sin\theta \tag{11.13}$$

where R is hole radius, r is radial position, v is Poisson's ratio, E is modulus of elasticity and θ is an angle measured from the x-axis. To find a profile of stress with depth, incremental micro-hole drilling can be perfomed within an SEM, and a computational methodology for implementation is available [86]. Since the displacements associated with micro-hole drilling are so small, SEM imaging artifacts can compromise determination of stresses; however, an approach to compensate for such artifacts has been developed [87]. Ring coring is yet another method that may be used. As illustrated in Figure 11.14, a ring is milled to release residual stresses. The resulting displacements (or strains) measured on the surface of the island can be used to determine residual stresses using available analytical approaches [89], including how stresses vary with depth [90].

A practical experimental difficulty that may arise with FIB milling is re-deposition [88,90], in which material removed by FIB is deposited near the slot, hole or ring core being formed. Since re-deposition can seriously disrupt image correlation, means to protect against it may be needed. For instance, deposition of a thin platinum ring has been used to protect such protection [90]. The micro-slotting, hole drilling or ring coring methods have used computational approaches that relate measured displacements to residual stresses based on continuum mechanics assumptions. It is important to be aware that microstructural features such as grain boundaries may influence the validity of such analyses.

11.3.4 Deflection

Digital image correlation may also be used to measure deflections that result from the generation of residual stresses, such as by deposition of a coating, or from the removal of layers of material containing residual stresses. For instance, the curvature of a thin stainless steel strip resulting from deposition of a thermal barrier coating was obtained by 3D digital image correlation and used to determine residual stresses [91].

11.4 Other Interferometric Approaches

11.4.1 Shearography

Unlike Moiré, holographic or electronic speckle pattern interferometry, shearography [92,93] produces fringe patterns that depend on displacement derivatives, rather than displacements. An advantage of shearography is reduced sensitivity to vibration and less stringent mechanical stability requirements relative to those other interferometric methods. The use of shearography combined with hole drilling has been explored for determination of residual stresses [94] but not developed to the extent of the methods in Sections 11.1 to 11.3.

11.4.2 Interferometric Strain Rosette

A tiny pattern of indentations (such as made with a hardness tester) illuminated with laser light forms an interference pattern by diffraction that can be used to measure strains in different directions [95]. The resulting optical strain rosette has been applied with hole drilling [96] and ring coring [97] to measure residual stresses vs. depth, but the approach has not been developed commercially.

11.5 Photoelasticity

Photoelasticity [98] has been a mainstay of experimental stress analysis for decades. Reflection photoelasticity, in which a coating of photoelastic material is applied to a surface, has been explored with hole drilling to determine residual stresses [94] but not developed to the extent of the methods in Sections 11.1 to 11.3. Reflection photoelasticity can be used with dissection to find residual stresses [99], but applications have been infrequent.

Transmission photoelasticity, in which light travels through transparent or translucent materials, can be used to find residual stresses in a variety of test objects from optical discs to tempered glass windshields. The method is well established and details of implementation are given in test standards for plastic materials [100] and glass [101–104]. Photoelasticity can even be used to measure residual stresses in glass objects with complicated geometries [105]. A variety of automated stress measuring systems for glass articles are available from providers such as StrainOptics Technologies, GlasStress, Stress Photonics and Ilis gmbh.

Transmission photoelasticity may also be applied to find residual stresses in materials opaque to visible light but transparent to electromagnetic radiation of other wavelengths such as infra-red (IR) light. Primary applications have been to measurement of residual stresses in silicon wafers and sheets [106].

11.6 Examples and Applications

This section presents practical examples of hole-drilling combined with optical methods for residual stress determination. The references cited in this chapter give additional examples. The hole-drilling device shown in Figures 11.8 and 11.15 has been applied to determine stresses at different locations on a gas pipeline in service in Brazil, illustrating that hole drilling with an interferometer can be applied in a challenging environment. Stresses were a combination of those from pipe fabrication, installation, ground movements over time as well as temperature gradients and internal pressure. Other hole drilling-ESPI devices have been applied in the Ukraine to investigate residual stresses in welded shells [39] and in Russia to determine residual stresses in and adjacent to pipe welds under field conditions [40]. Comprehensive holographic-hole drilling determinations of residual stresses in both thin and thick walled welded aluminum plates, and in tubular specimens have also been performed in Russia [107–109].

Hole drilling with ESPI or holographic interferometry can also be used to measure residual stresses from manufacturing processes. For instance, residual stresses from quenching of a stainless steel sample have been found [110] using a compact ESPI interferometer from the E.O. Paton Electric Welding Institute (Kiev). Profiles of residual stresses vs. depth in shot-peened aluminum and steel samples [37] have been found using an ESPI hole drilling device like that shown in Figure 11.7(a).

As another example, hole drilling with holographic interferometry has been used by the author to find the profile of biaxial residual stresses vs. depth in a rolled, undercut fillet of a crankshaft shown in Figure 11.16(a). A 0.8 mm diameter hole was drilled incrementally into the curved surfaces of various fillets each with a radius of 2 mm, producing fringe patterns like that in Figure 11.16(b). Corrections for the effects of curvature on the fringe pattern as well as on the relation between fringes and surface deformations caused by release of residual stresses enabled successful determination of the stresses from rolling.

As an example of digital image correlation used with hole drilling, Figure 11.17 shows a micro-hole (4 µm diameter) used to determine the profile of residual stresses vs. depth in a peened specimen of a metallic glass [86]. The hole drilling was perfomed by a focused ion beam technique within a scanning electron microscope. This example and the ones above demonstrate the versatility of the hole drilling method combined with optical methods for determining residual stresses over a size regime spanning orders of magnitude and in environments as vastly different as underground piping to the chamber of a scanning electron microscope.

11.7 Performance and Limitations

This chapter has summarized several ways for using optical methods to measure residual stresses. The measurement accuracies achieved by these methods will depend on factors such as the precision of the measurement equipment used, the computational methods used to process data, and the specific test conditions. The application of optical methods to measure residual stresses continues to develop, and performance is likely to improve over time. Table 11.1 summarizes typical ranges of accuracy for four methods. Accuracies of other residual stress measurement approaches are more difficult to generalize.

Table 11.1 Performance and limitations of hole drilling combined with optical methods for residual stress determination

Method	Accuracy*	Advantages	Limitations
Holographic interferometry or ESPI	5 to 15% [21–24,42, 111–115,117]	• Strain gages and milling guide avoided • Applicable to curved and/or rough surfaces • Equipment and software commercially available specifically for hole drilling as well as other measurement purposes	• Test setup needs interferometric stability
Moire interferometry	5 to 15% [57,116]	• Strain gages and milling guide avoided • Equipment and software commercially available for general purpose measurements	• Test setup needs interferometric stability • Requires installation of a grating
Digital image correlation (macro)	10 to 20% [76–78]	• Strain gages and milling guide avoided • Applicable to curved and/or rough surfaces • Interferometric stability not needed • Equipment and software commercially available for general purpose measurements	• Deformation measurements not as precise as for interferometric methods
Digital image correlation[†] (micro-hole or slot)	20 to 40% [81,85,117]	• Micro-scale measurements • Applicable to microscopically curved and/or rough surfaces • Software commercially available for general purpose measurements	• Requires scanning electron microscope with focused ion beam

* Accuracy estimates are based on residual stresses determined in carefully controlled lab conditions for stresses uniform with depth. Reduced accuracy can be expected for measurements of stresses vs. depth.
[†] References 81 and 85 report agreement of better than 10% between residual stress values determined by the use of micro-holes or micro-slots milled through the thickness of a thin membrane and values inferred from bulge testing. Reduced accuracy is expected when uncertainties in the depth of blind micro-holes/slots and inaccuracies associated with image data from a scanning instrument are taken into account [117].

Figure 11.15 Hole drilling system being used to find stresses in a gas pipeline in service. Reproduced with permission from [43], Copyright 2009 SPIE

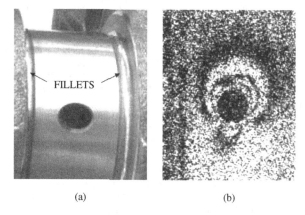

(a)   (b)

Figure 11.16 (a) Undercut fillets in crankshaft and (b) holographic-hole drilling fringe pattern caused by release of residual stresses by drilling a hole in the rolled fillet, with illumination directed parallel to the fillet

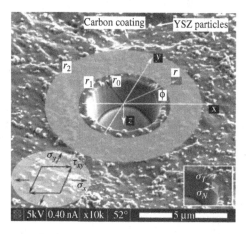

Figure 11.17 Determination of residual stresses using a micro-hole with an SEM. Reproduced with permission from [86], Copyright 2012 Elsevier

References

[1] Kreis, T. (2005) *Handbook of Holographic Interferometry*. Wiley-VCH: Weinheim, Germany.
[2] Pryputniewicz, R. (2008) "Holography." In: Sharpe, W. (ed) *Handbook of Experimental Solid Mechanics*. Springer: New York, pp 675–699.
[3] Vandenrijt, J., Georges, M. (2010) "Electronic Speckle Pattern Interferometry and Digital Holographic Interferometry With Microbolometer Arrays at 10.6 μm." *Appl Optics* 49:5067–5075.
[4] Toal, V. (2011) *Introduction to Holography*. CRC Press, Taylor & Francis Group: Boca Raton, Florida.
[5] Schnars, U., Jueptner, W. (2005) *Digital Holography*. Springer: New York.
[6] Asundi, A. (2011) *Digital Holography for MEMS and Microsystems Metrology*. Wiley: West Sussex, UK.
[7] Rastogi, P. (2001) *Digital Speckle Pattern Interferometry and Related Techniques*. Wiley: West Sussex, UK.
[8] Gan, Y., Steinchen, W. (2008) "Speckle Methods" In: Sharpe, J. (ed) *Handbook of Experimental Solid Mechanics*. Springer: New York, pp. 655–673.
[9] Yang, L., Ettemeyer, A. (2003) "Strain Measurement by Three-Dimensional Electronic Speckle Pattern Interferometry: Potentials, Limitations and Applications." *Opt Eng* 42:1257–1266.
[10] Huntley, J. (2001) "Automated Analysis of Speckle Interferograms." In: *Digital Speckle Pattern Interferometry and Related Techniques*. Wiley-VCH: West Sussex, UK, pp. 59–139.
[11] Soares, O. (1983) "Review of Resolution Factors in Holography." *Opt Eng* 22: SR-107 to SR-112.
[12] Georges, M., Scauflaire, V., Lemaire, P. (2001) "Compact Holographic Camera Based on Photorefractive Crystals and Applications in Interferometry." *Opt Mater* 18:49–52.
[13] Lobanov, L., Pivtorak, V. (2002) "Diagnostics of Residual Stresses State of Welded Structures Using the Methods of Holographic Interferometry and Electronic Speckle Interferometry." *Mater Sci Forum* 404–407:867–874.
[14] Thomas, B., Pillai, S. (2009) "High-Speed Generation of Digital Holographic Inteferograms and Shearograms for Non-Destructive Testing." *Insight* 51:252–256.
[15] Michalkiewicz, A., Kujawinski, M., Stasiewicz, K. (2008) "Digital Holographic Camera and Data Processing for Remote Monitoring and Measurement of Mechanical Parts." *Opto-Electon Rev* 16:68–75.
[16] McDonach, A., McKelvie, J., MacKenzie, P., Walker, C. A. (1983) "Improved Moire Interferometry and Applications in Fracture Mechanics, Residual Stress and Damaged Composites." *Exp. Tech* 7(6): 20–24.
[17] Nelson, D., McCrickerd, J. (1986) "Residual-Stress Determination Through Combined Use of Holographic Interferometry and Blind-Hole Drilling." *Exp Mech* 26:371–378.
[18] Bass, J., Schmitt, D., Ahrens, J. (1986) "Holographic in Situ Stress Measurements." *Geophysl J R Astr Soc* 85:13–41.
[19] Antonov, A. (1983) "Development of the Method and Equipment for Holographic Inspection of Residual Stresses in Welded Structures." *Weld Prod* 30(12): 41–43.
[20] Lobanov, L., Kasatkin, B., Pivtorak, V., Andrushchenko, S. (1983) "A Procedure for Investigating Residual Welding Stresses Using Holographic Interferometry." *Autom Weld* 36(3): 5–9.
[21] Makino, A., Nelson, D., Fuchs, E., Williams, D. (1996) "Determination of Biaxial Residual Stresses by a Holographic-Hole Drilling Technique." *J Eng Mater Technol* 118:583–588.
[22] Lin, S., Hsieh, C., Lee, C. (1998) "A General Form for Calculating Residual Stresses Detected by Using the Holographic Blind-Hole Method." *Exp Mech* 38:255–226.
[23] Andrushchenko, S., Krotenko, P. (2005) "Displacement Determination by Holographic Interferometry for Residual Stress Analysis in Elastic Bodies." *Int Appl Mech* 41:929–933.
[24] Apal'kov, A., Larkin, A., Osintev, A., Odintev, I., Shchepinov, V., Shchikanov, A., Fonatine, J. (2007) "Holographic Interference Method for Studying Residual Stresses." *Quantum Electron* 37:590–594.
[25] Zhang, J., Chong, J. (1998) "Fiber Electronic Speckle Pattern Interferometry and its Application in Residual Stress Measurement." *Appl Optics* 37:6707–6715.
[26] Baldi, A. (2005) "A New Analytical Approach for Hole Drilling Residual Stress Analysis by Full Field Method." *J Eng Mater Technol* 127:165–169.
[27] Schmitt, D., Hunt, R. (2000) "Inversion of Speckle Interferometer Fringes for Hole-Drilling Residual Stress Determinations." *Exp Mech* 40(2): 129–137.
[28] Diaz, F., Kaufmann, G., Moller, O. (2001) "Residual Stress Determination Using Blind-Hole Drilling and Digital Speckle Pattern Interferometry With Automated Data Processing." *Exp Mech* 41(4): 319–323.

[29] Focht, G., Schiffner, K. (2003) "Determination of Residual Stresses by an Optical Correlative Hole-Drilling Method." *Exp Mech* 43(1): 97–104.
[30] Steinzig, M., Ponslet, E. (2003) "Residual Stress Measurement Using the Hole Drilling Method and Laser Speckle Interferometry," part I. *Exp Tech* 27(3): 43–46.
[31] Ponslet, E., Steinzig, M. (2003) "Residual Stress Measurement Using the Hole Drilling Method and Laser Speckle Interferometry," part II: analysis technique. *Exp Tech* 27(4):17–21.
[32] Ponslet, E., Steinzig, M. (2003) "Residual Stress Measurement Using the Hole Drilling Method and Laser Speckle Interferometry," part III: analysis technique. *Exp Tech* 27(5): 45–48.
[33] Schajer, G., Steinzig, M. (2005) "Full-field calculation of hole drilling residual stresses from electronic speckle pattern interferometry data." *Exp Mech* 45:526–532.
[34] Dolinko, A., Kaufmann, G. (2006) "A Least-Squares Method to Cancel Rigid Body Displacements in Hole Drilling and DSPI Systems for Measuring Residual Stresses." *Opt Lasers Eng* 44(12): 1336–1347.
[35] Schajer, G., Rickert, T. (2011) "Incremental Computation Technique for Residual Stress Calculations Using the Integral Method." *Exp Mech* 51:1217–1222.
[36] Makino, A., Nelson, D. (1997) "Determination of Sub-Surface Distributions of Residual Stresses by a Holographic-Hole Drilling Technique." *J Eng Mater Technol* 119:95–103.
[37] Rickert, T. (2009) "ESPI Residual Stress Determination in Shot Peened Aluminum and Steels." In: Proc. SEM Annual Conf., vol. 2, pp. 998–1002.
[38] Sedivy, O., Krempaszky, C., Holy, S. (2007) "Residual Stress Measurement by Electronic Speckle Pattern Interferometry." In: Proc 5th Australasian Congr Appl Mech, pp. 342–347.
[39] Lobanov, L., Pivtorak, V., Savitskii, V., Tkachuk, G. (2010) "Using Electronic Speckle Interferometry for the Accurate Determination of the Residual Stresses in Welded Joints and Structural Members." *Welding Intl* 24:439.
[40] Antonov, A. (2011) "Operative Determination of the Stress-strain State of Welded Joints in Objects in Oil and Gas Industries." *Weld Int* 25:795–799.
[41] Viotti, M., Dolinko, A., Gallzzi, G., Kaufmann, G. (2006) "A Portable Digital Speckle Pattern Interferometry Device to Measure Residual Stresses Using the Hole Drilling Technique." *Opt Lasers Eng* 44:1052–1066.
[42] Viotti, M., Albertazzi, A., Kapp, M. (2008) "Experimental Comparison Between a Portable DSPI Device With Diffractive Optical Element and a Hole Drilling Strain Gage Combined System." *Opt Lasers Eng* 46:835–841.
[43] Viotti, M., Albertazzi, G. (2009) "Industrial Inspections by Speckle Interferometry: General Requirements and a Case Study." In *Optical Measurement Systems for Industrial Inspection* VI, Proc SPIE Int Soc Opt Eng, v. 7389, pp 73890 G-1 to G-15.
[44] Schajer, G., An, Y. (2010) "Residual Stress Determination Using Cross-Slitting and Dual-Axis ESPI." *Exp Mech* 50:169–177.
[45] Montay, G., Sicot, O., Maras, A., Rouhard, E., Francois, M. (2009) "Two Dimensions Residual Stresses Analysis Through Incremental Groove Machining Combined With Electronic Speckle Pattern Interferometry." *Exp Mech.* 49:459–469.
[46] Makino, A., Nelson, D., Hill, M. (2011) "Hole Within a Hole Method for Determining Residual Stresses." *J Eng Mater Technol* 113: 021020-1 to -8.
[47] Lira, I., Vial, C., Robinson, K. (1997) "The ESPI Measurement of the Residual Stress Distribution in Chemically Etched Cold-Rolled Metallic Sheets." *Meas Sci Technol* 8:1250–1257.
[48] Palma, J., Rivero, R., Lira, I., Franscois, M. (2009) "Measurement of the Residual Stress Tensor on the Surface of a Specimen by Layer Removal and Interferometry: Uncertainty Analysis." *Meas Sci Technol* 20: 115302+10.
[49] Kakunai, S., Hayahira, H., Sakamoto, T., Matsuda, H. (2005) "In-Situ Measurement of Internal Stress in Electroless Plating by Television Holographic Interferometry." In: *Appl Mech and Mater*, v.4, TransTech Publications, pp. 65–70.
[50] Coppola, G., Ferraro, P., Iodice, M., De Nicola, S., Finizio, A., Grilli, S. (2004) "A Digital Holographic Microscope for Complete Characterization of Micromechanical Systems." *Meas Sci Technol* 15:529–539.
[51] Kim, D., Huh, Y-H., Kee, C. (2006) "Out-Of-Plane Micro-ESPI System for Measurement of Mechanical Properties of Film Materials." *Key Eng Mater* 321–323:116–120.
[52] Post, D., Han, B. (2008) "Moiré Interferometry" In: Sharpe, W. (ed) *Handbook of Experimental Solid Mechanics*. Springer: New York, pp. 627–653.

[53] Han, B., Post, D., Ifju, P. (2001) "Moiré Interferometry for Engineering Mechanics: Current Practices and Future Developments." *J Strain Anal Eng Des* 36:101–117.
[54] Nicoletto, G. (1988) "Theoretical Fringe Analysis for a Coherent Optics Method or Residual Stress Measurement." *J Strain Anal* 23:169–178.
[55] Furgiuele, F. M., Pagnotta, L., Poggialini, A. (1991) "Measuring Residual Stresses by Hole-Drilling and Coherent Optics Techniques: A Numerical Calibration." *J Eng Mater & Technol* 113:41–50.
[56] Wu, Z., Lu, J., Han, B. (1998) "Study of Residual Stress Distribution by a Combined Method of Moire Interferometry and Incremental Hole Drilling," part I: theory and part II: implementation. *J. Appl Mech* 65:837–850.
[57] Riberio, J., Monteiro, J., Lopes, Vaz, M. (2011) "Moiré Interferometry Assessment of Residual Stress Variation in Depth on a Shot Peened Surface." *Strain* 47:e542–e550.
[58] Ya, M., Marquette, P., Belahcene, F., Lu, J. (2004) "Residual Stresses in Laser Welded Aluminium Plate by Use of Ultrasonic and Optical Methods." *Mater Sci Eng*, A 382:257–264.
[59] Shankar, K., Xie, H., Wei, R., Asundi, A., Boay, C. G. (2004) "A Study on Residual Stresses in Polymer Composites Using Moire Interferometry." *Adv Compos Mater* 13:237–253.
[60] Chen, J., Xin, Q., Yang, F. (2007) "Relationship Between the Depth of Drilling and Residual Strain Relief in Fiber Reinforced Composite Materials." *J Mater Eng Perform* 16:46–51.
[61] Cardenas, J., Ekwaro-Oise, S., Berg, J., Wilson, W. (2005) "Non-Linear Least Squares Solution to the Moire Hole Method Problem in Orthotropic Materials," part I: Residual Stresses. *Exp Mech* 45:314–324.
[62] Baldi, A. (2007) "Full Field Methods and Residual Stress Analysis in Orthotropic Material. (I) Linear Approach." *Int J Solids Struc* 44:8229–8243.
[63] Sicot, O., Gong, X., Cherouat, A., Lu, J. (2003) "Determination of Residual Stress in Composite Laminates Using the Incremental Hole-Drilling Method." *J Compos Mater* 37:831–841.
[64] Wang, Y., Chiang, F. (1997) "Experimental Study of Three-Dimensional Residual Stresses in Rails by Moiré Interferometry and Dissecting Methods." *Opt Lasers Eng* 27:89–100.
[65] Czarnek, R., Skrzat, A., Lin, S. (2011) "Application of Moiré Interferometry to Reconstruction of Residual Stresses in Cut Railroad Car Wheels." *Measurement* 44:569–579.
[66] Sciammarella, C., Sciammarella, F. (2012) "Shadow Moiré & Projection Moiré – The Basic Relationships," in *Experimental Mechanics of Solids*. John Wiley & Sons, Ltd, Chichester, UK. doi: 10.1002/9781119994091.ch15
[67] Daniel, I., Wang, T., Karalekas, D., Gotro, J. (1990) "Determination of Chemical Cure Shrinkage in Composite Laminates." *J Compos Tech Res* 12:172–176.
[68] Chen, K-S., Chen, T., Chuang, C-C., Lin, L-K. (2004) "Full-Field Wafer Level Thin Film Stress Measurement by Phase-Stepping Shadow Moiré." *IEEE Trans Compon Packag Technol* 27:594–601.
[69] Ifju, P., Han, B. (2010) "Recent Applications of Moiré Interferometry." *Exp Mech* 50:1129–1147.
[70] Kishimoto, S. (2012) "Electron Moiré Method." *Theo Appl Mech Lett* 2:011001.
[71] Li, B., Tang, X., Xie, H., Zhang, X. (2004) "Strain Analysis in MEMS/NEMS Structures and Devices by Using Focused Ion Beam System." *Sens Actuators*, A 111:57–62.
[72] Sutton, M. (2008) "Digital Image Correlation for Shape and Deformation Measurements," In: Sharpe, W. (ed) *Handbook of Experimental Solid Mechanics*. Springer: New York, pp. 565–600.
[73] Schmidt, T., Tyson, J., Galanulis, K. (2003) "Full-Field Dynamic Displacement and Strain Measurement Using Advanced 3D Image Correlation Photogrammetry," part I. *Exp Tech* 27(3):47–50.
[74] Moser, G., Lightner, J. (2007) "Using Three-Dimensional Digital Imaging Correlation Techniques to Validate Tire Finite Element Model." *Exp Tech* 31(4): 29–36.
[75] Reu, P. (2012) "Introduction to Digital Image Correlation: Best Practices and Applications." *Exp Tech* 36(1): 3–5.
[76] Nelson, D., Makino, A., Schmidt, T. (2006) "Residual Stress Determination Using Hole Drilling and 3D Image Correlation." *Exp Mech* 46:31–38.
[77] McGinnis, M., Pessiki, S., Turker, H. (2005) "Application of Three-Dimensional Digital Image Correlation to the Core-Drilling Method." *Exp Mech* 45:359–367.
[78] Gao, J., Shang, H. (2009) "Deformation-Pattern-Based Digital Image Correlation Method and its Application to Residual Stress Measurement." *Appl Opt* 48:1371–1381.
[79] Lord, J., Penn, D., Whitehead, P. (2008) "The Application of Digital Image Correlation for Measuring Residual Stress by Incremental Hole Drilling." *Appl Mech Mater* 13–14:65–73.
[80] Kang, K. J., Darzens, S., Choi, G. (2004) "Effect of Geometry and Materials on Residual Stress Measurement in Thin Films by Using the Focused Ion Beam." *J Eng Mater Technol* 126(4): 457–464.

[81] Sabate, N., Vogel, D., Gollhardt, A., Keller, J., Cane, C., Gracia, I., Morante, J. R., Michel, B. (2006) "Measurement of Residual Stress by Slot Milling with Focused Ion-Beam Equipment." *J Micromech Microeng* 16(2): 254–259.

[82] Yang, Y., Bae, J., Park, C. (2008) "Measurement of Residual Stress by Using Focused Ion Beam and Digital Image Correlation Method in Thin-Sized Wires Used for Steel Cords." *J Phys: Conf Ser* 100:012018.

[83] Winiarski, B., Langford, R., Tian, J., Yokoyama, Y., Liaw, P., Withers, P. (2010) "Mapping Residual Stress Distributions at the Micron Scale." *Metall Mater Trans*, A 41:1743–1751.

[84] Winiarski, B., Gholina, A., Tian, J., Yokoyama, Y., Liaw, P., Withers, P. (2012) "Submicron-Scale Depth Profiling of Residual Stress in Amorphous by Incremental Focused Ion Beam Slotting." *Acta Mater* 60:2237–2349.

[85] Sabate, N., Vogel, D., Gollhardt, A., Keller, J., Cane, C., Gracia, I., Morante, J. R., Michel, B. (2007) "Residual Stress Measurement on a MEMS Structure With High-Spatial Resolution." *J Micromech Syst* 16:365–372.

[86] Winiarski, B., Withers, P. (2012) "Micron-Scale Residual Stress Measurement by Micro-Hole Drilling and Digital Image Correlation." *Exp Mech* 52:417–428.

[87] Schajer, G., Winiarski, B., Withers, P. (2012) "Hole-Drilling Residual Stress Measurement With Artifact Correction Using Full-Field DIC." *Exp Mech* DOI: 10.1007/s11340-012-9626-0.

[88] Blair, A., Daynes, N., Hamilton, D., Horne, G., Heard, P., Hodgson, D., Scott, T., Shterenlikht, A. (2009) "Residual Stress Relaxation Measurements Across Interfaces at Macro- and Micro-Scales Using Slitting and DIC." *J Phys: Conf Ser* 181:012078.

[89] Korsunsky, A., Sebastiani, M., Bemporad, E. (2010) "Residual Stress Evaluation at the Micrometer Scale. Analysis of Thin Coatings by FIB Milling and Digital Image Correlation." *Surf Coat Tech* 205:2393–2403.

[90] Sebastiani, M., Eberl, C., Bemporad, E., Pharr, G. (2011) "Depth-Resolved Residual Stress Analysis of Thin Coatings by a New FIB-DIC Method." *Mater Sci Engr*, A 528:7901–7908.

[91] Zhu, J., Xie, H., Hu, Z., Chen, P., Zhang, Q. (2011) "Residual Stress in Thermal Spray Coatings Measured by Curvature Based on a 3D Digital Image Correlation Technique." *Surf Coat Tech* 206:1396–1402.

[92] Hung, Y. Y., Ho, H. (2005) "Shearography: An Optical Measurement Technique and Applications." *Mater Sci Eng*, R 49:61–87.

[93] Francis, D., Tatum, R., Groves, R. (2010) "Shearography Technology and Applications: A Review." *Meas Sci Technol* 21:102001+29.

[94] Nelson, D. (2010) "Residual Stress Determination by Hole Drilling Combined With Optical Methods." *Exp Mech* 50:145–158.

[95] Li, K. (1996) "The Interferometric Strain Rosette Technique." *Exp Mech* 36:199–203.

[96] Tjhung, T., Li, K. (2003) "Measurement of In-Plane Residual Stresses Varying With Depth by the Interferometric Strain/Slope Rosette and Incremental Hole-Drilling." *J Eng Mater Technol* 125:153–162.

[97] Li, K., Ren, W. (2007) "Application of Miniature Ring-Core and Interferometric Strain/Slope Rosette to Determine Residual Stress Distribution With Depth – Part I: Theories and Part II – Experiments." *J Appl Mech* 74:298–314.

[98] Ramesh, K. (2008) Photoelasticity in Handbook on Experimental Solid Mechanics (ed. W. Sharpe) *Springer*, pp. 701–742

[99] Corby, T., Nickola, W. (1997) "Residual Strain Measurement Using Photoelastic Coatings." *Opt Lasers Eng* 27:111–123.

[100] ASTM. Standard test method for photoelastic measurements of birefringence and residual strains in ransparent or translucent plastic materials, Standard D 4093–95. American Society for Testing and Materials, West Conshohocken, PA.

[101] ASTM. Standard test method for non-destructive photoelastic measurements of edge and surface stresses in annealed, heat-strengthened and fully tempered flat glass, Standard C 1279–09, American Society for Testing and Materials, West Conshohocken, PA.

[102] ASTM. Standard test method for photoelastic determination of residual stress in transparent glass matrix using a polarizing microscope and optical retardation compensator, Standard C 978–04, American Society for Testing and Materials, West Conshohocken, PA.

[103] ASTM. Standard test method for measurement of glass stress-optical coefficient. Standard C 770–98, American Society for Testing and Materials, West Conshohocken, PA.

[104] ASTM. Standard test method for measuring optical retardation and analyzing stress in glass. Standard F218-12, American Society for Testing and Materials, West Conshohocken, PA.

[105] Aben, H., Anton, J., Errapart, A. (2008) "Modern Photoelasticity for Residual Stress Measurement in Glass." *Strain* 44:40–48.
[106] He, S., Danyluk, S., Tarasov, I., Ostapenko, S. (2006) "Residual Stresses in Polycrystalline Silicon Sheet and Their Relation to Electron–hole Lifetime." *App Phys Lett* 89: 111909-1 to -3.
[107] Pisarev, V. S., Balalov, V., Aistov, V., Bondarenko, M., Yustus, M. (2001) "Reflection Hologram Interferometry Combined With Hole Drilling Technique as an Effective Tool for Residual Stress Fields Investigation in Thin-Walled Structures." *Opt Lasers Eng* 36:551–597.
[108] Pisarev, V., Aisotve, V., Balalov, V., Bondarenko, M., Chumak, S., Griogoriev, V., Yustus, M. G. (2004) "Metrological Justification of Reflection Hologram Interferometry with Respect to Residual Stresses Determination by Means of Blind Hole Drilling." *Opt Lasers Eng* 41:353–410.
[109] Balalov, V., Pisarev, V., Moshensky, V. (2007) "Combined Implementing the Hole Drilling Method and Reflection Hologram Interferometry for Residual Stress Determination in Cylindrical Shells and Tubes." *Opt Lasers Eng* 45:661–676.
[110] Cheng, J., Kwak, S., Choi, J. (2008) "ESPI Combined with Hole Drilling to Evaluate the Heat Treatment Induced Residual Stress." In Fourth International Symposium on Precision Mechanical Measurements. Proc SPIE *Int Soc Opt Eng* v. 7130, pp. 71302 C-1 to C-6.
[111] Steinzig, M., Takahashi, T. (2003) "Residual Stress Measurement Using the Hole Drilling Method and Laser Speckle Interferometry, Part IV: Measurement Accuracy." *Exp Tech* 27(6): 59–63.
[112] Viotti, M., Kaufmann, G. (2004) "Accuracy and Sensitivity of a Hole Drilling and Digital Speckle Interferometry Combined Technique to Measure Residual Stress." *Opt Lasers Eng* 41:297–305.
[113] Schmitt, D., Diallo, M., Weichman, F. (2006) "Quantitative Determination of Stress by Inversion of Speckle Interferometer Fringe Patterns: Experimental Laboratory Tests." *Geophys J Int* 167:1425–1438.
[114] Ribeiro, J., Monteiro, J., Vaz, M., Lopes, H., Piloto, P. (2009) "Measurement of Residual Stresses With Optical Techniques." *Strain* 45:123–130.
[115] Albertazzi, A., Viotti, M., Kapp, W. (2010) "Performance evaluation of a radial in-plane digital speckle pattern interferometer using a diffractive optical element for residual stresses measurement." In: Interferometry XV: Applications, Proc SPIE *Int Soc Opt Eng*, v. 7791, pp 77910G-1 to -8.
[116] Schwarz, R., Kutt, L., Papazian, J. (2000) "Measurement of Residual Stress Using Interferometric Moire:a New Insight." *Exp Mech* 40:271–281.
[117] Schajer, G. S. (2012) Private communication, Sept. 20, 2012.

Further Reading
Optical Methods Used in Experimental Mechanics

Cloud, G. (1995) *Optical Methods of Engineering Analysis*. Cambridge University Press: Cambridge, UK.
Vest, C. (1979) *Holographic Interferometry*. Wiley: New York.
Jones, R., Wykes, C. (1989) *Holographic and Speckle Interferometry*. Cambridge University Press: Cambridge UK.
Rastogi, P. (ed.) (2000) "Photomechanics," *Topics Appl. Phys.* 77. Spinger-Verlag: Berlin.
Sharpe, W. (ed.) (2008) *Handbook of experimental Solid Mechanics*. Springer: New York.
Sihori, R. (2009) *Optical Methods of Measurement: Whole Field Techniques*, 2d ed. CRC Press, Taylor & Francis Group: Boca Raton, Florida.
Hariharan, P. (1996) *Optical Holography: Principles, Techniques and Applications*, 2d ed. Cambridge University Press: Cambridge, UK.
Post, D., Han, B., Ifju, P. (1994) *High Sensitivity Moiré*. Springer: New York.
Steinchen, W., Yang, L. (2003) *Digital Shearography: Theory and Applications of Digital Speckle Pattern Interferometry*. SPIE Press: Bellingham, Washington.

Index

ABE, 17, 229–30, 233
Abrasion, 40
Abutments, 244
Accuracy
   estimates, 20
   of measurements, 7, 19, 24
Acoustic Barkhausen Emission. *See* ABE
Acoustic birefringence, 259, 262
Acoustic wave, 17
Acoustic wave speed, 259
Acoustoelastic coefficient, 264, 267–8, 272
Acoustoelastic effect, 259, 262
Air probe, 67
Analysis of data, 211
Analyzer crystal, 184
ASTM E837, 31–2, 35–6, 54, 58
ASTM standards
   hole-drilling method, 9
   X-ray measurements, 15
Attenuation length, 169

Basis function, 94
Bending magnet, 165, 167
Bimetallic sleeve, 82
Blocklength, 68
Boundary element method, 158
Bragg
   angle, 14–16, 172, 199
   diffraction, 164
   edge, 203
   law, 13, 141, 163–4, 173, 195, 210
   reflection, 191
Bueckner's principle, 110, 114

Calibration constant, 32, 36, 48, 51–2
Calibration standard, 274
Calibration, finite element, 8
Carbon fibre composite, 82
Coated cylinder bore, 57
Coating stress, 190
Coefficient of thermal expansion, 156
Coercivity, 247
Combination strains, 49
Combination stresses, 48
Compliance matrix, 70, 90–92, 95
Compliance tensor, 150, 152, 178
Composite laminate, 288–9
Concentricity, 46
Condition number, 94
Confocal ranging probe, 129
Conical slit, 184, 189
Contour method, 12, 109
   anti-symmetric errors, 113
   bulge error, 113, 115, 123, 131–2
   cut width, 130
   finite element model, 109
   fixturing, 114
   free surface, 110
   in-plane displacement, 110
   patent, 134

Contour method (*continued*)
    perimeter trace, 118
    symmetric errors, 113
Convolution, 104
Coordinate measurement machine, 116, 118, 121
Crack tip
    damage, 192
    opening displacement, 192
    plasticity, 192
    shielding, 192
    strain mapping, 189
    stress field, 192
    stress intensity, 190
Critical angle, 260
Cross-method validation, 103
CTE. *See* Coefficient of thermal expansion
Cure referencing method, 289
Curie temperature, 249
Curvature method. *See* Stoney's method
Curved surfaces, 252
Cylindrical parts, 134

Data averaging, 49
de Broglie wavelength, 199, 210
Debonding, 46
Debye-Scherrer, 203
Deep-hole drilling
    conventional method, 68
    incremental method, 68
    method, 11, 31, 65
    remote control, 84
Deformation effects, 5
Demagnetisation, 246
Destructive measurements, 6
Deviator stress, 156
DHD. *See* Deep-hole drilling
Diamond tipped hole saw, 73
DIC, 34, 63, 289, 292, 295
    camera, 290
    hole drilling, 291
    macro scale, 296
    micro/nano scale, 291, 296
Differential shrinkage, 3

Diffraction
    angle, 195
    angle-dispersive, 184
    elastic constants, 179, 201
    energy-dispersive, 188
    methods, 13, 139
    peak fitting, 173, 179
    spotty pattern, 183
    systematic errors, 180
    textured materials, 143, 170
digiMAPS, 238
Digital holographic
    interferometry, 279
    microscope, 285
    system, 280
Digital Image Correlation. *See* DIC
Digital photogrammetry, 98
Digital speckle pattern interferometry. *See* ESPI
Digitizing board, 271
Dimensional continuity, 1
Dispersion of waves, 17
Displacement boundary conditions, 120
Dissimilar metal weld, 80, 128
Dölle-Hauk, 155, 156
    analysis, 145
    method, 143
Domain wall, 227, 247
Drilling cutter selection, 42
Drilling machines, 43
Ductile cast iron, 273
Dynamic diffraction, 148

Eccentricity, 60
Eddy current, 244
Eddy current screening, 230
Edges, 244
EDM, 56, 67, 96, 109, 113, 116, 131
    finish cut setting, 116
    skim cut setting, 116
    wire, 116, 120
Eigenstrain method, 3, 68
Elastic anisotropy, 178
Electric Discharge Machine. *See* EDM
Electrical interference, 97
Electrical resistivity, 249

Electro-chemical machining, 73
Electromagnet, 225
Electron beam lithography, 289
Electronic Speckle Pattern Interferometry. *See* ESPI
Electroplated materials, 9
Electropolishing, 133
Entrance and exit face distortions, 70
Error propagation, 90
ESPI, 34, 279–82, 285, 294, 296
   hole-drilling, 283–4
   micro system, 285
   shearography, 292

Fast strain measurments, 170
Fatigue
   cracking, 100
   fracture, 4
Ferritic steels, 226
Ferromagnetic metals, 225
FIB, 295–6
   milling, 289, 291–2
   ring coring, 291
Finite element model, 68, 118, 120–121, 123, 133
Finnie, I, 89
FMD, 236
Focused Ion Beam. *See* FIB
Foreign object damage, 186
Fourier series, 120
Fracture mechanics, 193
Fracture toughness, 100, 106
Frame overlap, 216
Friction stir welding, 272
Friction welding, 122
Fringe pattern, 281–3
Fusion welding, 122

Gauge volume, 164, 170–171, 180, 196, 206, 209
   instrumental, 171, 180
   partly filled, 213
   sampled, 173, 180
Goniometer
   ψ type, 148
   Ω type, 148

Grain boundaries, 247
Grain size, 17
Guide block, 243

Heat-affected zone, 3, 248
Heterogeneous elasticity, 114
Hole-drilling method, 9, 29, 32, 281, 286, 291
   ASTM E837, 9
   DIC, 291, 295
   distance to a step feature, 37
   distance to an edge feature, 37
   drilled hole depth, 36
   ESPI, 283, 293
   FIB, 295
   gauge spacing, 38
   hole diameter, 36
   holographic interferometry, 281, 294
   incremental drilling, 45, 54, 79, 283, 287
   Instrumentation, 41
   Moiré, 287–8
   radius of curvature, 38
   specimen thickness, 37
   stress data depth, 36
Holographic interferometry, 33, 278, 282, 285, 294, 296
   hole-drilling, 281, 294
Hooke's Law, 140, 145, 178, 201
Hoop stress, 123
Hysteresis loop, 225

Ill-conditioning, 53
Incorrect use of slits, 214
Incremental drilling, 45, 54, 79, 283, 287
Incremental slitting, 130
Indentation method, 18
Indented disks, 122, 215
Influence function, 96
Insertion devices, 166
In-situ welded component, 83
Integral method, 47, 51, 54, 57, 59–60, 94, 101
Interference fringe pattern, 281

Interferometric strain rosette, 293
Intergranular effects, 122, 215
Intergranular stress, 170, 179
Intra-laboratory repeatability, 102, 106
Inverse problem, 51, 90

Kerf, 115
Kernel function, 51
Kinematical diffraction theory, 148
Kröner
    limit, 153
    model, 152
    X-ray elastic constant, 153

Laboratory coordinate system, 141, 150, 152
Large grain effect, 214
Laser scanner, 123
Laser shock peening, 99, 125, 254, 267
Lattice spacing, 14–16, 173
Layer removal method, 9, 126, 285
L_{CR} wave, 259, 262, 265, 274
Least-squares method, 71, 143, 145, 155–6
Legendre polynomial, 91–2
Lift-off, 233, 239
Linear friction weld, 123
Los Alamos National Laboratory, 134

Machining-induced stresses, 39
Macroscopic cross section, 205
MAE, 230
Magnetic 'easy' axis, 226
Magnetic Anisotropy and Permeability System. *See* MAPS
Magnetic Barkhausen Noise. *See* MBN
Magnetic conditioning, 246
Magnetic domain alignment, 228
Magnetic domains, 227
Magnetic field
    frequency, 230
    penetration, 245
Magnetic loading history, 245
Magnetic method, 16, 225
    calibration, 240

Magnetising solenoid, 232
Magneto-Acoustic Emission. *See* MAE
Magneto-dynamic distortion, 252
Magneto-elastic energy, 227
Magnetostriction, 226, 230, 256
MAPS, 229, 235, 248, 250
Material Delta Value
    flux leakage sensor. *See* FMD
    linkage sensor. *See* PMD
MBE, 229
MBN, 16, 229, 232–3
Measurement
    depth, 237
    environment, 22
    objective, 22
    uncertainties, 59
    volume, 5, 147
Mechanical relaxation, 89, 106
MEMS, 289
Microbeam diffraction, 156
Micromagnetic methods, 233
Micromechanical device, 285
Miller indices, 196
Misfit, 95, 101
Moiré
    fringe pattern, 286–7, 289
    hole-drilling, 287–8
    interferometer, 286
    interferometry, 33, 285, 288–9, 296
    micro scale, 289
    shadow method, 289
Monochromator, 207
Morozov criterion, 54, 95
Multiaxial stress, 125
Multiple cut method, 133
Multiple stress components, 133

Near-surface stresses, 114
Neerfeld-Hill
    limit, 153, 156
    model, 152
    X-ray elastic constant, 153
Neutron
    absorption, 203
    absorption cross section, 205
    beam wavelength, 195

coherent scattering, 203
diffracted beam, 196
diffraction, 13, 15, 79, 122–3, 133, 195–6, 201
diffractometer, 206
flux, 208
incident beam, 196
incoherent scattering, 203
magnetic moment, 204
path length, 198
scattering cross-section, 202
small angle scattering, 191
time-of-flight, 199
transmission, 205
wavelets, 196
Non-destructive measurements, 7, 13–14, 16
Nozzle-to-nozzle intersection, 83
Nuclear reactor, 127, 196
Numerical validation, 75

ODF. *See* Orientation distribution function
Optical measurements, 33, 50
Optical scanner, 117
Orbit eccentricity, 59
Orbital drilling, 42, 44
Orientation distribution function, 151
Orthotropic materials, 50
Overall uncertainty, 75
Over-coring methods, 65

Path length, 198
Penetration depth, 7, 148–50, 190, 238, 273
  neutron, 169
  synchrotron X-ray, 169
  X-ray, 169
Performance and limitations of measurements, 19
Permeability
  differential, 228
  maximum, 228
Personnel qualifications, 274
Phase
  composition, 17
  transformation, 2, 192
Phenomenological correlation, 103–4, 106
Photoelasticity, 18
  reflection, 293
  transmission, 293
Pilot borehole, 65
Pinning, 247
Plane normals <hkl>, 197
Plane stress, 14
Plastic anisotropy, 178
Plastic strain, 248
Plastic zone, 102
Plasticity, 50, 71, 102, 112, 125, 132
PMD, 236, 240
Polariscope, 18
Polarized light, 18
Polycrystal, 142
Polymer (PMMA) wedges, 260
Polynomial
  basis, 101
  series, 90
Powder diffraction pattern, 164
Practical considerations, 57
Precision of measurements, 213
Pressure vessels, 272
Pressuriser nozzle, 80
Principal axes, 200
Principal direction, 68
Principle of superposition, 69, 92, 110, 133
Prior elastic-plastic loading, 78
Probe
  rotation, 236
  size, 249
  standoff, 249
Probe calibration, 73
Prong test, 8
Pseudoinverse, 92, 94

Quasi-isotropic
  elastic aggregate, 155
  polycrystalline materials, 145
  XEC, 153
Quenching, 79

Radial collimator, 206, 209
Railway rail, 12, 129, 130
Rayleigh wave, 17
Reconstruction of stress, 72
Reference
    bush, 67
    coupon, 217
    frame, 73
    hole, 67, 73
    lattice spacing, 220
    samples, 211
    stress-free, 7
Regularization, 53
    parameter, 95
    Tikhonov, 95
Relative permeability, 230
Relaxation methods, 6–7, 65
Relief nozzle, 127
Remnant magnetic field, 246
Repeatability, 89
Residual stress
    definition, 1
    equibiaxial, 70
    gradients, 4
    in-situ, 84
    intensity factor, 96
    length scale, 2
    measurement characteristics, 23
    measurement choice, 19
    profile calculation, 90
    Type I, 2, 197
    Type II, 2, 13, 197
    Type III, 2, 13, 198
    uniaxial, 70
Reuss limit, 151
Reversible hydrogen attack, 273
Rietveld
    method, 212
    refinement, 217
Ring and plug specimen, 254
Ring-core method, 9, 29, 50
    FIB, 291

Sachs' method, 9
Sacrificial layer, 131
Safe-end nozzle, 80

Sample coordinate system, 141, 199
Sampled gauge volume SGV, 173, 180
Sampling errors, 159
Scanning electron microscope, 291, 295–6
Sectioning method, 9, 288–9
Semi-destructive measurements, 16, 83
Series expansion, 90
Shadow Moiré method, 289
Shear stresses, 111
Shear wave, 259
Shear wave birefringence, 17, 259, 262, 264
Shearography, 292
Shot-peened alloy steel plate, 54
Shot-peening, 2, 9, 56, 245, 273
Shrink fitting, 77
Signal to noise ratio, 180
Silicon-on-insulator, 156–7
$Sin^2\psi$ method, 146, 176–7, 190
Single crystal, 142, 148
    coordinate system, 152
    strain, 156
Slitting method, 11, 89, 126
    FIB, 291
Slitting plane, 96
SMA, 236
Snell's Law, 260
SOI. See Silicon-on-insulator
Solidification, 3
Spallation source, 196, 209
SPATE, 17
Spatial resolution, 237, 239
    of measurements, 7, 19, 24
Specimen
    damage, 22
    dimensions, 22
    material, 22
    shape, 22
    table, 73
Spline smoothing, 120
Splitting method, 8
Spurious strain, 181
SSCANS, 211
St. Venant's Principle, 53, 112
Stainless steel 316L, 122

Statistical error, 159
Steel cylinder, 76
Steel rolling, 82
Stereoscopic imaging, 34
Stiffness tensor, 178
Stoney's method, 9
   optical measurements, 289
   X-ray, 140
Strain cell, 65
Strain gauge
   backing material, 46
   installation, 40
   rosette, 31, 46, 57
   selection, 35
   three-wire connection, 41
Strain incompatibility, 78
Strain mapping, 189
Strain measurment
   deconvolution, 172
   gauge volume, 170
   precision, 179
Strain tensor, 178, 199
Strain-free lattice spacing, 170
   global, 174
   mapping, 176
   stress balance, 177
Stranded wire, 244
Stress corrosion cracking, 4
Stress profile, 47, 190, 237, 273
Stress resolution, 270
Stress tensor, 178
Stress uncertainties, 61
Stress-free reference, 198
Stress-free reference powder, 175
Stress-free test cut, 122
Stress-induced Magnetic Anisotropy. *See* SMA
Surface
   coating, 244
   height map, 111
   preparation, 38
Surface-skimming longitudinal waves, 262
Swab-etching, 39
Synchrotron, 165
   diffraction, 122, 163

   penetration depth, 169
   surface effects, 170
   X-rays, 15
Systematic error, 213

Thermal guide, 207
Thermal output, 41
Thermodynamic models, 240
Thermoelastic measurements, 17
Thick specimen, 47
Thin specimen, 47
Thompson scattering, 205
Three-pass bead-in-slot weld, 218
Ti-6Al-4V, 103, 123, 125
Tikhonov regularization, 54, 95
Titanium, 56
Toughened glass, 1, 3
T-R arrangement, 267
Triaxial constraint, 79
Triaxial stress, 146, 154, 156
Truth data, 103
Type A rosette, 32, 35
Type B rosette, 32, 35, 60
Type C rosette, 32, 35, 60

Ultrasonic method, 17, 259
   couplant, 262, 265, 267
   dual probe setup, 260
   dual receivers, 271
   fixed probes, 267
   freely rotating probes, 270
   grain size, 265
   history, 264
   immersion, 267, 272
   impedance matching, 262
   instrumentation, 266
   probe clamping, 267
   repeatability, 267
   surface roughness, 265
   temperature effects, 265
   texture analysis, 264
   travel time, 265, 267, 269
   wave speed, 260, 262, 265–6
Uncertainty sources, 58, 270
Undulator, 165, 167
Uniform stresses, 47

Unit pulse method. *See* Integral method

Voigt limit, 152
Volterra equation, 51

Wave velocity, 17
Wave-particle duality, 202
Weighting factor, 95
Weld cap, 244
Weld overlay, 80
Weld stresses, 272
Welded metal component, 66
Wiggler, 165, 167
Wire EDM, 90, 115
Wire electrode, 97

XEC. *See* X-ray elastic constant
X-ray
 diffraction, 14, 104, 126, 139
 diffraction angle oscillation, 142
 diffraction angle splitting, 142
 diffraction peak breadth, 140
 elastic constant, 145, 153–5
 instrument errors, 159
 small angle scattering, 191
 strain equation, 142

Yield strength, 132
Yielded region, 71